Advanced Discrete Mathematics

Uday Singh Rajput

Assistant Professor

Department of Mathematics and Astronomy

Lucknow University

Lucknow

PHI Learning Private Limited

New Delhi-110001

2012

₹ 350.00

ADVANCED DISCRETE MATHEMATICS
Uday Singh Rajput

ISBN-978-81-203-4589-8

Published by Asoke K. Ghosh, PHI Learning Private Limited, M-97, Connaught Circus, New Delhi-110001 and Printed by Baba Barkha Nath Printers, Bahadurgarh, Haryana-124507.

To

The Almighty who keeps me alive

and

My wife and daughter who keep me inspired

Contents

Part E AUTOMATA THEORY AND GRAMMAR

Preface

The "Advanced Discrete Mathematics" course aims to provide a comprehensive collection of useful topics of mathematics and computer science to scientists, engineers, students and teachers engaged in research and higher education in applied mathematics and computer science disciplines. Discrete mathematics is a compulsory course in these disciplines in most of the engineering institutions/universities. This book is designed for the postgraduate students of mathematics, computer science and computer applications as well as undergraduate students of computer science and engineering.

The whole text is divided into six parts namely, algebraic structure, mathematical logic, lattice and Boolean structure, graph theory, automaton, and recurrence relations. The symbols used in this book are usual and standard.

Part A (Chapters 1 to 6) discusses algebraic systems. Most of the modern algebra texts contain material only on group and ring theory. Semigroups and monoids are generally ignored despite their applications in computer science. Therefore, these topics have also been covered along with groups and rings. **Part B** (Chapters 7 and 8) deals with mathematical logic. This part sets the tone for the study of mathematical modelling of human intelligence. It will definitely find applications in developing software packages for knowledge representation, storage and its analysis. Chapter 7 is devoted to statement calculus. It contains theory related with symbolization of English language sentences. Argument and its proof models are also discussed. Chapter 8 is an extension of Chapter 7. It contains more realistic proof models incorporating the details of quantifiers. **Part C** (Chapters 9 to 12) gives details of lattices and Boolean algebra. Lattice theory is considered as the foundation of Boolean algebra. In Chapter 9, partial order relation, Hasse diagram, lattice and related theory are discussed. Chapters 10, 11 and 12 are related with logic circuits. **Part D** (Chapters 13 to 15) is dedicated to graph theory. It is a great mathematical tool with its applications in computer science, communication, physical science, networking and so on. Efforts have been made to provide details of graphs and trees along with related algorithms. Flow networks have also been discussed. **Part E** (Chapters 16 and 17) covers automata theory and grammars. Some useful automata models have been described in Chapter 16. Language theory plays a crucial role in developing compilers. Therefore, the parser, which is one of the most important parts of a compiler, is discussed in Chapter 17 along with grammars. **Part F** (Chapter 18) deals with the study of discrete numeric functions and recurrence relations. The textbook contains a large number of examples and diagrams to support the theoretical explanations and to make learning easy. Some solved problems are also given at the end of each chapter.

I would like to thank my wife Dr. Debangana Rajput and daughter Preksha who helped me a lot in the completion of this project.

Suggestions and comments from the readers are welcome.

Uday Singh Rajput

Symbols Used

Symbol	*Description*
$\mathbb{R}$	Set of real numbers
$\mathbb{N}$	Set of natural numbers
$\mathbb{Q}$	Set of rational numbers
$\mathbb{I}$, or $\mathbb{Z}$	Set of integers
$\forall$	For all
$\exists$	There exists
$K_{m,n}$	Bipartite graph
#	Hash
$\rho(X)$	Power set of X

Part A
ALGEBRAIC STRUCTURE

1

Preliminaries of Set Theory

1.1 Introduction

The use of set theory is so wide that the word **set** can be found in almost every course of modern mathematics. The actual work on set theory began in the year 1873 with a letter written by Georg Cantor* to Dedekind, although the word 'Menge' (meaning 'set') had been coined by Bernhard Bolzano much earlier. Georg Cantor, in his letter, informed Dedekind of his work related with the uncountability of the real line, rather uncountability of sets. Cantor's work between 1874 and 1884 is considered as the origin of set theory. Prior to this work, the concept of a set was very elementary. In fact, before Cantor's development, no one had imagined the intricacies involved with infinity. It is an amazing fact that the set of all positive integers has a proper subset (like the set of their squares, $\{1, 4, 9, 16, 25, ...\}$) of the same cardinality. This was

---

***Georg Ferdinand Ludwig Philip Cantor** (1845–1918) was a German mathematician. He is known as the inventor of set theory. Cantor was born in 1845 in the western merchant colony in Saint Petersburg, Russia. Cantor defined infinite and well-ordered sets, and proved that the real numbers are "more numerous" than the natural numbers. He established the importance of one-to-one correspondence between sets. However, Cantor's theory got a lot of resistance and criticism from a number of mathematicians, particularly Leopold Kronecker. Kronecker was a well-established figure within the mathematical community and Cantor's former professor. He disagreed fundamentally with the Cantor's work. Kronecker's public opposition and personal attacks were fierce. Kronecker even objected to Cantor's proofs that the algebraic numbers are countable, and that the transcendental numbers are uncountable. It is said that Kronecker's criticisms left Cantor quite depressed and ultimately led to his health deterioration. In 1904, Cantor's contribution was recognized by the Royal Society, and it awarded Cantor its Sylvester Medal, the highest honour in mathematics. Cantor did his PhD, on number theory, at the University of Berlin. He taught briefly at a Berlin girls' school, before taking up a position at the University of Halle, where he spent his entire career. Cantor died in 1918.

considered somewhat ambiguous by the mathematicians; though it had been discussed by several other researchers, including the great Galileo. Nevertheless, it really seems absurd.

Cantor's discovery of set theory came from his work on Fourier series and on **transcendental** numbers (a number, possibly a complex number, which is not algebraic, i.e., not a root of a polynomial equation with rational coefficients). The most prominent examples of transcendental numbers are π and e. All real transcendental numbers are **irrational** (a number that cannot be written in p/q form, where p and q are integers, and $q \neq 0$), but the converse is not true. For instance, the square root of two ($\sqrt{2}$) is an irrational number but it is an algebraic number, and therefore, not transcendental.

Cantor declared that the **real numbers** (consisting of rational and irrational numbers) are uncountable. According to him an infinite set X is said to be **countable** if its elements can be arranged in a sequence indexed by positive integers, like $x_1, x_2, x_3, \ldots, x_n, \ldots$ etc, and hence many infinite sets of numbers are countable, such as the set of all integers, the set of all rational numbers and so on. However, due to the presence of irrational numbers, the set of real numbers cannot be arranged in a sequence indexed by positive integers, and hence it is not countable. A set, which is not countable, is called an **uncountable set**. Cantor supported his claim of uncountability of irrational numbers by his famous proof known as **diagonal argument*** or proof by contradiction.

This proved that the set of real numbers is **more numerous** than the set of natural numbers. Thus, he showed, for the first time that there exist infinite sets of different sizes, i.e., some infinite sets being larger than others. He was also the first person to appreciate the importance of one-to-one correspondences in set theory. He used this concept to define finite and infinite sets, subdividing the latter into denumerable (or countably infinite) sets and uncountable sets. Cantor also introduced fundamental constructions in set theory, such as the power set of a set A, which is the set of all possible subsets of A. He later proved that the size of the power set of A is strictly larger than the size of A.

1.2 Basic Definitions

The elements of set theory are **sets**. The sets are basic objects, which are used to define all other structures or concepts in mathematics and, therefore, it will be improper to limit the scope of set theory to pure mathematics.

Formally, a **set is a well-defined collection of distinct objects.** Why do we say well defined? Think. The answer can be found when we will discuss Russell's paradox (Sec. 1.13.2). The set theory is based on a relation called membership. One can say that x is a member of X (or $x \in X$), or that the set X contains x as its element. A set is determined by its elements. In formal notation form, $X = \{x, y, z\}$, where x, y and z are the members of the set X.

A set may be **finite** or **infinite** depending on the number of elements in it. The number of elements in a set is called the **cardinality** of the set. It is denoted by $|A|$. If the number of elements in a set is infinite, then it is not possible to list all the elements of the set. Therefore, a better representation called **tabular or set builder form** is used for the set. For instance, $X = \{x \mid x \text{ is an integer}\}$. Here, X is a set having infinite number of elements, which are integers.

*For diagonal argument, readers may refer any good text on real analysis.

Two sets are said to be **equal** if they have exactly same elements. A set with no element is called **null** or **empty** set and is written as **{ }** or ϕ, symbolically. A set that has exactly one member is called a **singleton set,** for example, $\{x\}$.

In set theory problems, we consider sets of points, sets of numbers, sets of functions, sets of sets and so on. Using membership relation, other concepts of set theory such as union of sets, intersection of sets and so on can be developed. For instance, a set Z is the **union** of two sets X and Y if its members are exactly those elements which are either the members of X or members of Y. The set Z is uniquely determined because we have specified what its elements are.

Using the basic construction principles, and assuming the existence of infinite sets, one can define numbers, for example, integers, real numbers, complex numbers etc. Functions, functionals and topological concepts can also be defined. In this sense, set theory can be seen as the foundation of Mathematics.

1.3 Operations on Sets

Set theory operations are defined using membership relation. Some important operations are discussed in the following subsections:

1.3.1 Union of Sets

The union of any two sets results in a set that contains elements of both the sets. Let X and Y be two sets, then

$$X \cup Y = \{x \mid x \in X \text{ or } x \in Y\}.$$

In general, union of an arbitrary number of sets will contain elements of all the sets. Symbolically, it is written as

$$A = \bigcup_{i=1}^{n} X_i.$$

1.3.2 Intersection of Sets

The intersection of two or more number of sets forms a set that contains common elements of the sets. Suppose X and Y are two sets, then

$$X \cap Y = \{x \mid x \in X \text{ and } x \in Y\}.$$

In general, the intersection of n sets is written as

$$A = \bigcap_{i=1}^{n} X_i.$$

If $X \cap Y = \phi$, then X and Y are said to be **disjoint sets,** or mutually disjoint.

1.3.3 Difference of Sets

The difference of two sets $(X - Y)$ is a set containing exactly those elements of X that are not in Y. Hence,

$$X - Y = \{x \mid x \in X \text{ and } x \notin Y\}.$$

EXAMPLE 1.1

$$\{p, q, r, s, t\} - \{p, q, r\} = \{s, t\}$$
$$\{p, q, r, s, t\} - \{p, q, r, u\} = \{s, t\}$$
$$\{p, q, r\} - \{s, t\} = \{p, q, r\}$$

1.3.4 Symmetric Difference of Sets

The symmetric difference of two sets X and Y, denoted by $X \oplus Y$, contains exactly those elements of X and Y, which are in X or in Y but not in both. Symbolically,

$$X \oplus Y = (X \cup Y) - (X \cap Y).$$

EXAMPLE 1.2

$$\{p, q, r, s, t\} \oplus \{p, q, r, t, u\} = \{s, u\}$$
$$\{p, q, r, s, t\} \oplus \{p, q, r, s, t\} = \{\ \} = \phi$$
$$\{p, q, r\} \oplus \phi = \{p, q, r\}$$

1.3.5 Complement of a Set

Let $\mathcal{U}$ be a set. Further, let $A \subseteq \mathcal{U}$. Then the set $\mathcal{U} - A$ is called the **complement** of A. It is denoted by $\bar{A}$. Set $\mathcal{U}$ in the context of A, is called the **universal set**.

For instance, if $\mathcal{U} = \{2, 3, 4, 5, 6, 7, 8, 9\}$ and $A = \{2, 4, 7\}$, then $\bar{A} = \{3, 5, 6, 8, 9\}$.

1.4 Subset of a Set

Suppose X and Y are two sets. Set X is said to be a subset of Y if every element of X is also an element of Y. Symbolically, $X \subseteq Y$.

Let $X = \{p, q, r, s, t\}$ and $Y = \{p, q, r, s, t, u, v\}$. Then $X \subseteq Y$.

Note that, every set is a subset of itself. However, if two sets are not equal then one set will be the **proper subset** of other. For instance, if $X = \{p, q, r\}$ and $Y = \{p, q, r, s\}$, then X is said to be proper subset of Y. Notation used for proper subset is $\subset$, i.e., $X \subset Y$.

1.5 Principle of Inclusion and Exclusion

In this section, we shall list some obvious results of set theory related to cardinality of finite set.

(i) $|X \cup Y| \le |X| + |Y|$

(ii) $|X \cap Y| \le \min\{|X|, |Y|\}$

(iii) $|X \oplus Y| = |X| + |Y| - 2|(X \cap Y)|$

(iv) $|X - Y| \ge |X| - |Y|$

(v) $|X \cup Y| = |X| + |Y| - |(X \cap Y)|$

Equality (v) is known as **principle of inclusion and exclusion**. It can be written in a generalized form as:

$$|X_1 \cup X_2 \cup \ldots \cup X_n| = \Sigma_i|X_i| - \Sigma_{1\le i<j\le n}|X_i \cap X_j| + \Sigma_{1\le i<j<k\le n}|X_i \cap X_j \cap X_k| + \cdots + (-1)^{n-1}\,|X_1 \cap X_2 \cap \ldots \cap X_n|.$$

1.6 Laws of Set Theory

A set theory **expression** is an expression that consists of sets and usual set theory operators. A set theory expression is called a set theory law, if it holds always true. Most of the laws occur in pairs and if one law is known, then the other can be obtained by principle of **duality**. If a set theory expression is a law, then by principle of **duality** its dual will also be a law. Operations union and intersection are said to be **duals** of each other. **Dual** of an expression is obtained by replacing all occurrences of $\cup$ by $\cap$ and vice versa, and all occurrences of $\mathcal{U}$ by φ and φ by $\mathcal{U}$. Some important laws that govern the set theory are listed in Table 1.1. The proof of these laws can be constructed easily.

Table 1.1 Laws of Set Theory

S. No.	*Set theory expression*	*Dual of expression*	*Known as*
1.	$A \cup A \equiv A$	$A \cap A \equiv A$	Idempotent laws
2.	$(A \cup B) \cup C \equiv A \cup (B \cup C)$	$(A \cap B) \cap C \equiv A \cap (B \cap C)$	Associative laws
3.	$A \cup (B \cap C) \equiv (A \cup B) \cap (A \cup C)$	$A \cap (B \cup C) \equiv (A \cap B) \cup (A \cap C)$	Distributive laws
4.	$A \cup \varphi \equiv A$ (φ is empty set)	$A \cap \mathcal{U} \equiv A$ ($\mathcal{U}$ is universal set)	Identity laws
5.	$A \cup (A \cap B) \equiv A$	$A \cap (A \cup B) \equiv A$	Absorption laws
6.	$\overline{\overline{A}} \equiv A$	—	Double complement
7.	$\overline{A \cup B} \equiv \overline{A} \cap \overline{B}$	$\overline{A \cap B} \equiv \overline{A} \cup \overline{B}$	De Morgan's laws
8.	$A \cup B \equiv B \cup A$	$A \cap B \equiv B \cap A$	Commutative laws

1.7 Ordered Pairs

Suppose $X = \{x_1, x_2\}$. Since set is simply a collection, its elements are **unordered**. Therefore, set X can be written as $\{x_2, x_1\}$. Also, $\{x_1, x_2\} = \{x_2, x_1\}$, i.e., the elements of the set $\{x_1, x_2\}$ can appear in any order. We call $\{a, b\}$ to be an unordered pair of x_1 and x_2. In contrast, for ordered pairs, the order of the elements matters. We denote an **ordered pair** involving x_1 and x_2 as (x_1, x_2). Also, $(x_1, x_2) \ne (x_2, x_1)$. If $(x_1, x_2) = (x_3, x_4)$, then $x_1 = x_3$ and $x_2 = x_4$. Similarly, an ordered n-tuple involving $x_1, x_2, \ldots, x_n$ is denoted as $(x_1, x_2, \ldots, x_n)$.

1.8 Cartesian Product

The Cartesian product $(X \times Y)$ of two sets X and Y is the set of all ordered pairs (x, y), where $x \in X$ and $y \in Y$. In set-theoretic notations, $X \times Y = \{(x, y) | x \in X \text{ and } y \in Y\}$. If X and Y are finite sets, then the cardinality of $X \times Y$ is $|X| \cdot |Y|$.

For example, let $A = \{a, b, c\}$ and $B = \{c, d\}$. Then the Cartesian product of A and B is given by

$$A \times B = \{(a, c), (a, d), (b, c), (b, d), (c, c), (c, d)\}$$

or in set builder form $A \times B = \{(x, y) | x \in A \text{ and } y \in B\}$.

Since the Cartesian product is ordered, therefore, $A \times B \neq B \times A$, in general. Cartesian product of more than two sets is defined in a similar way.

That is, $A \times B \times C = \{(a, b, c) | a \in A, b \in B \text{ and } c \in C\}$.

In general,

$$A_1 \times A_2 \times A_3 \times \cdots \times A_n = \{(a_1, a_2, a_3, \ldots, a_n) | a_1 \in A_1, a_2 \in A_2, \ldots, \text{and } a_n \in A_n\} \quad (1.1)$$

Here, element $(a_1, a_2, a_3, \ldots, a_n)$ is called ***n*-tuple**.

If $A_1 = A_2 = A_3 = \cdots = A_n$, then Eq. (1.1) can be written as:

$$A^n = \{(a_1, a_2, a_3, \ldots, a_n) | a_i \in A \text{ for all } i = 1, 2, 3, \ldots, n\}.$$

1.9 Power Set

Suppose X is a set. Power set of X, denoted as $\rho(X)$ is said to be a set that contains exactly all its subsets. Let $X = \{p, q, r\}$, then

$$\rho(X) = \{\phi, \{p\}, \{q\}, \{r\}, \{p, q\}, \{p, r\}, \{q, r\}, \{p, q, r\}\}.$$

The cardinality of power set is given by $|\rho(X)| = 2^n$, where n is the number of elements in X.

1.10 The Continuum Hypothesis

Continuum hypothesis has been the most famous problem in set theory that deals with cardinality. Let the cardinality of an infinite countable set is the smallest infinite cardinal number C. The set of all integers is countable, and so is the set of all rational numbers. However, the set of all real numbers is uncountable, and hence its cardinality is greater than the least infinite cardinal. A question arises: is this cardinal (the **continuum**), the next cardinal? In other words, is it the case that there are no cardinals between the countable and the continuum?

Cantor was unable to find any subset of real numbers whose cardinal lies strictly between the countable and the continuum, therefore, he conjectured that the continuum is the next cardinal **(the Continuum Hypothesis)**. Cantor himself spent a lot of time trying to prove the Continuum Hypothesis. Later on, many other mathematicians too, tried to find a proof of it. David Hilbert, at the World Congress of Mathematicians in Paris in 1900, presented a list of major unsolved problems of the time, and the continuum hypothesis was the very first problem on Hilbert's list. Except some efforts by researchers, the problem is still unsolved.

1.11 Axiomatic Set Theory

After Cantor's development, the work on set theory continued without any concern about how exactly sets should be defined. However, Cantor's informal 'definition' was sufficient for proofs in the new concepts. After the discovery of paradoxes, mathematicians came up with the idea to formalize the informal definitions as a system of axioms. A set theory developed with axioms to avoid paradoxes is known as axiomatic set theory. Among many axiomatic constructs, Zermelo's axiomatic set theory is more popular. Zermelo introduced axioms to define properly, all the concepts of set theory. Among other axioms, his axiom of choice became the most discussed construct of set theory.

1.12 The Axiom of Choice

According to this axiom, for every set of mutually disjoint nonempty sets, there exists a set that has exactly one member common with each of these sets. For example, let S be a set whose members are mutually disjoint finite sets of real numbers. Now, choose from each $X \in S$, the smallest number, and thus, form a set A that has exactly one member in common with each $X \in S$. The axiom of choice postulates the existence of a certain set (the choice set) without giving specific instructions on how to construct such a set. Due to this non-constructive nature of this axiom, it went through controversies and discussions.

1.13 Some Paradoxes

Paradox is a logical demonstration, which exhibits flaws in some existing definitions, concepts, and models through contradiction. Some famous paradoxes are discussed here for enhancing students' awareness of such contradictions.

1.13.1 Paradox of Enumeration

Before the development of set theory, the concept of the size of a collection (or set) was ambiguous. Several mathematicians like Galileo and Bolzano discussed it. For instance, are there as many natural numbers as the squares of natural numbers when counted by the method of enumeration? The answer is both, yes and no.

1. The answer is yes because for every natural number n, there is a square number n^2, and likewise the other way around.
2. The answer is no because the set of the squares of natural numbers is a proper subset of the set of natural numbers; since every square number is a natural number but there are natural numbers, like 2, 3, 5 and so on, which are not squares of any natural numbers.

The issue can be settled by defining the size of a set in terms of its cardinality. Two sets are said to have same cardinality if there is a one-to-one correspondence between the elements of both the sets.

1.13.2 Russell's Paradox

Russell recognized that the statement $x = x$ is true for every set, and thus the set of all sets may be defined as $\{x \mid x = x\}$. In 1906, he constructed several paradox sets, the most famous of which is the set of all sets, which do not contain themselves. Russell himself explained this idea by means of some very concrete examples. One such example, known as the Barber paradox, states: The male barber of a town shaves each male of the town who does not shave himself. Since barber also belongs to the same town, therefore, the question arises whether he shaves himself. But, he may shave himself only if he does not shave himself.

In terms of set theory, let us say that there is a set M of males which contain those males of the town who do not shave themselves, and thus, are shaved by the barber. Now, can we say that the barber belongs to the set M? Answer is both, yes and no. The reason is that the set M is not well defined, and therefore, it suggests an improvement in the very definition of the set itself. Thus, a set must be a **well-defined** collection of objects.

1.13.3 Ross–Littlewood Paradox

It is also known as the balls and vase (vessel made of glass and without handles) problem or the ping-pong ball problem. It is a hypothetical problem in abstract mathematics designed to illustrate the non-intuitive nature of infinity. More specifically, the Ross–Littlewood paradox tries to illustrate the fundamental difficulties involved with the execution of a super task, in which an infinite number of tasks are completed sequentially. The problem starts with an empty vase and an infinite number of balls. An infinite number of steps are then performed in sequence, in such a way that at each step, balls are added as well as removed from the vase.

It is assumed that the vase is empty at two minutes before noon, and that the following steps are performed:

1. The first step is performed at one minute before noon.
2. The second step is performed at 30 seconds before noon.
3. The third step is performed at 15 seconds before noon.
4. Each subsequent step n is performed at 2^{1-n} minutes before noon.

This guarantees that an infinite number of steps are performed by noon; since each subsequent step takes half as much time as the previous step.

At each step, ten balls are added to the vase, and one ball is removed from the vase. The question is: How many balls are in the vase at noon? The most intuitive answer seems to be that the vase contains an infinite number of balls by noon; since at every step more balls are being added than removed.

However, suppose that the balls are numbered. At step 1, balls numbering 1 to 10 are inserted into the vase, and ball number 1 is removed. At step 2, balls numbering 11 to 20 are inserted, and ball 2 is removed. Since, infinite number of steps can be performed by noon, hence, every ball labelled n that is inserted into the vase is eventually removed in a subsequent step (namely, at step n). Therefore, the vase is empty at noon.

1.13.4 Skolem's Paradox

Skolem's paradox says that every countable axiomatisation of set theory has a consistent model of first order logic (logic is discussed in Chapters 7 and 8) that contains sets, which are all countable. This appears contradictory because it is possible to prove, from those same axioms, that there exist sets that are not countable. Thus, there seems a contradiction that a model which is countable, and which contains only countable sets, satisfies the first order sentence that intuitively states, "there are uncountable sets".

Many more paradoxes in mathematics arise due to insufficient and faulty formulations of definitions and concepts. Improper use of language sentences plays an important role in creating paradoxes. In most of the cases, terms are not defined properly and a lot is left to rely on intuitionist notion. To get rid of paradoxes, set theory has been recast in formal terms where all properties of sets are built from some axioms. The discussion about some paradoxes is just to make the readers aware of noxious effects of poorly defined concepts.

1.14 Solved Problems

PROBLEM 1.1 If $A = \{\{\ \}, a, b, \{a, b\}\}$ and $B = \{a, b, \{a, b, c\}\}$, then find $A \cup B$ and $A \cap B$.

Solution

$$A \cup B = \{\{\ \}, a, b, \{a, b\}, \{a, b, c\}\}.$$
$$A \cap B = \{a, b\}.$$

PROBLEM 1.2 If $A = \{\phi, a, b\}$, then find $\rho(X)$.

Solution

$$\rho(X) = \{\phi, \{\phi\}, \{a\}, \{b\}, \{\phi, a\}, \{\phi, b\}, \{a, b\}, \{\phi, a, b\}\}.$$

PROBLEM 1.3 If $A = \{a, b, c, \{a, b, c, d\}\}$ and $B = \{a, b, \{a, b, c, d\}\}$, then find $A - B$ and $(A \cap B) - B$.

Solution

$$A - B = \{c\}$$

Also,
$$(A \cap B) - B = \phi.$$

PROBLEM 1.4 If $A = \{\{\ \}, \{a\}, \{a, b\}\}$ and $B = \{a, \{a, b\}\}$, find $A \times B$.

Solution

Here, $A \times B = \{(\{\ \}, a), (\{\ \}, \{a, b\}), (\{a\}, a), (\{a\}, \{a, b\}), (\{a, b\}, a), (\{a, b\}, \{a, b\})\}$.

PROBLEM 1.5 If $A = \{a, b, c\}$ and $B = \{a, b\}$, then find $(A \times B) \cap (B \times A)$ and $(A \times B) - (B \times A)$.

Solution

Here,
$$A \times B = \{(a, a), (a, b), (b, a), (b, b), (c, a), (c, b)\}.$$
$$B \times A = \{(a, a), (a, b), (a, c), (b, a), (b, b), (b, c)\}.$$

Therefore,
$$(A \times B) \cap (B \times A) = \{(a, a), (a, b), (b, a), (b, b)\}.$$
$$(A \times B) - (B \times A) = \{(c, a), (c, b)\}.$$

PROBLEM 1.6 If $A = \{a, b, c\}$, $B = \{d, e\}$, and $C = \{g, h\}$, find $A \times B \times C$.

Solution

Here, $A \times B = \{(a, d), (a, e), (b, d), (b, e), (c, d), (c, e)\}$.

Therefore, $(A \times B) \times C = \{((a, d), g), ((a, e), g), ((b, d), g), ((b, e), g), ((c, d), g), ((c, e), g), ((a, d), h), ((a, e), h), ((b, d), h), ((b, e), h), ((c, d), h), ((c, e), h)\}$.

Hence, $A \times B \times C = \{(a, d, g), (a, e, g), (b, d, g), (b, e, g), (c, d, g), (c, e, g), (a, d, h), (a, e, h), (b, d, h), (b, e, h), (c, d, h), (c, e, h)\}$.

PROBLEM 1.7 If $A = \{a, b, c, \{a, b\}\}$ and $B = \{a, b, \{a, b, c, d\}\}$, then find $(A \cup B) - A$, and $(A \cap B) - B$.

Solution

Here,
$$A \cup B = \{a, b, c, \{a, b\}, \{a, b, c, d\}\},$$
$$A \cap B = \{a, b\}.$$

Therefore,
$$A \cup B - A = \{\{a, b, c, d\}\},$$
$$A \cap B - B = \phi.$$

PROBLEM 1.8 If $A = \{a, b, c\}$ and $B = \{g, h\}$, then find
$$(A \times B) \cup (B \times A) - (A \times B) \cap (B \times A).$$

Solution

Here,
$$A \times B = \{(a, g), (a, h), (b, g), (b, h), (c, g), (c, h)\},$$
$$B \times A = \{(g, a), (g, b), (g, c), (h, a), (h, b), (h, c)\}.$$

Therefore,
$$(A \times B) \cup (B \times A) = \{(a, g), (a, h), (b, g), (b, h), (c, g), (c, h), (g, a), (g, b), (g, c), (h, a), (h, b), (h, c)\}$$
$$(A \times B) \cap (B \times A) = \{\ \}.$$

Hence, $(A \times B) \cup (B \times A) - (A \times B) \cap (B \times A) = \{(a, g), (a, h), (b, g), (b, h), (c, g), (c, h), (g, a), (g, b), (g, c), (h, a), (h, b), (h, c)\}$.

Exercises

1.1 Prove that the set of real numbers is uncountable.

1.2 Find the cardinalities of the following sets:

(i) ϕ
(ii) $\{\phi, \{\phi\}\}$
(iii) $\{\phi, \{a, b, c\}, \{\phi\}, \{\phi, \{\phi\}\}\}$
(iv) $\{a, (a, b), (a, b, c)\}$

1.3 What is the role of set theoretic paradoxes? Discuss in brief.

1.4 Let $A = \{a, b, c, d\}$ and $B = \{a, b, c, d, e\}$. Find $(A \times B) - (B \times A)$.

1.5 Let $A = \{a, b, (c, d)\}$ and $B = \{a, b\}$. Find $(A \times B) \cup (B \times A)$.

1.6 Given $A = \{0, 1\}$ and $B = \{0, 1, a, b\}$. Find $(A \times B) \oplus (B \times A)$.

2

Relations and Functions

2.1 Introduction

In Chapter 1, we have discussed several preliminaries of set theory including Cartesian product.

In this chapter, we will discuss two important concepts of mathematics: functions and relations. Relation is more general as far as constraints are concerned. Relation is defined as a subset of Cartesian product. Function is too defined as the subset of Cartesian product but with restrictions.

Therefore, in some sense function is a special type of relation.

2.2 Relation

A **relation** is a subset of the Cartesian product of the sets. Let A and B be two sets, then a subset of $A \times B$ is called a binary relation, or simply a relation R. Of course, the entire Cartesian product consisting of all ordered pairs of $A \times B$ is a trivial binary relation, and it is called universal relation. Also, ϕ is a subset of $A \times B$, or in general, it is a subset of each Cartesian product, therefore, it is also trivial. It is termed an empty relation. The study of binary relations is of interest in set theory, although there could be any n-ary relation defined over the Cartesian product of n number of sets. Such n-ary relations are subsets of Cartesian products consisting of n-tuples.

A popular notation used for relation is $a\mathcal{R}b$, which denotes $(a, b) \in \mathcal{R}$.

EXAMPLE 2.1 Given, sets $A = \{1, 2, 3, 4\}$ and $B = \{3, 5\}$. Then $R = \{(a, b)|\ a \leq b\}$ is a relation, such that

$$R = \{(1, 3), (1, 5), (2, 3), (2, 5), (3, 3), (3, 5), (4, 5)\}.$$

EXAMPLE 2.2 The relation defined by the equation of the circle $x^2 + y^2 = 16$ is given by $C = \{(x, y) \mid x \text{ and } y \text{ satisfy } x^2 + y^2 = 16\}$.

2.3 Inverse Relation

Let A and B be two sets, and R be a relation from A to B. Then R^{-1} is called the inverse relation of R, and it is the set of those ordered pairs which, when reversed, belong to R.

Formally,

$$R^{-1} = \{(b, a) \mid (a, b) \in R\}.$$

For example, let $A = \{1, 2, 3, 4, 5\}$ and $B = \{3, 5\}$. Also, let $R = \{(1, 3), (1, 5), (2, 3)\}$, then R^{-1} will be $\{(3, 1), (5, 1), (3, 2)\}$.

2.4 Composition of Relations

Suppose R and S are two relations defined over sets A, B and C in such a way that,

$$R \subseteq A \times B \text{ and } S \subseteq B \times C.$$

Then $R \circ S$ gives rise to a relation from A to C, and is defined by

$$R \circ S = \{(a, c) \mid (a, b) \in R \text{ and } (b, c) \in S \text{ for some } b \in B\}.$$

The relation $R \circ S$ is called the composition of R and S.

EXAMPLE 2.3 Let $A = \{1, 2, 3, 4, 5, 6\}$, $B = \{a, b, c, d, e\}$ and $C = \{x, y, z\}$. Consider relations $R = \{(1, b), (2, c), (3, d), (6, e)\}$ and $S = \{(a, x), (b, y), (c, z)\}$ defined on $A \times B$ and $B \times C$, respectively. Then using definition of composition given in Section 2.4, we get

$$R \circ S = \{(1, y), (2, z)\}.$$

The pictorial representation of this relation is given in Figure 2.1(a) and (b).

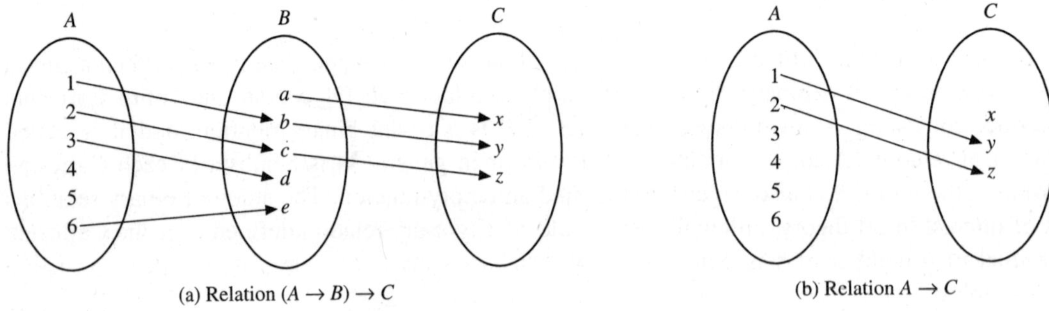

(a) Relation $(A \to B) \to C$

(b) Relation $A \to C$

Figure 2.1 Composition of functions.

Observe the complete paths from set A to set C in Figure 2.1(a). There are only two complete paths, $1 \to b \to y$ and $2 \to c \to z$, and thus, $R \circ S$ has two ordered pairs corresponding to these paths: $(1, y)$ and $(2, z)$.

2.5 Types of Relations

Some relations having special properties are discussed here. These relations are defined on a single set (say A); i.e., $R \subseteq A \times A$.

2.5.1 Reflexive

Let R be a relation on a set A. Then R is reflexive if aRa for every $a \in A$; i.e. $(a, a) \in R$ for every $a \in A$.

EXAMPLE 2.4 Consider the following relations defined on $A = \{1, 2, 4, 7\}$:

1. $R_1 = \{(1, 2), (2, 4), (4, 7)\}$
2. $R_2 = \{(1, 4), (2, 4), (4, 7)\}$
3. $R_3 = \{(1, 1), (1, 2), (2, 2), (4, 4), (7, 7), (2, 4), (4, 7)\}$
4. R_4 = Universal relation ($A \times A$)
5. R_5 = Empty relation(ϕ)
6. $R_6 = \{(1, 1), (1, 2), (2, 2), (7, 7), (2, 4), (4, 7)\}$

Only R_3 and R_4 are reflexive.

2.5.2 Symmetric

A relation R on a set A is said to be **symmetric**, if whenever $(a, b) \in R$, then $(b, a) \in R$. Hence, if R is not symmetric, then there exists $a, b \in A$ such that $(a, b) \in R$ but $(b, a) \notin R$. A relation R is said to be **antisymmetric** if whenever $(a, b) \in R$ and $(b, a) \in R$, then $a = b$. Moreover, if a relation is not symmetric, then it does not mean that it is antisymmetric. Sometimes, a relation may be both symmetric and antisymmetric.

EXAMPLE 2.5 Consider the following relations defined on $A = \{1, 2, 3, 4, 5, 7\}$:

1. $R_1 = \{(1, 1), (2, 4), (4, 7)\}$
2. $R_2 = \{(1, 4), (4, 1), (4, 7), (7, 4)\}$
3. $R_3 = \{(1, 1), (1, 2), (2, 2), (4, 4), (7, 7), (2, 4), (4, 7), (7, 4)\}$
4. $R_4 = \{(1, 1), (2, 2), (3, 3), (4, 4), (5, 5), (7, 7)\}$
5. R_5 = Universal relation ($A \times A$)

Relation R_1 is antisymmetric and R_2 is symmetric. Further, relation R_3 is neither symmetric nor antisymmetric, but R_4 is both symmetric and antisymmetric. R_5 is symmetric but not antisymmetric.

2.5.3 Transitive

A relation R on a set A is said to be transitive if whenever $(a, b) \in R$, $(b, c) \in R$, then $(a, c) \in R$ for some $a, b, c \in A$.

EXAMPLE 2.6 Consider a set $A = \{1, 2, 3, 4, 7, 8, 9\}$. The relations defined on it are:

1. $R_1 = \{(1, 1), (2, 2), (2, 7)\}$
2. $R_2 = \{(1, 4), (4, 1), (4, 7), (7, 4)\}$
3. $R_3 = \{(1, 1), (1, 2), (2, 3), (1, 3), (3, 4), (1, 4), (2, 4)\}$
4. R_4 = Universal relation ($A \times A$)

All relations except R_2 are transitive.

2.5.4 Equivalence Relation

A relation R on a set A is said to be equivalence if R is reflexive, symmetric and transitive. Formally, R is an equivalence relation on A, if the following properties are satisfied:

1. $\forall\, a \in A, (a, a) \in R$.
2. If $(a, b) \in R$, then $(b, a) \in R$.
3. If $(a, b) \in R$ and $(b, c) \in R$, then $(a, c) \in R$.

EXAMPLE 2.7

1. Relation of equality, i.e., '$a = a$', is an equivalence relation.
2. Relation, set inclusion ($\subset$) is not equivalence, since it is not symmetric.

Remark: An equivalence relation R on a set A, partitions the set into disjoint equivalence classes. The collection of all equivalence classes of elements of A under an equivalence relation R is denoted by A/R, which is given by

$$A/R = \{[a] \mid a \in A\}, \text{ where } [a] = \{x \mid (a, x) \in R\}.$$

2.6 Functions

Suppose A and B are two sets. Then, a function f from A to B, denoted by $f : A \to B$ is a subset of Cartesian product $A \times B$, such that each $a \in A$ appears exactly once as the first component of an ordered pair in f. In other words, for each $a \in A$, there is a unique $b \in B$ such that $(a, b) \in f$. In fact, function is a relation from A to B such that each element of A is associated with exactly one element of B. In $(a, b) \in f$, element b is called the **image** of element a under the function f, whereas, a is a pre-image of b. Set A is called the **domain,** and set B, the **co-domain** of the function f. The set of elements, $y \in B$ for each $(a, y) \in f$, is called the **range** of the function f. Let R be the range of f, then $R \subseteq B$.

Sometimes, the **domain** and the **co-domain** of a function may be identical. For instance, $f : A \to A$.

EXAMPLE 2.8 Let $A = \{3, 4, 5\}$ and $B = \{a, b, c\}$.

Then $A \times B = \{(3, a), (3, b), (3, c), (4, a), (4, b), (4, c), (5, a), (5, b), (5, c)\}$.

Further, let f_1 be the subset of $A \times B$ such that $f_1 = \{(3, a), (4, a), (5, b)\}$.

Then f_1 is called a function from A to B, i.e., $f_1 : A \to B$.

Here, **domain** of $f_1 = A$,

co-domain of $f_1 = B$,

and **range** of $f_1 = \{a, b\}$.

EXAMPLE 2.9 The following are real valued functions, i.e., $f : \mathbb{R} \rightarrow \mathbb{R}$.

1. $f : \mathbb{R} \rightarrow \mathbb{R}$ such that $f(x) = x^2$
2. $f : \mathbb{R} \rightarrow \mathbb{R}$ such that $f(x) = 1 + x + x^2$

A function is also called a **mapping**. But, the word mapping is more appropriately used when the elements of domain and co-domain are non-numeric. A pictorial representation of a function $f : A \rightarrow B$ is given in Figure 2.2. Here $A = \{-2, 2, 3, 5\}$, and $B = \{4, 9, 25\}$.

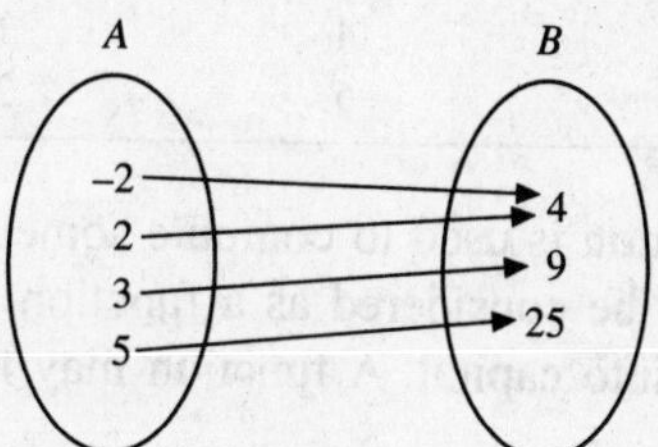

Figure 2.2 Function $f: A \rightarrow B$.

Another notation for the function is

$$y = f(x).$$

Here, x and y may be some quantities. Quantity x is called the **argument**, and y, the **value** of the function. A function assigns exactly one value to each argument of a specified type. The variable used for argument of a function is called **independent variable**, and the variable representing the value of the function is called **dependent variable**. For example, let $y = f(x)$ be a function with the set of real numbers as both its domain and co-domain. Then, x is called an independent variable, and y is called a dependent variable.

There are different ways to represent a function. Some of these ways are: a formula, an algorithm that computes it, a graph, and a table. The table, listing arguments versus function values is a common way to represent a function. The functions with numerical values may be operated (added, multiplied, etc.), yielding new functions. An important operation on functions is the **composition** of functions. The collections of functions with certain properties (like continuity, differentiability, etc.) are called **function spaces**. As discussed, the usual notation used for the function is $y = f(x)$. However, this is just a convention and not a rule. In fact, we may take any letters in place of f, x and y depending on the context of the function. For instance, $u = s(t)$. We may also visualize a function (Figure 2.3) as a computing device that takes an input x and returns an output $f(x)$.

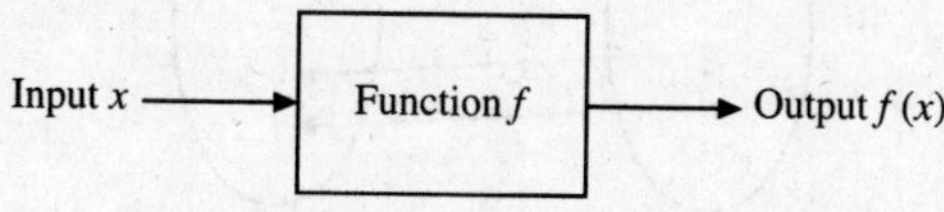

Figure 2.3 Function.

Here, input x may be from a set of values, which is domain of the function. The set of output values $f(x)$ is the image set, or range of the function. For instance, if the domain of the function $f(x) = x^2$ is the set of all real numbers, then its **range** will be the set of all positive real numbers. Here, the **co-domain** may be the set of all real numbers. This type of function is called real-valued function. Some values of this function are listed in Table 2.1.

Table 2.1 Function $f(x) = x^2$

S. No.	x	$f(x) = x^2$
1	1	1
2	2	4
3	3	9
4	4	16
5	5	25

In computer science, a function is used to compute some value, or process some data. For example, v = uppercase (w), may be considered as a function that takes an English word w as input and converts all its letters into capital. A function may have more than one independent variable.

For example, add $(x, y, z) = x + y + z$.

Here, **add** is a function that takes three arguments and returns their sum.

Sometimes, a function is written in a compact form that may contain the whole information.

For example, $((x, x^2) \mid x \in \mathbb{R})$ means $f(x) = x^2$, and x belongs to the set of real numbers.

2.7 Types of Functions

There are following three types of functions:

One-to-one function

A function $f: A \to B$ is said to be one-to-one if the elements of set A have distinct images in set B. Formally, f is one-to-one if

$$f(a) = f(b) \Rightarrow a = b.$$

For example, the function f given by Figure 2.4 is one-to-one.

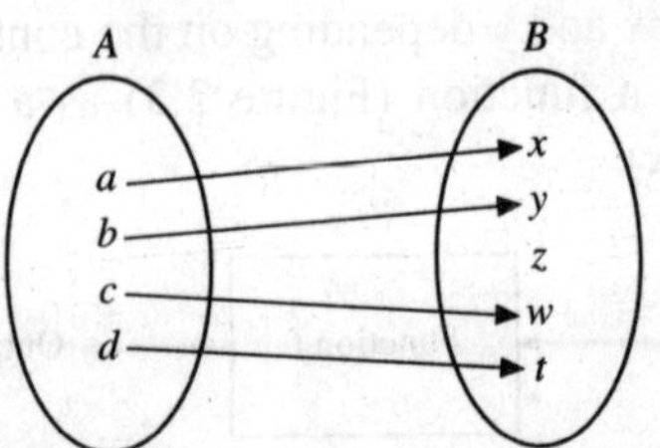

Figure 2.4 One-to-one function.

Onto function

A function $f : A \to B$ is said to be an onto function if each element of set B is the image of some elements in set A. In fact, if a function is onto, then its image is the entire co-domain. If a function is not onto, then it is called into function.

For example, the function f given by Figure 2.5 is onto.

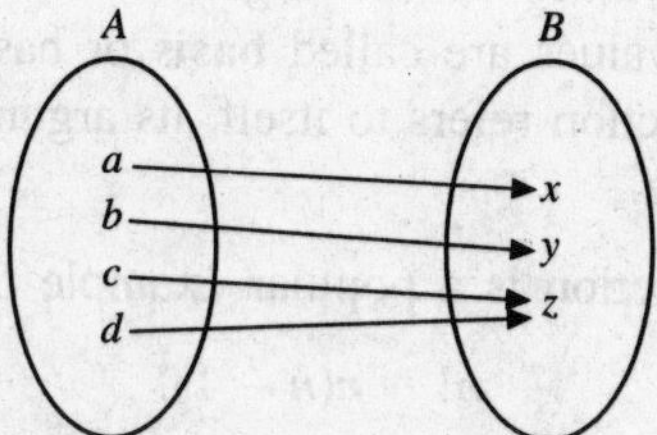

Figure 2.5 Onto function.

One-to-one and onto functions

The functions, which are both, one-to-one and onto are called one-to-one and onto functions. For instance, function $f(x) = x$ is one-to-one and onto. A function, which is one-to-one and onto is also **invertible**.

Formally, a function $f : A \to B$ is said to be **invertible** if its inverse f^{-1} is a function from B to A, i.e., $f^{-1} : B \to A$.

For example, the function f given by Figure 2.6 is one-to-one and onto.

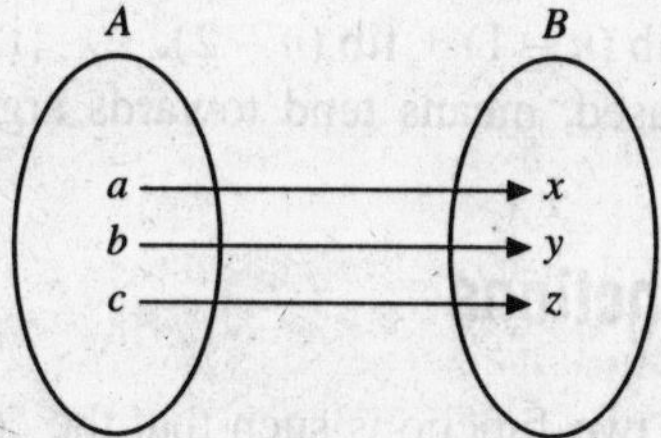

Figure 2.6 One-to-one and onto function.

EXAMPLE 2.10 Consider the diagrams given in Figure 2.7. These diagrams represent relations between x and y. Note that these relations are not functions.

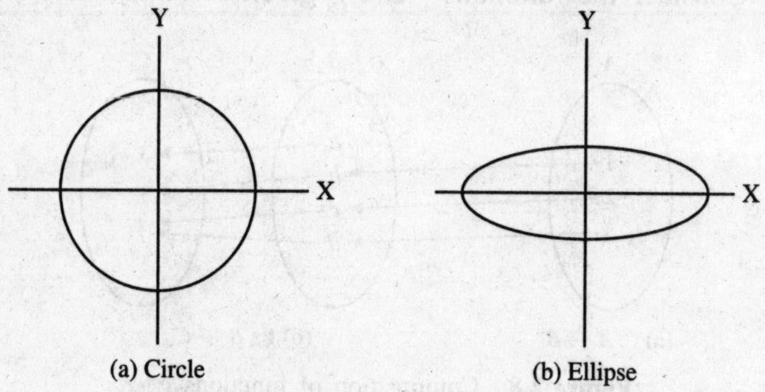

Figure 2.7 Relations but not functions.

2.8 Recursive Function

There are some special types of functions which are defined recursively. A function is said to be recursive if it is defined in terms of itself. In fact, recursive function must be **well-defined**. There are two conditions for a recursive function to be well-defined. These are:

1. There must be some values of the argument for which the function is defined non-recursively. These values are called **basis** or base values of the function.
2. Each time when the function refers to itself, its arguments must become closure to the arguments of base value.

EXAMPLE 2.11 Factorial function is a popular example of recursive function.

$$n! = n(n-1)!$$

It is defined as:

1. If $n = 0$, then $n! = 1$. {i.e., basis value is $0! = 1$, with argument $n = 0$}
2. If $n > 0$, then $n! = n(n-1)!$. {i.e., each time function is referred, its argument n is decreased by 1; means its value tends towards $n = 0$, which is the argument of base value.}

This function is well defined, since it satisfies both the conditions.

EXAMPLE 2.12 Fibonacci sequence is another example of a recursive function. It is defined as:

1. If $n = 0$, or $n = 1$, then fib $(n) = n$. {i.e., fib(0) = 0 and fib(1) = 1. These are the base values.}
2. If $n > 1$, then fib (n) = fib $(n - 1)$ + fib $(n - 2)$. {i.e., each time function is referred, its arguments are decreased, means tend towards arguments of base values.}

2.9 Composition of Functions

Let $f: A \to B$ and $g: B \to C$ be two functions such that the co-domain of f be the domain of g. Then a function $(g \circ f)(a)$ defined from A to C is called the composition of functions.

Formally, $(g \circ f)(a) = g(f(a))$.

The condition for $g \circ f$ to exist is that the range of f should be a subset of domain of g.

EXAMPLE 2.13 Consider the functions f and g given in Figure 2.8.

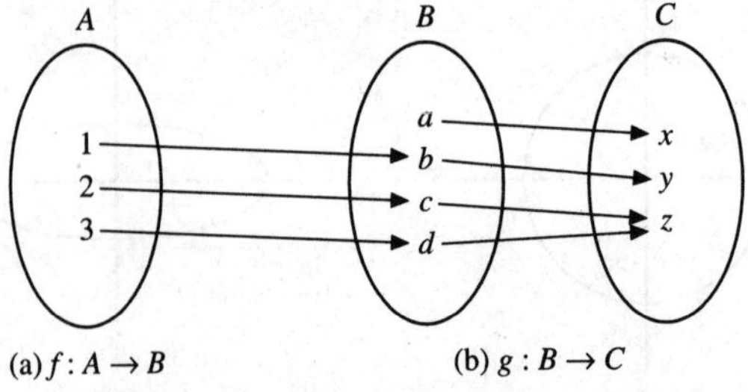

Figure 2.8 Composition of functions $g \circ f$.

Then the composition of g and f will be given by

$$g \circ f = \{(1, y), (2, z), (3, z)\}.$$

2.10 Pigeonhole Principle

Pigeonhole principle is an obvious idea, but sometimes, it becomes very useful in solving the problems. According to this principle, "if m pigeons occupy n pigeonholes and $m > n$, then at least two pigeons roost in the same pigeonhole."

EXAMPLE 2.14 Consider a set of positive integers X such that $|X| = 29$. With the help of pigeonhole principle, we can establish that X contains at least two elements, which have the same remainder on division by 28.

By Division Algorithm, we know that when an integer n, positive or negative, is divided by another positive integer m, there exists a unique quotient q, and a unique remainder r, such that

$$n = mq + r.$$

For $m = 28$, we get

$$n = 28q + r.$$

Obviously, $0 \le r < 28$. Here, the number of pigeons may be taken as 29 (the given numbers), and the possible pigeonholes 28.

EXAMPLE 2.15 Suppose, you have ten pairs of socks kept in a bag. Each pair of socks is of different colour. The problem is, 'how many socks do you have to draw randomly from the bag to get at least one matched pair?'. The answer is 11. This problem may be interpreted as having 10 pairs of pigeons, i.e., 20 pigeons, and 10 pigeonholes. If each of ten pigeons go to different pigeonholes (at worst, these may be ten socks each from different pairs), and the remaining ten pigeons all share one pigeonhole, already occupied by one, the total number of pigeons in that pigeonhole becomes 11. This ensures that there will be at least one matched pair.

2.11 Solved Problems

PROBLEM 2.1 Find relation $R = \{(x, y) | \ x \ge y\}$, defined on the sets $X = \{3, 5\}$ and $Y = \{1, 2, 3, 4\}$.

Solution

Given

$$X = \{3, 5\}$$

and

$$Y = \{1, 2, 3, 4\}.$$

Then Cartesian product,

$$X \times Y = \{(x, y) | \ x \in X \text{ and } y \in Y\}.$$

Therefore, relation $R = \{(3, 1), (5, 1), (3, 2), (5, 2), (3, 3), (5, 3), (5, 4)\}$.

PROBLEM 2.2 What is the relation defined by the equation $x^3 + y^3 = 8$?

Solution

The relation defined by the equation $x^3 + y^3 = 8$ is given by

$$C = \{(x, y) \mid x \text{ and } y \text{ satisfy } x^3 + y^3 = 8\}.$$

PROBLEM 2.3 Discuss the functions given in Figure 2.9. The functions are from $\mathbb{R}$ to $\mathbb{R}$.

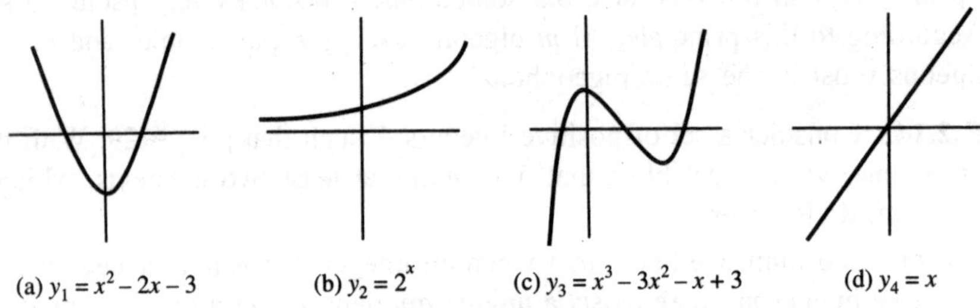

Figure 2.9 Functions.

Solution

(i) $y_1 = x^2 - 2x - 3$

Function y_1 is neither one-to-one nor onto.

(ii) $y_2 = 2^x$

Function y_2 is one-to-one, but not onto.

(iii) $y_3 = x^3 - 3x^2 - x + 3$

Function y_3 is onto, but not one-to-one.

(iv) $y_4 = x$

Function y_4 is both, one-to-one and onto.

Note: Here $y_4 = x$ is the only function which is invertible.

PROBLEM 2.4 Prove that the function $f: \mathbb{R} \to \mathbb{R}$, given by $f(x) = 5x + 8$ is one-to-one.

Solution

Let $x_1, x_2 \in f$, then

$$f(x_1) = 5x_1 + 8,$$
$$f(x_2) = 5x_2 + 8.$$

Therefore,

$$f(x_1) = f(x_2).$$

$\Rightarrow$ $$5x_1 + 8 = 5x_2 + 8,$$

$\Rightarrow$ $$x_1 = x_2.$$

Hence, function is one-to-one.

PROBLEM 2.5 Prove that the function $f : \mathbb{R} \to \mathbb{R}$, given by $f(x) = x^3$ is onto.

Solution

The range of f is $\mathbb{R}$. Let r be an arbitrary real number in the co-domain of f, then we can always find a real number $r^{1/3}$ in the domain of f, such that

$$f(r^{1/3}) = r.$$

Therefore, the co-domain of f is equal to the range of f. Thus, f is onto.

PROBLEM 2.6 Prove that the function $f : \mathbb{Z} \to \mathbb{Z}$, given by $f(x) = 2x + 1$ is not onto. Here $\mathbb{Z}$ is the set of integers.

Solution

The range of $f = \{\ldots, -7, -5, -3, -1, 1, 3, 5, \ldots\}$. Obviously, range of f is not equal to the co-domain of f. Therefore, f is not onto. However, f is one-to-one.

PROBLEM 2.7 Two functions f and g are defined as follows:

$$f : \mathbb{R} \to \mathbb{R}, \text{ such that } f(x) = x^2,$$

$$g : \mathbb{R} \to \mathbb{R}, \text{ such that } g(x) = x + 7.$$

Find $g \circ f$ and $f \circ g$.

Solution

(i) $$(g \circ f)(x) = g(f(x)) = g(x^2) = x^2 + 7$$

(ii) $$(f \circ g)(x) = f(g(x)) = f(x + 7) = (x + 7)^2 = x^2 + 14x + 49$$

PROBLEM 2.8 Consider $f : \mathbb{R} \to \mathbb{R}$, given by $f = \{(x, y) | \; y = mx + c\}$. Find $f^{-1}(x)$.

Solution

Given

$$y = mx + c$$

$$\Rightarrow \quad x = \frac{y - c}{m}$$

$$\Rightarrow \quad f^{-1}(y) = \frac{y - c}{m}. \qquad \text{(i)}$$

Now, for $f^{-1}(x)$, replace y by x in Eq. (i)

Therefore,

$$f^{-1}(x) = \frac{x - c}{m} = \frac{1}{m}(x - c).$$

Exercises

2.1 If $A = \{1, 3, 4, 5\}$ and $B = \{3, 6\}$, find $a\mathcal{R}b$ with $a \in A$ and $b \in B$, such that

(i) $a \leq b$

(ii) $a \geq b$

(iii) $a^2 \leq b$

(iv) $\sqrt{a} > b$

2.2 Determine whether each of the following relations is a function.

(i) $\{(x, y)|\ x, y \in \mathbb{R}, y^2 = x\}$, relation defined from $\mathbb{R}$ to $\mathbb{R}$.

(ii) $\{(x, y)|\ x, y \in \mathbb{Z}, y = x^2 + 6\}$, relation defined from $\mathbb{Z}$ to $\mathbb{Z}$.

(iii) $\{(x, y)|\ x, y \in \mathbb{R}, y = 2x + 3\}$, relation defined from $\mathbb{R}$ to $\mathbb{R}$.

(iv) $\{(x, y)|\ x, y \in \mathbb{Q}, x^2 + y^2 = 4\}$, relation defined from $\mathbb{Q}$ to $\mathbb{Q}$.

2.3 Give an example of finite sets A and B with $|A|, |B| \geq 5$ and a function $f: A \rightarrow B$, such that

(i) f is one-to-one but not onto.

(ii) f is onto but not one-to-one.

(iii) f is one-to-one and onto.

(iv) f is neither one-to-one nor onto.

2.4 Determine whether each of the following diagrams in Figure 2.10 defines a function from $A = \{1, 2, 4, 6\}$ to $B = \{x, y, z\}$. State the reason.

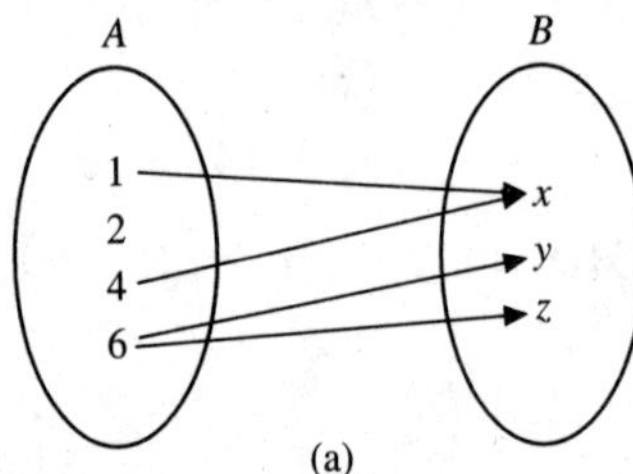

(a)

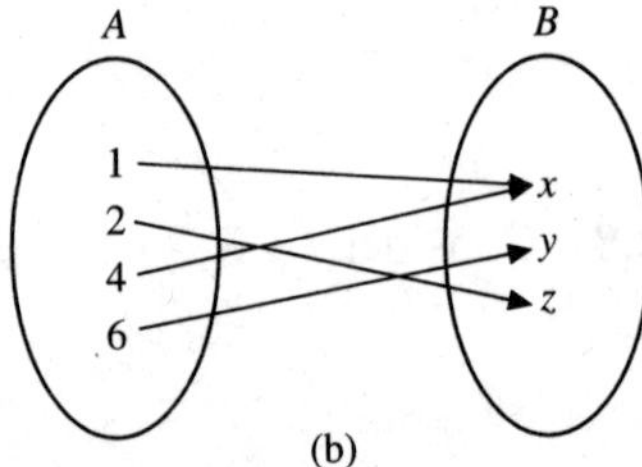

(b)

Figure 2.10

2.5 Prove that the relation given by $a \equiv b(\text{mod } n)$, where $a, b \in \mathbb{I}$, and $n \in \mathrm{N}$, is equivalence. Here $\mathbb{I}$ is the set of integers and N, the set of natural numbers.

2.6 Find the equivalence classes of the relation $a \equiv b(\text{mod } 5)$, where $a, b \in \mathbb{I}$, the set of integers.

2.7 Two functions f and g are defined as follows:

$f: \mathbb{R} \rightarrow \mathbb{R}$ such that $f(x) = x^3$,

$g: \mathbb{R} \rightarrow \mathbb{R}$ such that $g(x) = x^2 + 3x + 7$. Find $f \circ g$ and $g \circ f$.

2.8 Let $f: \mathbb{R} \rightarrow \mathbb{R}$ be given by $f = \{(x, y)|\ y = 3x + 7\}$. Find $f^{-1}(x)$.

3

Semigroup

3.1 Introduction

The semigroup is an important algebraic structure. Its formal study began in the early 20th century. The theory of finite semigroups has been of particular importance in theoretical computer science because of its natural link with finite automata and language theory.

3.2 Definition of Semigroup

A semigroup is a set S, together with a binary operation '$*$' that satisfies the following axioms:

1. **Closure**

   For all $a, b \in S$

$$a * b \in S.$$

2. **Associative**

   For all $a, b, c \in S$

$$(a * b) * c = a * (b * c).$$

A semigroup is denoted by $< S, * >$, where S is a set and $*$ a binary operation. The Cayley table represents binary operation of an algebraic structure. It was introduced by Cayley in 1854 through his paper on group theory.

Let $< S, * >$ be a semigroup such that $S = \{1, -1, i, -i\}$ and $*$ be multiplication operation. Then the Cayley table for this semigroup is given by Table 3.1.

Table 3.1 Cayley Table for Semigroup S {1, –1, i, –i}

*	1	–1	i	$-i$
1	1	–1	i	$-i$
–1	–1	1	$-i$	i
i	i	$-i$	–1	1
$-i$	$-i$	i	1	–1

Following are some of the examples of semigroup. These examples will help the readers understand the semigroup structure well.

EXAMPLE 3.1 Semigroup with one element: Let $S = \{a\}$ be a set. Then, the algebraic structure $< \{a\}, * >$ with operation defined as:

$$a * a = a$$

is a semigroup. A set with natural number 1 and operation as multiplication is a semigroup.

EXAMPLE 3.2 Empty semigroup: Let $S = \phi$ be an empty set. Then $< \varphi, * >$ forms a semigroup with operation defined as:

$$\varphi * \varphi = \varphi.$$

EXAMPLE 3.3 Semigroup of positive integers: The set of positive integers with operation addition, i.e., $<\mathbb{Z}^+, +>$ is a semigroup. Similarly, the set of positive integers with operation multiplication, i.e., $< \mathbb{Z}^+, * >$ is also a semigroup.

EXAMPLE 3.4 Semigroup of matrices: Let $S = \phi$ be the set of all square matrices of order n. Then it will form a semigroup with respect to the operation, matrix addition. It will also form a semigroup if the operation is matrix multiplication.

EXAMPLE 3.5 Semigroup of strings: A string is an ordered sequence of symbols. The set of all finite strings over a fixed alphabet Σ with operation defined as **concatenation** forms a semigroup. It is called **free semigroup** over Σ. Operation **concatenation**, or simply **catenation** is defined as

$$\alpha \circ \beta = \alpha\beta.$$

For instance, if $\alpha = ab$, and $\beta = cd$, then

$$\alpha \circ \beta = abcd.$$

Further, let $\alpha = ab^3$, and $\beta = bc^2d$, then

$$\alpha \circ \beta = ab^4c^2d.$$

Note that in strings

$$(ab)^n \neq a^nb^n,$$

since $(ab)^n = abab \ldots ab$, n times, but

$$a^nb^n = aaa \ldots abbb \ldots b,$$

i.e., n times a followed by n times b.

EXAMPLE 3.6 Consider a set $S = \{m, n, o, p\}$. Two operations $*$ and Δ are defined by two Cayley tables, Table 3.2(a) and (b), respectively.

Table 3.2 Cayley Tables for Set S $\{m, n, o, p\}$

(a) Cayley Table for $*$

$*$	m	n	o	p
m	m	m	m	m
n	n	n	n	n
o	o	o	o	o
p	p	p	p	p

(b) Cayley Table for Δ

Δ	m	n	o	p
m	m	n	o	p
n	n	m	m	n
o	o	n	m	m
p	p	m	m	m

Note that operation $*$ may be defined as

$$a * b = a \text{ for any } a, b \text{ in } S$$

Therefore,

$$(m * n) * o = m * o = m$$

and

$$m * (n * o) = m * n = m.$$

Hence,

$$(m * n) * o = m * (n * o).$$

Therefore, $< S, * >$ is a semigroup.

Further,

$$(n \,\Delta\, o) \,\Delta\, o = m \,\Delta\, o = o$$

and

$$n \,\Delta\, (o \,\Delta\, o) = n \,\Delta\, m = n.$$

Hence, $< S, \Delta >$ is not a semigroup.

3.3 Subsemigroup and Ideal

Let $< S, * >$ be a semigroup, and T be the subset of S. Then $< T, * >$ will be called a subsemigroup, if it is a semigroup itself.
Formally,
Let $< S, * >$ be a semigroup and $T \subseteq S$, then T will be called a subsemigroup, if

1. for all $a, b \in T$, $a * b \in T$,
2. for all $a, b, c \in T$, $(a * b) * c = a * (b * c)$.

Again, let $< S, * >$ be a semigroup and $A, B \subseteq S$. Also, let operation '$*$' induced over subsets be defined as:

$$A * B = \{a * b \mid a \in A \text{ and } b \in B\}.$$

Then, with respect to this operation, a subset A is called

1. a **left ideal** of $< S, * >$ if $S * A \subseteq A$, and
2. a **right ideal** if $A * S \subseteq A$.

If A is both a left ideal and a right ideal, then it is called an **ideal** (or a **two-sided ideal**).

EXAMPLE 3.7 Let < F, ∘ > be a free semigroup defined over alphabet $\Sigma = \{a, b\}$. Also, let E be the set of strings of length even (i.e. $|\alpha|$ is even, $\alpha \in E$ is any string over alphabet Σ).

Then, < E, ∘ > will be a free subsemigroup of < F, ∘ >.

EXAMPLE 3.8 Consider a semigroup $S = \{m, n, o, p, q\}$ with the binary operation defined by the following Cayley table (Table 3.3):

Table 3.3 Cayley Table for the Given Semigroup

$*$	m	n	o	p	q
m	m	m	m	p	p
n	m	n	o	p	p
o	m	o	n	p	p
p	p	p	p	m	m
q	p	q	q	m	m

Let $I = \{m, p\} \subseteq S$. Then,

$S * I = \{m * m, n * m, o * m, p * m, q * m, m * p, n * p, o * p, p * p, q * p\} = \{m, p\} \subseteq I,$

$I * S = \{m * m, p * m, m * n, p * n, m * o, p * o, m * p, p * p, m * q, p * q\} = \{m, p\} \subseteq I.$

Therefore, set I is an ideal of S.

EXAMPLE 3.9 Let $\mathbb{N}$ be the set of natural numbers and < $\mathbb{N}$, $*$ > be a semigroup with operation multiplication, then < $\mathbb{E}$, $*$ > will be an ideal of $\mathbb{N}$; where $\mathbb{E}$ is the set of all even numbers.

3.4 Semigroup Homomorphism

Let < S, $*$ > and < T, Δ > be any two semigroups. A mapping $f : S \to T$, such that, for all $a, b \in S$, $f(a * b) = f(a)\ \Delta\ f(b)$, is called semigroup homomorphism.

A semigroup homomorphism is a function that preserves semigroup structure. Further, if mapping f is also one-to-one and onto, then it is called isomorphism.

In other words, two semigroups S and T are said to be isomorphic, if f: $S \to T$ is a one-to-one and onto mapping. Symbolically, it is written as $S \approx T$.

A semigroup isomorphism onto itself is called semigroup automorphism.

EXAMPLE 3.10 Consider a free semigroup < F, ∘ > defined over an alphabet Σ. Also, let < $\mathbb{Z}$, + > be the semigroup of set of integers with respect to operation addition.

Then a mapping, $f : F \to \mathbb{Z}$ defined by $f(w) = |w|$, where $|w|$ is the length of w, is a semigroup homomorphism. For instance,

$$|u \circ v| = |u| + |v|.$$

However, it will not be an isomorphism.

EXAMPLE 3.11 Let $< \mathbb{Z}, +_4 >$ be a semigroup defined over integers with respect to addition modulo 4. Also, consider a semigroup of set $A = \{1, 3, 7, 9\}$ over the operation multiplication modulo 10 (i.e., $\times_{10}$).

Then a mapping $f : \mathbb{Z} \rightarrow A$ defined by

$$f(0) = 1, f(1) = 3, f(2) = 9 \text{ and } f(3) = 7$$

will be a semigroup homomorphism. Cayley tables for the two operations are given in Tables 3.4 and 3.5.

Table 3.4 Cayley Table of Addition Modulo 4

$+_4$	0	1	2	3
0	0	1	2	3
1	1	2	3	0
2	2	3	0	1
3	3	0	1	2

Table 3.5 Cayley Table of Multiplication Modulo 10

$\times_{10}$	1	3	7	9
1	1	3	7	9
3	3	9	1	7
7	7	1	9	3
9	9	7	3	1

Remark: Let a, b, r and n (> 0) belong to the set of integers, $\mathbb{Z}$. Then operations addition and multiplication modulo are defined as:

1. **Addition modulo over integers $\mathbb{Z}$:** It is defined as:

$$a +_n b = r$$

i.e., if $(a + b)$ is divided by n, it leaves remainder r.

2. **Multiplication modulo over integers $\mathbb{Z}$:** It is defined as:

$$a \times_n b = r$$

i.e., if $(a \times b)$ is divided by n, it leaves remainder r.

Now, if $f: \mathbb{Z} \rightarrow A$ is a semigroup homomorphism, then it should satisfy the homomorphism property, i.e., $f(a * b) = f(a) \Delta f(b)$.

Let us consider

$$f(2 +_4 3) = f(2) \times_{10} f(3).$$

$$\Rightarrow \quad f(1) = 9 \times_{10} 7.$$

$$\Rightarrow \quad f(1) = 3.$$

Therefore, it is valid. In fact, readers may try for other numbers too. Further, since this mapping is one-to-one and onto, hence it is isomorphism.

THEOREM 3.1 Let, $< S, * >$, $< T, \Delta >$ and $< U, \odot >$ be three semigroups, and $g: S \to T$ and $h: T \to U$ be two semigroup homomorphisms. Then $(h \circ g): S \to U$ is a semigroup homomorphism from $< S, * >$ to $< U, \odot >$.

Proof: Let $a, b \in S$.

Consider,
$$\begin{aligned}(h \circ g)(a * b) &= h(g(a * b)) \\ &= h(g(a) \Delta g(b)) \\ &= h(g(a)) \odot h(g(b)) \\ &= (h \circ g)(a) \odot (h \circ g)(b).\end{aligned}$$

Therefore, $(h \circ g)$ is a homomorphism from S to U.

3.5 Quotient Semigroup and Congruence Relation

Let $< S, * >$ be a semigroup and R be an equivalence relation defined on S. Since an equivalence relation partitions the set into disjoint equivalence classes, therefore, R will partition S. The collection of all equivalence classes is denoted by S/R.

Further, define an operation on the equivalence classes by

$$[a] * [b] = [a * b]$$

or

$$[a][b] = [ab].$$

Since this operation is associative, therefore, S/R will form a semigroup with respect to operation '$*$'. It is called quotient semigroup.

In addition, let the equivalence relation R on S has the following property:

$$(a, a'), (b, b') \in R \Rightarrow (ab, a'b') \in R, \text{ where } a, a', b, \text{ and } b' \in S.$$

Then R is called the **congruence relation** on S.

EXAMPLE 3.12 Let $< F, \circ >$ be a free semigroup defined over an alphabet Σ. Also, let $u, u', v, v' \in F$. Then a relation R on F given by ${}_uR_{u'}$, if $| u | = | u' |$ is an equivalence relation.

Further, let, $(u, u'), (v, v') \in R$, such that

$$| u | = | u' | = m$$

and

$$| v | = | v' | = n.$$

Then

$$| u \circ v | = | u' \circ v' | = m + n$$

and therefore, $(u \circ v, u' \circ v') \in R$. Thus, R is a congruence relation on F.

THEOREM 3.2 Fundamental theorem of semigroup homomorphism

Let, $f: S \to S$ be a semigroup homomorphism. Let $(a, b) \in R$ iff

$$f(a) = f(b).$$

Then,

(i) R is a congruence relation on S, and
(ii) S/R is isomorphic to $f(S)$.

Proof:

(i) Let us show that R is congruence relation on S.

Since, $(a, b) \in R$, iff $f(a) = f(b)$. (given)

Therefore, $\forall\, a, f(a) = f(a)$.

$\Rightarrow$ $(a, a) \in R$.

$\Rightarrow$ R is reflexive.

If $f(a) = f(b)$ then

$f(b) = f(a)$.

$\Rightarrow$ if $(a, b) \in R$, then $(b, a) \in R$.

$\Rightarrow$ R is symmetric.

If $f(a) = f(b)$

and $f(b) = f(c)$

then $f(a) = f(c)$.

$\Rightarrow$ If $(a, b) \in R$ and $(b, c) \in R$ then $(a, c) \in R$.

$\Rightarrow$ R is transitive.

Thus, R is an equivalence relation.

Now, let $(a, a') \in R$ and $(b, b') \in R$, where $a, a', b, b' \in S$.

$\Rightarrow$ $f(a) = f(a')$.

and $f(b) = f(b')$.

Also,

$$f(a \cdot b) = f(a) \cdot f(b) \qquad (f \text{ is homomorphism})$$
$$= f(a') \cdot f(b')$$
$$= f(a' \cdot b').$$

$\Rightarrow$ $(ab, a'b') \in R$.

Hence, R is a congruence relation on S.

(ii) Let us show that S/R is isomorphic to $f(S)$.

Define a function $g : S/R \to f(S)$ by $g([a]) = f(a)$, $a \in S$. (A)

Then we have to prove that, (a) g is well defined and (b) g is an isomorphism.

(a) g is well defined:

Since $a \in S$, therefore,

$f(a) \in f(S)$.

Also,

$g([a]) = f(a)$. [From (A)]

$\Rightarrow$ $g([a]) \in f(S)$.

Again let $[a] = [b]$, i.e., a and b are in same equivalence class.

Then $(a, b) \in R$.

$\Rightarrow \qquad f(a) = f(b).$

Therefore,
$$\begin{aligned} g([a]) &= f(a) \\ &= f(b) \\ &= g([b]). \end{aligned}$$

$\Rightarrow$ g is well defined.

(b) g is an isomorphism:

$$\begin{aligned} g([a][b]) &= g([ab]) \\ &= f(ab) \\ &= f(a)f(b) \qquad (f \text{ is homomorphism}) \\ &= g([a])g([b]) \end{aligned}$$

$\Rightarrow$ g is homomorphism.

Now, let

$$g([a]) = g([b]).$$

$\Rightarrow \qquad f(a) = f(b).$

$\Rightarrow \qquad (a, b) \in R.$

$\Rightarrow \qquad [a] = [b].$

$\Rightarrow$ g is one-to-one.

Also, let $y \in f(S)$.

$\Rightarrow \qquad f(a) = y$ for some $a \in S$

$\Rightarrow \qquad g([a]) = f(a) = y.$ [From (A)]

$\Rightarrow$ g is onto.

Since g is homomorphism, one-to-one and onto, therefore, it is an isomorphism.

Thus, the theorem is proved.

3.6 Direct Product of Semigroups

Suppose $< S, * >$ and $< T, \Delta >$ are two semigroups. The direct product of S and T is an algebraic structure $< S \times T, \lozenge >$ with operation '$\lozenge$' defined as:

$$(s_1, t_1) \lozenge (s_2, t_2) = (s_1 * s_2, t_1 \Delta t_2)$$

where $(s_1, t_1), (s_2, t_2) \in S \times T$, $s_1, s_2 \in S$ and $t_1, t_2 \in T$.

It is easy to show that the direct product of semigroups is closed and associative. Therefore, it is a semigroup.

3.7 Cyclic Semigroup

A semigroup $< S, * >$ is said to be cyclic if it can be generated by a single element. For instance, set of positive even integers with respect to addition, $< \mathbb{E}, + >$ is a cyclic semigroup generated by 2. The element that generates the semigroup is called the generator. If the cyclic semigroup is finite, then generator is said to be of **finite order**, otherwise, it is of **infinite order**. A semigroup is said to be **periodic** if all of its elements are of finite order. If a cyclic semigroup is infinite then it is isomorphic to the semigroup of positive integers with the operation of addition.

3.8 Solved Problems

PROBLEM 3.1 Let $< F, \circ >$ be a free semigroup defined over alphabet $\Sigma = \{0, 1\}$. Prove that a subset of F containing strings of only odd length will not form a subsemigroup.

Solution

Let N be the set of strings of length odd (i.e. $|\alpha|$ is odd, $\alpha \in N$ is any string over alphabet Σ). Also, let α_1 and α_2 be two strings of N, then string $\alpha_1 \circ \alpha_2$ will be of length even.

That is, $< N, \circ >$ is not closed. Therefore, it will not be a free subsemigroup of $< F, \circ >$.

PROBLEM 3.2 Consider the set $\mathbb{Q}$ of rational numbers, and let $*$ be the operation on $\mathbb{Q}$ defined by $a * b = a + b + ab$. Prove that $< \mathbb{Q}, * >$ is a semigroup.

Solution

Algebraic structure $< \mathbb{Q}, * >$ will be closed, since $a * b = a + b + ab$ will be rational.

Further, for $a, b, c \in \mathbb{Q}$,

$$b * c = b + c + bc$$

and
$$\begin{aligned} a * (b * c) &= a + b + c + bc + a(b + c + bc) \\ &= a + b + c + ab + ac + bc + abc. \end{aligned}$$

Also,
$$(a * b) * c = a + b + c + ab + ac + bc + abc.$$

Therefore, it is a semigroup.

PROBLEM 3.3 Consider a free semigroup $< F, \circ >$ defined over an alphabet Σ. Also, let E be a subset of F such that elements of E are of length even. Then $< E, \circ >$ will be a subsemigroup of $< F, \circ >$. Prove that a mapping, $f: F \to E$ defined by $f(u) = v$, such that $|v| = 2|u|$, is a semigroup homomorphism.

Solution

Let a and b be strings of F. Then,

$$f(a \circ b) = c$$

where,
$$\begin{aligned} |c| &= 2|a \circ b| \\ &= 2(|a| + |b|) \\ &= 2|a| + 2|b|. \end{aligned}$$

Therefore, operation is preserved. Hence, it is a semigroup homomorphism.

PROBLEM 3.4 Let E be a set of strings of 1's of length even over an alphabet $\Sigma = \{0, 1\}$. Prove that $< E, \circ >$ is a free semigroup.

Solution

Let E be the set of strings of 1's of length even. Let α_1 and α_2 be two strings of E. Then string $\alpha_1 \circ \alpha_2$ will also be a string of 1's of length even. Therefore, E is closed. Further, let $\alpha_1, \alpha_2, \alpha_3 \in E$. Then

$$(\alpha_1 * \alpha_2) * \alpha_3 = \alpha_1 * (\alpha_2 * \alpha_3).$$

Therefore, E is associative. Thus, $< E, \circ >$ is a free semigroup.

PROBLEM 3.5 Let $< S, +_3 >$ and $< P, \times_4 >$ be two semigroups, such that $S = \{0, 1, 2\}$ and $P = \{1, 2, 3\}$. Operations $+_3$ and $\times_4$ are addition modulo 3 and multiplication modulo 4, respectively. Find the direct product of S and P, and show that it is a semigroup.

Solution

The direct product of $< S, +_3 >$ and $< P, \times_4 >$ is an algebraic structure $< S \times P, \Diamond >$ with operation '$\Diamond$' defined as:

$$(s_1, p_1) \Diamond (s_2, p_2) = (s_1 +_3 s_2, p_1 \times_4 p_2)$$

where, $\quad (s_1, p_1), (s_2, p_2) \in S \times P,$

$$s_1, s_2 \in S \text{ and } p_1, p_2 \in P.$$

Further, $\quad S = \{0, 1, 2\}$ and $P = \{1, 2, 3\}$.

Therefore, $S \times P = \{(0, 1), (0, 2), (0, 3), (1, 1), (1, 2), (1, 3), (2, 1), (2, 2), (2, 3)\}$.

Now, to prove that $< S \times P, \Diamond >$ is a semigroup, we have to show that it is closed and associative. To prove closure, we can construct a Cayley table. In the Cayley table, if all the elements are from the set $S \times P$, then $S \times P$ will be closed. Table 3.6 is a Cayley table of $S \times P$.

Table 3.6 Cayley Table of $S \times P$

$\Diamond$	(0, 1)	(0, 2)	(0, 3)	(1, 1)	(1, 2)	(1, 3)	(2, 1)	(2, 2)	(2, 3)
(0, 1)	(0, 1)	(0, 2)	(0, 3)	(1, 1)	(1, 2)	(1, 3)	(2, 1)	(2, 2)	(2, 3)
(0, 2)	(0, 2)	(0, 1)	(0, 2)	(1, 2)	(1, 1)	(1, 2)	(2, 2)	(2, 1)	(2, 2)
(0, 3)	(0, 3)	(0, 2)	(0, 1)	(1, 3)	(1, 2)	(1, 1)	(2, 3)	(2, 2)	(2, 1)
(1, 1)	(1, 1)	(1, 2)	(1, 3)	(2, 1)	(2, 2)	(2, 3)	(0, 1)	(0, 2)	(0, 3)
(1, 2)	(1, 2)	(1, 1)	(1, 2)	(2, 2)	(2, 1)	(2, 2)	(0, 2)	(0, 1)	(0, 2)
(1, 3)	(1, 3)	(1, 2)	(1, 1)	(2, 3)	(2, 2)	(2, 1)	(0, 3)	(0, 2)	(0, 1)
(2, 1)	(2, 1)	(2, 2)	(2, 3)	(0, 1)	(0, 2)	(0, 3)	(1, 1)	(1, 2)	(1, 3)
(2, 2)	(2, 2)	(2, 1)	(2, 2)	(0, 2)	(0, 1)	(0, 2)	(1, 2)	(1, 1)	(1, 2)
(2, 3)	(2, 3)	(2, 2)	(2, 1)	(0, 3)	(0, 2)	(0, 1)	(1, 3)	(1, 2)	(1, 1)

We can observe that each element of Cayley table belongs to $S \times P$. Therefore, $S \times P$ is closed. Further, since the operation $\Diamond$ is defined in terms of $+_3$ and $\times_4$ which are associative, therefore, we can deduce that $\Diamond$ will be associative. Thus, $< S \times P, \Diamond >$ is a semigroup.

Exercises

3.1 Let $\mathbb{Z}$ be the set of integers and $\mathbb{E}$ be the set of all even integers. Prove that the semigroup $< \mathbb{Z}, + >$ and $< \mathbb{E}, + >$ are isomorphic with respect to mapping $f: \mathbb{Z} \to \mathbb{E}$ given by $f(a) = 2a$.

3.2 Let S be the set of ordered pairs of natural numbers. An operation '$*$' is defined on the elements of S by $< a_1, b_1 > * < a_2, b_2 > = < a_1 \cdot a_2, b_1 \cdot b_2 >$. Show that $< S, * >$ is a commutative semigroup.

Also, a mapping $f: < S, * > \to < \mathbb{Q}, \cdot >$ defined by $f(< a, b >) = a/b$ is a semigroup homomorphism.

(In a commutative semigroup, $a * b = b * a\}$

3.3 Consider the set N of positive integers, and let $*$ denote the operation of least common multiple on N. Show that $< N, * >$ is a commutative semigroup.

3.4 Consider the set $\mathbb{Q}$ of rational numbers, and let $*$ be the operation on $\mathbb{Q}$ defined by $a * b = a + b - ab$. Prove that $< \mathbb{Q}, * >$ is a semigroup.

3.5 Let $< S, +_2 >$ and $< P, \times_{10} >$ be two semigroups such that $S = \{0, 1\}$ and $P = \{1, 3, 7, 9\}$. Operations $+_2$ and $\times_{10}$ are addition modulo 2 and multiplication modulo 10, respectively. Find the direct product of S and P, and show that it is a semigroup.

4

Monoid

4.1 Introduction

In this chapter we will discuss another algebraic structure called **monoid**. Monoids are special types of semigroups. If a semigroup S, possesses an element e which when operated with any other element $x \in S$, leaves x unchanged, then e is called identity of the semigroup S. A semigroup with identity is called a monoid. Like semigroup, monoid is also a useful algebraic structure.

4.2 Definition of Monoid

A monoid is a set M, together with a binary operation '$*$' that satisfies:

1. **Closure**
   For all $a, b \in M$
   $$a * b \in M.$$
2. **Associative**
   For all $a, b, c \in M$,
   $$(a * b) * c = a * (b * c).$$
3. **Identity**
   There exists an element e, such that $\forall\ a \in M$
   $$e * a = a.$$
   Notation used for monoid is $< M, *, e >$.

EXAMPLE 4.1 Empty monoid: Let $M = \varphi$ be an empty set. Then $< \varphi, *, e >$ forms a monoid with operation defined as:

$$\varphi * \varphi = \varphi.$$

Identity of this will be φ, i.e., $e = \varphi$.

EXAMPLE 4.2 Monoid with one element: Let M be a set that contains only one element, say e. Then, the set M together with an operation '$*$' defined as:

$$e * e = e$$

will form a monoid. A singleton set with natural number 1 and operation as multiplication is a monoid.

EXAMPLE 4.3 Monoid of positive integers: The set of positive integers including zero, with operation addition, i.e., $< \mathbb{Z}^+, +, 0 >$ is a monoid with 0 as identity. Similarly, the set of positive integers with operation multiplication, i.e., $< \mathbb{Z}^+, \bullet, 1 >$ is also a monoid, where identity is 1.

EXAMPLE 4.4 Monoid of matrices: Let M be the set of all square matrices of order n. Then, it will form a monoid with respect to matrix addition. Identity, in this case will be null matrix. It will also form a monoid if operation is matrix multiplication with identity as unit matrix.

EXAMPLE 4.5 Monoid of strings: The set of all strings over a finite alphabet Σ with operation defined as concatenation forms a monoid. The identity will be empty string, λ. Let α be an element of monoid then

$$\alpha\lambda = \lambda\alpha = \alpha.$$

The monoid of strings is called **free monoid.**

EXAMPLE 4.6 Consider a set $M = \{m, n, o, p\}$ with operator '$*$' defined by Cayley Table 4.1.

Table 4.1 Cayley Table for $M = \{m, n, o, p\}$

$*$	m	n	o	p
m	m	n	o	p
n	n	n	n	n
o	o	o	o	o
p	p	p	p	p

Obviously operation $*$ is closed

Further, operation $*$ is associative.

For instance, $(m * n) * o = n * o = n$

and $m * (n * o) = m * n = n.$

Hence,

$$(m * n) * o = m * (n * o).$$

This can be observed with other combination of elements too. Thus, $< M, *, m >$ is closed and associative. Also identity is m, therefore, $< M, *, m >$ is a monoid.

4.3 Submonoid

Let $< M, *, e >$ be a monoid, and T be the subset of M. Then, $< T, *, e >$ will be called a submonoid, if it is a monoid itself.

Formally, let $< M, *, e >$ be a monoid and $T \subseteq M$, then T will be called a submonoid, if

1. for all $a, b \in T$,
$$a * b \in T;$$
2. for all $a, b, c \in T$,
$$(a * b) * c = a * (b * c);$$
3. identity of M belongs to T.

EXAMPLE 4.7 Let $< M, \circ, e >$ be a monoid defined over alphabet $\Sigma = \{a, b\}$. Also, let E be the set of strings of length even (i.e. $|\alpha|$ is even, $\alpha \in E$ is any string over alphabet Σ).

Then, $< E, \circ, e >$ will be a submonoid of $< M, \circ, e >$. The identity e will be empty string (λ) with length $|\lambda| = 0$.

4.4 Monoid Homomorphism

Let $< M, *, e_1 >$ and $< T, \Delta, e_2 >$ be any two monoids. Then the mapping $f : M \to T$, such that, for all $a, b \in M$,

1. $f(a * b) = f(a) \, \Delta \, f(b)$ and
2. $f(e_1) = e_2$

is called monoid homomorphism.

A monoid homomorphism is a function that preserves monoid structure. Further, if mapping f is also one-to-one and onto, then it is called the monoid isomorphism. Therefore, two monoids M and T are said to be isomorphic, if $f : M \to T$ is a one-to-one and onto mapping. Symbolically, it is written as $M \approx T$.

A monoid isomorphism onto itself is called monoid automorphism.

EXAMPLE 4.8 Consider a monoid $< M, \circ, \lambda >$ defined over an alphabet Σ. Also, let $< Z, +, 0 >$ be the monoid of set of integers with respect to operation addition.

Then a mapping, $f : M \to Z$ defined by $f(w) = |w|$, where $|w|$ is the length of w, is a monoid homomorphism. For instance,
$$|u \circ v| = |u| + |v|.$$
It will not be an isomorphism.

EXAMPLE 4.9 Let $< \mathbb{Z}, +_4, 0 >$ be a monoid defined over $\mathbb{Z} = \{0, 1, 2, 3\}$ with respect to addition modulo 4. Also, consider a monoid of set $A = \{1, 3, 7, 9\}$ over the operation multiplication modulo 10 (i.e. $\times_{10}$) with multiplicative identity, 1.

Then a mapping $f : \mathbb{Z} \to A$ defined by
$$f(0) = 1,$$
$$f(1) = 3,$$

$$f(2) = 9,$$

and $$f(3) = 7,$$

will be a monoid homomorphism. Cayley tables for the two operations are given in Tables 4.2 and 4.3.

Table 4.2 Cayley Table for Addition Modulo 4

$+_4$	0	1	2	3
0	0	1	2	3
1	1	2	3	0
2	2	3	0	1
3	3	0	1	2

Table 4.3 Cayley Table for Multiplication Modulo 10

$\times_{10}$	1	3	7	9
1	1	3	7	9
3	3	9	1	7
7	7	1	9	3
9	9	7	3	1

Now, if $f: \mathbb{Z} \to A$ is a monoid homomorphism, then it should satisfy the monoid homomorphism properties, i.e.,

1. $f(a * b) = f(a)\ \Delta\ f(b)$ and
2. $f(e_1) \to e_2$.

Let us consider,

$$f(2 +_4 3) = f(2) \times_{10} f(3).$$

$$\Rightarrow \quad f(1) = 9 \times_{10} 7.$$

$$\Rightarrow \quad f(1) = 3.$$

Readers may try for other numbers too.

Also, $f(0) \to 1$.

Therefore, it satisfies the properties of monoid homomorphism. Further, since this mapping is one-to-one and onto, hence it is monoid isomorphism.

THEOREM 4.1 Let $< M, *, e_1 >$, $< T, \Delta, e_2 >$ and $< U, \odot, e_3 >$ be three monoids, and $g: M \to T$ and $h: T \to U$ be two monoid homomorphisms. Then

$$(h \circ g): M \to U$$

is a monoid homomorphism from $< M, *, e_1 >$ to $< U, \odot, e_3 >$.

Proof: Let $a, b \in M$.

Consider,

$$\begin{aligned}(h \circ g)(a * b) &= h(g(a * b)) \\ &= h(g(a)\ \Delta\ g(b)) \\ &= h(g(a)) \odot h(g(b)) \\ &= (h \circ g)(a) \odot (h \circ g)(b).\end{aligned}$$

Also, since

$$g(e_1) = e_2 \text{ and } h(e_2) = e_3,$$

therefore, $$(h \circ g)(e_1) = e_3.$$

Thus, $(h \circ g)$ is a homomorphism from M to U.

4.5 Quotient Monoid

Let $< M, *, e >$ be a monoid and R be an equivalence relation defined on M. Since an equivalence relation partitions the set into disjoint equivalence classes, therefore, R will partition M. The collection of all equivalence classes is denoted by M/R.

Further, define an operation on the equivalence classes by,

$$[a] * [b] = [a * b]$$

or

$$[a][b] = [ab].$$

Also,

$$[e][a] = [ea] = [a].$$

Since this operation is associative, therefore, M/R will form a monoid with respect to operation '$*$'. It is called quotient monoid with identity $[e]$.

4.6 Direct Product of Monoids

Let $< M, *, e_1 >$ and $< T, \Delta, e_2 >$ be any two monoids. The direct product of M and T is an algebraic structure $< M \times T, \lozenge, e_3 >$, with operation '$\lozenge$' defined as:

$$(m_1, t_1) \lozenge (m_2, t_2) = (m_1 * m_2, t_1 \Delta t_2),$$

where $(m_1, t_1), (m_2, t_2) \in M \times T$, $m_1, m_2 \in M$, and $t_1, t_2 \in T$.

Also,

$$e_3 = (e_1, e_2).$$

It is easy to show that the direct product of monoids is closed and associative. Further, if we consider (e_1, e_2) as the identity of the set of ordered pairs, then direct product will form a monoid.

4.7 Commutative Monoid

Suppose $< M, *, e >$ is a monoid. If operation $*$ is commutative, then it is called commutative monoid. For instance, set of integers with 0 as identity is a monoid with respect to operation addition.

4.8 Cyclic Monoid

A monoid $< M, *, e >$ is said to be cyclic if it can be generated by a single element. The element that generates the monoid is called the generator. For instance, set of positive integers with respect to addition modulo 4, i.e., $< I^+, +_4 >$ is a cyclic monoid generated by 1.

THEOREM 4.2 Let $< M, *, e >$ be a commutative monoid. Then, the set of idempotent elements of M forms a submonoid.

Proof: An element $a \in M$ is said to be idempotent if

$$a * a = a.$$

Let S be the set of idempotent elements of M.
Since,

$$e * e = e,$$

therefore, $e \in S$.
Let $a, b \in S$
$\Rightarrow$ $$a * a = a$$
and $$b * b = b.$$

Consider $(a * b) * (a * b) = (a * b) * (b * a)$ {since $a, b \in M$ and M is commutative}
$= (a * (b * b) * a)$ {since M is associative}
$= a * b * a$ {b is idempotent}
$= a * a * b$
$= (a * a) * b$
$= a * b.$

$\Rightarrow$ $a * b$ is idempotent.
$\Rightarrow$ S is closed.
Now, since operation $*$ is associative, therefore, $< S, *, e >$ will be a submonoid.

4.9 Solved Problems

PROBLEM 4.1 Let $< M, \circ, \lambda >$ be a free monoid defined over alphabet $\Sigma = \{0, 1\}$. Prove that a subset of M containing all strings of length odd will not form a monoid; operation being catenation.

Solution
Let O be the set of strings of length odd (i.e. $|\alpha|$ is odd, $\alpha \in O$ is any string over alphabet Σ).
Also, let α_1 and α_2 be any two strings of O.
Then string $\alpha_1 \circ \alpha_2$ will be of length even.
That is, $< O, \circ >$ is not closed.
Therefore, it will not be a monoid.

PROBLEM 4.2 Consider the set $\mathbb{Q}$ of rational numbers, and let $*$ be the operation on $\mathbb{Q}$ defined by

$$a * b = a + b + ab, \text{ where } a, b \in \mathbb{Q}.$$

Prove that $< \mathbb{Q}, *, 0 >$ is a monoid.

Solution
Algebraic structure $< \mathbb{Q}, *, 0 >$ will be closed, since

$$a * b = a + b + ab$$

will be rational.

Further, for $a, b, c \in \mathbb{Q}$,

$$b * c = b + c + bc$$

and $$a * (b * c) = a + b + c + bc + a(b + c + bc),$$
$$= (a * b) * c.$$

Therefore, it is a semigroup. Also, the identity e will be 0, since $0 + a = 0 + a + 0 \cdot a = a$. Thus, it is a monoid.

PROBLEM 4.3 Consider a free monoid $< M, \circ, \lambda >$ defined over an alphabet Σ. Also, let E be a subset of M, such that elements of E are of length even. Then $< E, \circ, \lambda >$ will be a submonoid of $< M, \circ, \lambda >$. Prove that a mapping, $f : M \rightarrow E$ defined by $f(u) = v$, such that $| v | = 2| u |$, is a monoid homomorphism.

Solution

$$f(a \circ b) = c$$

where, $$| c | = 2| a \circ b |$$
$$= 2(| a | + | b |)$$
$$= 2| a | + 2| b |.$$

Therefore, operation is preserved. Further, $f(\lambda) = \lambda$ because $| \lambda | = 2| \lambda |$. Hence, it is a monoid homomorphism.

PROBLEM 4.4 Let $< M, \circ, \lambda >$ be a free monoid. Let $E \subseteq M$ be a set of strings of 1's of length even over an alphabet $\Sigma = \{0, 1\}$. Prove that $< E, \circ, \lambda >$ is a free monoid.

Solution

Let E be the set of strings of 1's of length even. Let α_1 and α_2 be two strings of E.
Then string $\alpha_1 \circ \alpha_2$ will also be a string of length even. Therefore, E is closed.
Further, let $\alpha_1, \alpha_2, \alpha_3 \in E$. Then

$$(\alpha_1 * \alpha_2) * \alpha_3 = \alpha_1 * (\alpha_2 * \alpha_3).$$

Therefore, E is associative.
Thus, $< E, \circ >$ is a free semigroup. Further, empty string λ is considered to belong to E. Therefore, E contains identity. Hence, $< E, \circ, \lambda >$ is a free monoid.

PROBLEM 4.5 Let $< M, +_2, 0 >$ and $< P, \times_{10}, 1 >$ be two monoids, such that $M = \{0, 1\}$ and $P = \{1, 3, 7, 9\}$. Operations $+_2$ and $\times_{10}$ are addition modulo 2 and multiplication modulo 10, respectively. Find the direct product of M and P, and show that it is a monoid.

Solution

The direct product of $< M, +_2, 0 >$ and $< P, \times_{10}, 1 >$ is an algebraic structure $< M \times P, \lozenge, e >$, with operation '$\lozenge$' defined as:

$(m_1, p_1) \lozenge (m_2, p_2) = (m_1 +_2 m_2, p_1 \times_{10} p_2)$, where $(m_1, p_1), (m_2, p_2) \in M \times P$, $m_1, m_2 \in M$ and $p_1, p_2 \in P$.

Also, e may be the possible identity, if $< M \times P, \lozenge, e >$ happens to be a monoid.

Since, $M = \{0, 1\}$ and $P = \{1, 3, 7, 9\}$.

Hence, $M \times P = \{(0, 1), (0, 3), (0, 7), (0, 9), (1, 1), (1, 3), (1, 7), (1, 9)\}$.

Now, to prove that $< M \times P, \lozenge, e >$ is a monoid, we have to show that it is closed and associative. In addition, we have to prove that e is the identity. To prove closure, we can construct a Cayley table. In the Cayley table, if all the elements are from the set $M \times P$, then $M \times P$ will be closed. Table 4.4 is the Cayley table of $M \times P$.

Table 4.4 Cayley Table of $M \times P$

$\lozenge$	(0, 1)	(0, 3)	(0, 7)	(0, 9)	(1, 1)	(1, 3)	(1, 7)	(1, 9)
(0, 1)	(0, 1)	(0, 3)	(0, 7)	(0, 9)	(1, 1)	(1, 3)	(1, 7)	(1, 9)
(0, 3)	(0, 3)	(0, 9)	(0, 1)	(0, 7)	(1, 3)	(1, 9)	(1, 1)	(1, 7)
(0, 7)	(0, 7)	(0, 1)	(0, 9)	(0, 3)	(1, 7)	(1, 1)	(1, 9)	(1, 3)
(0, 9)	(0, 9)	(0, 7)	(0, 3)	(0, 1)	(1, 9)	(1, 7)	(1, 3)	(1, 1)
(1, 1)	(1, 1)	(1, 3)	(1, 7)	(1, 9)	(0, 1)	(0, 3)	(0, 7)	(0, 9)
(1, 3)	(1, 3)	(1, 9)	(1, 1)	(1, 7)	(0, 3)	(0, 9)	(0, 1)	(0, 7)
(1, 7)	(1, 7)	(1, 1)	(1, 9)	(1, 3)	(0, 7)	(0, 1)	(0, 9)	(0, 3)
(1, 9)	(1, 9)	(1, 7)	(1, 3)	(1, 1)	(0, 9)	(0, 7)	(0, 3)	(0, 1)

We can observe that each element of Cayley table belongs to $S \times P$. Therefore, $S \times P$ is closed. Further, since the operation $\lozenge$ is defined in terms of $+_2$ and $\times_{10}$, which are associative, therefore, we can deduce that $\lozenge$ will be associative.

From the Cayley table, it is obvious that (0, 1) is the identity. Thus, $< M \times P, \lozenge, (0, 1) >$ is a monoid.

Exercises

4.1 Show that $< \rho(X), \cup, \varphi >$ and $< \rho(X), \cap, X >$ are monoids. What are the identities?

4.2 Show that the set of natural numbers forms a monoid under the operation defined as $x * y = \max\{x, y\}$. What is the identity?

4.3 Let $\mathbb{Z}$ be the set of integers and E be the set of all even integers. Prove that the monoid $< \mathbb{Z}, +, 0 >$ and $< E, +, 0 >$ are isomorphic under the mapping $f : \mathbb{Z} \to E$ given by $f(a) = 2a$.

4.4 Let M be the set of ordered pairs of natural numbers. An operation '$*$' is defined on the elements of M by $< a_1, b_1 > * < a_2, b_2 > = < a_1 \cdot a_2, b_1 \cdot b_2 >$. Show that M is a commutative monoid. What is the identity?

4.5 Consider the set N of positive integers, and let $*$ denote the operation of least common multiple on N. Show that $< N, * >$ is a commutative monoid.

4.6 Consider the set $\mathbb{Q}$ of rational numbers, and let $*$ be the operation on $\mathbb{Q}$ defined by

$$a * b = a + b - a \cdot b.$$

Prove that $\mathbb{Q}$ is a monoid. Find the identity.

4.7 Let $< M, +_3, 0 >$ and $< P, \times_4, 1 >$ be two monoids, such that $M = \{0, 1, 2\}$ and $P = \{1, 2, 3\}$. Operations $+_3$ and $\times_4$ are addition modulo 3 and multiplication modulo 4, respectively. Find the direct product of M and P, and show that it is a monoid.

5

Group

5.1 Introduction

In Chapter 4, we have studied monoid. Here, in this chapter, we will discuss another algebraic structure called group. If each element in a monoid possesses its inverse then it is called a group. The study of groups is very vast as compared to other algebraic structures. The group structure is an important part of abstract algebra. There are different types of group structures like symmetry groups, lie groups, and matrix groups etc., which play the important roles in many academic disciplines.

5.2 Definition of Group

A group is a nonempty set G together with a binary operation '$*$' that satisfies:

1. **Closure**
   For all $a, b \in G$,
   $$a * b \in G.$$
2. **Associative**
   For all $a, b, c \in G$,
   $$(a * b) * c = a * (b * c).$$
3. **Identity**
   There exists an element e, such that,
   $$\forall a \in G, e * a = a.$$

4. **Inverse**
   For each element $a \in G$, $\exists\, a^{-1}$, such that,
$$a * a^{-1} = a^{-1} * a = e.$$

If operation $*$ is commutative, i.e., $a * b = b * a$, then the group is called **Abelian** group. The number of elements in a group G is called the **order** of G and is denoted as $|G|$. If for a group G, $|G|$ is finite then group G is called **finite**, otherwise **infinite**. Further, if for a least positive integer n, $a^n = e$, then n is called the **order of element** a. It is denoted as $o(a)$.

Here $a^n = a * a * a * \cdots * a$ (n times).

EXAMPLE 5.1 Group with one element: Let G be a set and it contains only one element, say e. Then, the set G together with an operation '$*$' defined as:
$$e * e = e$$
will form a group. The inverse of e will be e itself. A singleton set with natural number 1 and operation as multiplication is a group.

EXAMPLE 5.2 Group of roots of $(1)^{1/3}$: Let $G = \{1, \omega, \omega^2\}$. Then $< G, \cdot >$ is a finite group with respect to multiplication. It is a commutative group with identity 1. Inverse of ω is ω^2 and inverse of ω^2 is ω.

EXAMPLE 5.3 Group of integers: The set of integers including zero, with operation addition, i.e., $< \mathbb{Z}, + >$ is a group with 0 as identity. Similarly, the set of rational numbers, excluding zero, with operation multiplication, i.e., $< \mathbb{Q} - \{0\}, \cdot >$ is a group, with identity being 1. These are infinite groups.

EXAMPLE 5.4 Group of matrices: Let G be the set of all square matrices of order n. Then it will form an infinite group with respect to operation, matrix addition. The identity, in this case will be the null matrix.

5.3 Subgroup and Coset

Let $< G, * >$ be a group, and H be the subset of G. Then $< H, * >$ will be called a subgroup of G, if it is a group itself.

Formally,

Let $< G, * >$ be a group and $H \subseteq G$, then H will be called a subgroup of G, if

1. for all $h_1, h_2, \in H$, $h_1 * h_2 \in H$;
2. for all $h_1, h_2, h_3 \in H$, $(h_1 * h_2) * h_3 = h_1 * (h_2 * h_3)$;
3. identity of G belongs to H;
4. for each element $h \in H\ \exists\ h^{-1} \in H$, such that
$$h * h^{-1} = h^{-1} * h = e.$$

Again, let $< G, * >$ be a group and H be its subgroup. Then for $g \in G$, set
$$g * H = \{g * h \mid h \in H\},$$

or $gH = \{gh \mid h \in H\}$, is called the **left coset** of H in G.

In a similar way, Hg is called the **right coset**.

The cosets of any subgroup H form a **partition** of G, i.e., the union of all left cosets is equal to G, and two left cosets are either equal or disjoint. Similar considerations apply to the right cosets of H, too. The left and right cosets of H may or may not be equal. If they are equal, i.e., for all $g \in G$, $gH = Hg$, then H is said to be a **normal subgroup,** N of G. The number of left cosets of H in G is called the **index of** H in G, i.e., $(G : H)$.

EXAMPLE 5.5 Let $< \mathbb{Z}, +_4 >$ be a group with $\mathbb{Z} = \{0, 1, 2, 3\}$, and operation be the addition modulo 4. Consider a subgroup $H = \{0, 2\}$.

The left cosets are:

$$0 +_4 H = 2 +_4 H = \{0, 2\}$$

$$1 +_4 H = 3 +_4 H = \{1, 3\}.$$

THEOREM 5.1 (Lagrange) The order of a subgroup H of a group G divides the order of G.

Proof: Let G be a group of finite order n, and H be its subgroup.

Let $|H| = m$, i.e., say $H = \{h_1, h_2, h_3, \ldots, h_m\}$. Therefore, each left coset of H will have m distinct members. Let there are r such left cosets.

That is, $$g_1H, g_2H, g_3H, \ldots, g_rH.$$

Also, $$G = g_1H \cup g_2H \cup g_3H \cup, \ldots, \cup g_rH.$$

{Since union of all left cosets is G}

Therefore, $$|G| = |g_1H| + |g_2H| + |g_3H| + \cdots + |g_rH|.$$

{Since left cosets are disjoint}

$$\Rightarrow \quad n = m + m + m + \cdots + m \ \{r \text{ times}\}$$

$$\Rightarrow \quad n = rm$$

$$\Rightarrow \quad r = \frac{n}{m}$$

Hence, the theorem is true.

5.4 Normal Subgroup and Quotient Group

A subgroup N of a group $< G, \cdot >$ is said to be normal if and only if, for all $a \in G$, $a \cdot N = N \cdot a$,

or, in short,

$$aN = Na$$

where aN and Na are the left and right cosets of N in G, respectively. Normal subgroup of G is denoted by $N \lhd G$. Further, let the set G/N be the set of all left cosets of N in G, i.e., $G/N = \{aN \mid a \in G\}$, then it is easy to see that $< G/N, \cdot >$ forms a group.

Since,

1. for each aN and bN in G/N:

$$(aN)(bN) = a(Nb)N$$
$$= a(bN)N$$
$$= (ab)NN$$
$$= (ab)N.$$

Therefore, G/N is closed.

2. the operation is derived from the group G and elements of G/N contain elements of G, therefore, the operation is well-defined, associative, and has identity element N. The inverse of an element aN of G/N is $a^{-1}N$.

The group G/N is called **quotient group** or **factor group.** Obviously, group G, itself and its subgroup containing $\{e\}$ are its normal subgroups. These are called **trivial** normal subgroups.

Note: The motivation for calling G/N quotient group comes from division of integers. For instance, when dividing 15 by 3 one obtains the answer 5 because one can regroup 15 objects into 5 sub collections of 3 objects. In G/N with N as a normal subgroup of G, the group structure is used to form a natural 'regrouping'. The elements in the regrouping are the cosets of N in G. Since, we started with a group and its normal subgroup, the quotient G/N contains more information than just the number of cosets (which is what regular division yields). In fact, G/N has a group structure itself.

EXAMPLE 5.6 Consider the group G with respect to operation, addition modulo 6. Let $G = \{0, 1, 2, 3, 4, 5\}$ and $N = \{0, 3\}$.
Then the quotient group is:

$$G/N = \{aN \mid a \in G\}$$
$$= \{a\{0, 3\} \mid a \in \{0, 1, 2, 3, 4, 5\}\}$$
$$= \{0\{0, 3\}, 1\{0, 3\}, 2\{0, 3\}, 3\{0, 3\}, 4\{0, 3\}, 5\{0, 3\}\}$$
$$= \{\{(0 + 0) \bmod 6, (0 + 3) \bmod 6\}, \{(1 + 0) \bmod 6, (1 + 3) \bmod 6\},$$
$$\{(2 + 0) \bmod 6, (2 + 3) \bmod 6\}, \{(3 + 0) \bmod 6, (3 + 3) \bmod 6\},$$
$$\{(4 + 0) \bmod 6, (4 + 3) \bmod 6\}, \{(5 + 0) \bmod 6, (5 + 3) \bmod 6\}\}$$
$$= \{\{0, 3\}, \{1, 4\}, \{2, 5\}, \{3, 0\}, \{4, 1\}, \{5, 2\}\}$$
$$= \{\{0, 3\}, \{1, 4\}, \{2, 5\}\}.$$

$\{$Since $\{3, 0\} = \{0, 3\}, \{1, 4\} = \{4, 1\}$ and $\{5, 2\} = \{2, 5\}\}$

Identity of $G/N = \{0, 3\}$.

Also, $\{1, 4\}$ and $\{2, 5\}$ are inverses of each other.

EXAMPLE 5.7 Consider the group of integers $\mathbb{Z}$ (under addition) and the subgroup $2\mathbb{Z}$ consisting of all even integers. Now, $2\mathbb{Z}$ is a normal subgroup because $\mathbb{Z}$ is Abelian. There are only two cosets: the set of even integers and the set of odd integers. These cosets will form a quotient group containing two elements. The set of even integers is the identity of the quotient group.

5.5 Group Homomorphism

Let < G_1, $*$ > and < G_2, Δ > be any two groups. A mapping $f : G_1 \to G_2$, such that for all $a, b \in G$, and

$$f(a * b) = f(a) \,\Delta\, f(b)$$

is called group homomorphism. Also, for identities $e_1 \in G_1$ and $e_2 \in G_2$, $f(e_1) = e_2$. Further,

$$f(g)^{-1} = f(g^{-1})$$

for all g in G.

A group homomorphism is a function that preserves group structure. Further, if mapping f is also one-to-one and onto, then it is called group **isomorphism**. Therefore, two groups G_1 and G_2 are said to be isomorphic, if f: $G_1 \to G_2$ is a one-to-one and onto mapping. Symbolically, it is written as $G_1 \approx G_2$.

A group isomorphism onto itself is called a group **automorphism**.

EXAMPLE 5.8 Group < $\mathbb{R}$, + > is isomorphic to the group < $\mathbb{R}^+$, · > under the mapping defined as $f(x) = e^x$.

For showing isomorphism, we have to prove that mapping f is (1) one-to-one, (2) onto, and (3) preserves homomorphism.

1. Consider $f(x) = e^x$ and $f(y) = e^y$.

   Now, if $f(x) = f(y)$ then $e^x = e^y$

   $\Rightarrow$ $x = y$.

   $\Rightarrow$ f is one-to-one.

2. Let $y \in \mathbb{R}^+$, then

   $$e^{\log y} = f(\log y)$$

   $\Rightarrow$ $\log y \in \mathbb{R}$.

   Therefore, f is onto.

3. For $x, y \in \mathbb{R}$, we have

   $$\begin{aligned} f(x + y) &= e^{x+y} \\ &= e^x e^y \\ &= f(x)\, f(y). \end{aligned}$$

   $\Rightarrow$ f preserves homomorphism.

   Therefore, f is isomorphism.

EXAMPLE 5.9 Consider the group of integers $\mathbb{Z}$ (under addition), and the subgroup $2\mathbb{Z}$ consisting of all even integers. The quotient group $\mathbb{Z}/2\mathbb{Z}$ is isomorphic to the group {0, 1} with operation addition modulo 2.

EXAMPLE 5.10 Let < $\mathbb{Z}$, $+_4$ > be a group defined over $\mathbb{Z}$ = {0, 1, 2, 3} with respect to addition modulo 4. Also, consider a group of set A = {1, 3, 7, 9} over the operation multiplication modulo 10 (i.e., $\times_{10}$).

Then a mapping $f : \mathbb{Z} \to A$ defined by

$$f(0) = 1$$
$$f(1) = 3$$
$$f(2) = 9$$

and
$$f(3) = 7$$

will be a group homomorphism. Cayley tables for the two operations are given in Tables 5.1 and 5.2.

Table 5.1 Cayley Table for Addition Modulo

$+_4$	0	1	2	3
0	0	1	2	3
1	1	2	3	0
2	2	3	0	1
3	3	0	1	2

Table 5.2 Cayley Table for Multiplication Modulo

$\times_{10}$	1	3	7	9
1	1	3	7	9
3	3	9	1	7
7	7	1	9	3
9	9	7	3	1

Obviously, if $f : \mathbb{Z} \to A$ is a group homomorphism, then it should satisfy the homomorphism property, i.e.,

$$f(a +_4 b) = f(a) \times_{10} f(b).$$

Let us consider,

$$f(1 +_4 2) = f(1) \times_{10} f(2).$$
$$= 3 \times_{10} 9.$$

$\Rightarrow$ $f(3) = 7.$ {given}

Also, $f(0) = 1.$ {given}

That is, identity is mapped on identity.

Further, it is not difficult to see that the homomorphism property is also satisfied by other elements of the two groups. Therefore, f is homomorphism.

It is also one-to-one and onto, hence it is isomorphism.

5.6 Cyclic Group

A group G is said to be cyclic if all its elements can be generated by a single element. In other words, a group G is called cyclic, if for some $a \in G$, every element $x \in G$ is of the form a^n, where n is some integer. The element a is called the **generator** of G. Further, there may be more than one generator of a group. If a^n is the identity element of the group with n, the least positive integer, then n is called the order of a. It is written as $o(n)$. An element will be the generator of the group if its order is equal to the order of the group.

EXAMPLE 5.11 The group $G = \{1, \omega, \omega^2\}$ is cyclic with generators ω and ω^2. Similarly, $G = \{1, -1, i, -i\}$ is cyclic with i and $-i$ as its generators.

EXAMPLE 5.12 The group $G = \{0, 1, 2, 3, 4, 5\}$ with respect to operation $+_6$ is cyclic. The generators are 1 and 5. For instance,

$$1^1 = 1$$
$$1^2 = 1 +_6 1 = 2$$
$$1^3 = 1 +_6 1 +_6 1 = 3$$
$$1^4 = 1 +_6 1 +_6 1 +_6 1 = 4$$
$$1^5 = 1 +_6 1 +_6 1 +_6 1 +_6 1 = 5$$
$$1^6 = 1 +_6 1 +_6 1 +_6 1 +_6 1 +_6 1 = 0.$$

Therefore,

$$G = \{1^1, 1^2, 1^3, 1^4, 1^5, 1^6\}$$
$$= \{1, 2, 3, 4, 5, 0\}$$
$$= \{0, 1, 2, 3, 4, 5\}.$$

Similarly,

$$5^1 = 5$$
$$5^2 = 5 +_6 5 = 4$$
$$5^3 = 5 +_6 5 +_6 5 = 3$$
$$5^4 = 5 +_6 5 +_6 5 +_6 5 = 2$$
$$5^5 = 5 +_6 5 +_6 5 +_6 5 +_6 5 = 1$$
$$5^6 = 5 +_6 5 +_6 5 +_6 5 +_6 5 +_6 5 = 0.$$
$$G = \{5^1, 5^2, 5^3, 5^4, 5^5, 5^6\}.$$

5.7 Some Properties of Cyclic Group

There are some important theorems, which are frequently used in the study of cyclic groups. These are stated below without proof. The proof of these theorems can be found in any good text on abstract algebra.

1. Every cyclic group is abelian.
2. Every group of prime order is cyclic.
3. If a is a generator of a cyclic group G, then a^{-1} is also a generator of G.
4. If a cyclic group G is generated by an element a of order n, then a^m is also a generator of G if and only if, the GCD (greatest common divisor) of m and n is 1, i.e., m and n are relatively prime.
5. Every subgroup of a cyclic group is cyclic.
6. If G is an infinite cyclic group, then G has exactly two generators, and is isomorphic to the additive group of integers.
7. A cyclic group G with generator of finite order n is isomorphic to the multiplicative group of n nth roots of unity.

5.8 Permutation Group

A permutation of a set A is a function $f : A \to A$, that is both, one-to-one and onto. It is written in matrix form with two parallel rows.

For example, let $A = \{1, 2, 3, 4, 5\}$, then one of the permutations of A is

$$P_1 = \begin{pmatrix} 1\,2\,3\,4\,5 \\ 4\,2\,5\,3\,1 \end{pmatrix}.$$

The mapping defined here is given by

$$f(1) = 4$$
$$f(2) = 2$$
$$f(3) = 5$$
$$f(4) = 3$$

and $$f(5) = 1.$$

Similarly, $$P_2 = \begin{pmatrix} 1\,2\,3\,4\,5 \\ 3\,5\,4\,2\,1 \end{pmatrix}.$$

The mapping is given by

$$f(1) = 3$$
$$f(2) = 5$$
$$f(3) = 4$$
$$f(4) = 2$$

and $$f(5) = 1.$$

The composition operation on two permutations, say P_1 and P_2 (also called the product of the permutations) maps elements of P_1 to elements of P_2. For example:

$$P_1 \cdot P_2 = \begin{pmatrix} 1\,2\,3\,4\,5 \\ 4\,2\,5\,3\,1 \end{pmatrix}\begin{pmatrix} 1\,2\,3\,4\,5 \\ 3\,5\,4\,2\,1 \end{pmatrix}$$

$$= \begin{pmatrix} 1\,2\,3\,4\,5 \\ 4\,2\,5\,3\,1 \end{pmatrix}\begin{pmatrix} 4\,2\,5\,3\,1 \\ 2\,5\,1\,4\,3 \end{pmatrix}$$

$$= \begin{pmatrix} 1\,2\,3\,4\,5 \\ 2\,5\,1\,4\,3 \end{pmatrix}$$

It is easy to verify that the collection of all permutations of a non empty set A forms a group under the operation composition. This group is called the **permutation group** and is denoted by S_n.

Identity permutation on n symbols is given by

$$I = \begin{pmatrix} 1\,2\,3 \ldots n \\ 1\,2\,3 \ldots n \end{pmatrix}.$$

5.9 Cyclic Permutation

A permutation is said to be cyclic if it can be represented by a cycle. For example,

$$P_1 = \begin{pmatrix} 1\,2\,3\,4\,5 \\ 2\,3\,4\,5\,1 \end{pmatrix} = (1\,2\,3\,4\,5)$$

Here, (1 2 3 4 5) represents a cyclic permutation on 5 symbols with a cycle of length 5. In the cycle, the image of an element follows the element. In P_1, image of 1 is 2, image of 2 is 3, image of 3 is 4, image of 4 is 5, and image of 5 is 1.

Another cyclic permutation is

$$P_2 = \begin{pmatrix} 1\,2\,3\,4\,5 \\ 1\,3\,5\,4\,2 \end{pmatrix} = (2\,3\,5).$$

Its length is three. The element, which is the image of itself, is not mentioned in the cycle and its presence is considered as obvious. Since, image of 1 is 1 and image of 4 is 4, therefore, these are not mentioned in the cycle. A cycle of length 2 is called a **transposition**.

For example, $\begin{pmatrix} 1\,2\,3\,4\,5 \\ 4\,2\,3\,1\,5 \end{pmatrix} = (1\ 4).$

5.10 Even and Odd Permutations

A permutation is said to be even if it can be expressed as a product of an even number of transpositions. Similarly, a permutation is odd if it can be expressed as a product of odd number of transpositions. Of the $n!$ (n factorial) permutations on n symbols, $n!/2$ are even, and $n!/2$ are odd.

Remarks:

1. Identity permutation is always even.
2. The product of two even permutations is even.
3. The product of two odd permutations is even.
4. The product of one even and one odd permutation is odd.
5. The set S_n of all permutations on n symbols forms a finite group of order $n!$ with respect to composition of permutations.
6. The set A_n of all even permutations of degree n forms a finite group of order $n!/2$ with respect to permutation multiplication. It is called **alternating group of permutations.**

EXAMPLE 5.13 Consider a permutation

$$P = \begin{pmatrix} 1\,2\,3\,4\,5 \\ 1\,2\,4\,3\,5 \end{pmatrix}.$$

Then P can be written as (1)(2)(3 4)(5) or simply (3 4). It is the product of four cycles having one transposition. Therefore, it is an odd permutation.

EXAMPLE 5.14 Consider a permutation

$$P = \begin{pmatrix} 1\,2\,3\,4\,5\,6\,7 \\ 1\,2\,4\,3\,5\,7\,6 \end{pmatrix}.$$

Then P can be written as (3 4)(6 7). It is the product of two transpositions. Therefore, it is an even permutation.

EXAMPLE 5.15 Consider a permutation

$$\begin{aligned} P &= \begin{pmatrix} 1\,2\,3\,4\,5\,6 \\ 6\,5\,2\,4\,3\,1 \end{pmatrix} \\ &= (1\ 6)(2\ 5\ 3) \\ &= (1\ 6)(2\ 5)(2\ 3). \end{aligned}$$

It is the product of three transpositions. Therefore, it is an odd permutation.

EXAMPLE 5.16 Set S_3 of all permutations on three symbols is a finite non Abelian group of order 6 with respect to operation 'permutation composition'.

Let $A = \{1, 2, 3\}$. Then there will be 3! = 6, permutations.

$$P_1 = \begin{pmatrix} 1\,2\,3 \\ 1\,2\,3 \end{pmatrix}, P_2 = \begin{pmatrix} 1\,2\,3 \\ 1\,3\,2 \end{pmatrix}, P_3 = \begin{pmatrix} 1\,2\,3 \\ 3\,2\,1 \end{pmatrix}$$

$$P_4 = \begin{pmatrix} 1\,2\,3 \\ 2\,1\,3 \end{pmatrix}, P_5 = \begin{pmatrix} 1\,2\,3 \\ 2\,3\,1 \end{pmatrix}, P_6 = \begin{pmatrix} 1\,2\,3 \\ 3\,1\,2 \end{pmatrix}$$

Note that, identity permutation

$$I = P_1 = \begin{pmatrix} 1\,2\,3 \\ 1\,2\,3 \end{pmatrix}.$$

Further,

$$\begin{aligned} P_2P_2 &= I \\ P_3P_3 &= I \\ P_4P_4 &= I \\ P_6P_5 &= I. \end{aligned}$$

Therefore,

$$\begin{aligned} (P_2)^{-1} &= P_2 \\ (P_3)^{-1} &= P_3 \\ (P_4)^{-1} &= P_4 \\ (P_5)^{-1} &= P_6 \\ (P_6)^{-1} &= P_5. \end{aligned}$$

Table 5.3 gives Cayley table of S_3.

Table 5.3 Cayley Table of S_3

$*$	P_1	P_2	P_3	P_4	P_5	P_6
P_1	P_1	P_2	P_3	P_4	P_5	P_6
P_2	P_2	P_1	P_6	P_5	P_4	P_3
P_3	P_3	P_5	P_1	P_6	P_2	P_4
P_4	P_4	P_6	P_5	P_1	P_3	P_2
P_5	P_5	P_3	P_4	P_2	P_6	P_1
P_6	P_6	P_4	P_2	P_3	P_1	P_5

5.11 Simple Group

A group G is said to be **simple** if it has no proper nontrivial normal subgroups. That is, a simple group G has only normal subgroups G, itself and $\{e\}$.

EXAMPLE 5.17 The alternating group A_n is simple for $n \geq 5$.

5.12 Direct Product of Groups

In this section, we shall discuss two types of direct products, external direct product and internal direct product. These are given below.

5.12.1 External Direct Product

Let $< G, * >$ and $< T, \Delta >$ be two groups. Then the direct product of G and T is an algebraic structure $< G \times T, \lozenge >$, with operation '$\lozenge$' defined as:

$(g_1, t_1) \lozenge (g_2, t_2) = (g_1 * g_2, t_1 \Delta t_2)$, where $(g_1, t_1), (g_2, t_2) \in G \times T$; $g_1, g_2 \in G$ and $t_1, t_2 \in T$. It is called **external direct product** of groups.

It is easy to show that direct product of groups is closed and associative with identity (e_1, e_2), where $e_1 \in G_1$ and $e_2 \in G_2$. It will also possess inverse of each element. Therefore, it is a group.

5.12.2 Internal Direct Product

Suppose $< G, * >$ is a group. Also, let H and K be two normal subgroups of G, then G is said to be the internal direct product of H and K if

(i) $G = HK$, i.e., $G = \{g \mid g = hk$, where $h \in H$ and $k \in K\}$.

(ii) $H \cap K = \{e\}$.

In general, a group may be the internal direct product of its n normal subgroups.

Alternatively, let G has subgroups H_i for $i = 1$ to n. Then G is called internal direct product of the subgroups H_i if the mapping

$$f: \prod_{i=1}^{n} H_i \to G$$

given by,

$$f(h_1, h_2, h_3, \ldots, h_n) = h_1 h_2 h_3 \ldots h_n$$

is an isomorphism.

5.13 Coding Theory

Group theory has an important application in signal transmission. When information in electronic form is transmitted from one place to the other, it may get distorted due to interference with some other electronic signals. Information, which is coded into binary digits, contains a sequence of words having 0s and 1s. During transmission, it is possible that some 0s may change into 1s and vice versa. Therefore, it is desirable that at the receiving end, such errors should be detected and corrected. A code is a set of words used to represent information or message. A word in a code is generally called a codeword. Information is translated into a sequence of codewords before transmission. A **block code** is a special code that consists of words of same length. In this section, we will discuss a coding mechanism, called **group code** that is used for error correction in the transmitted word. Before discussing group code, we will study a simple mechanism for error detection. It is called parity check. It enables the detection of single bit error.

5.13.1 Parity Check

A **parity bit** is an additional bit that is added to each codeword to ensure that the number of 1s in the codeword is even or odd. Parity bits are used as the simplest form of error detecting code. There are two variants of parity bits: **even parity** and **odd parity**. In even parity, the parity bit is set to 1 if the number of 1s in a given codeword (excluding the parity bit) is odd, thus, making the total number of 1s in the codeword (including the parity bit) even. If the number of 1s is already even, parity bit is set to a 0. In odd parity, the parity bit is set to 1 if the number of 1s in a given codeword (excluding the parity bit) is even, keeping the number of 1s in the codeword (including the parity bit) odd. If the number of 1s is already odd, the odd parity bit is set to 0.

In even parity, if a codeword (including the parity bit) is transmitted incorrectly, the parity bit will be incorrect and thus, indicates that an error occurred in transmission. The parity bit is only suitable for detecting errors, it cannot correct any error, as there is no way to determine which particular bit is corrupted. The data must be discarded entirely, and re-transmitted from the scratch. Therefore, on a noisy transmission medium, successful transmission can take a long time. However, parity has the advantage that it uses only a single bit and requires only XOR gates to generate the code. Parity bit checking is used occasionally for transmitting ASCII characters, which have 7 bits, leaving the 8th bit as a parity bit.

Let us understand the even parity with the help of an example.

1. Transmission using even parity
   Codeword to be transmitted (7-bit word) = 1001001
   Parity bit value = 1
   Effective codeword (8-bit word) = 1001001$\underline{1}$
   Codeword received = 10010001
   Computed parity = 0 (odd number of 1s)
   Error is reported.
2. Transmission using odd parity
   Codeword to be transmitted (7-bit word) = 1001001
   Parity bit value = 0
   Effective codeword (8-bit word) = 1001001$\underline{0}$
   Codeword received = 10010000
   Computed parity = 1 (even number of 1s)
   Error is reported.

5.13.2 Group Codes

As stated earlier, parity mechanism can only detect error, but cannot correct it, and that too of one bit only. Group code is a scheme that does both, i.e., detects as well as corrects the error. There may be implementation of group code in many ways. We will discuss one mechanism that works on maximum likelihood principle, i.e., a received word is decoded as the word which is closest to a codeword in a given list of codewords.

Let G be the set of all words of length n. **Length** of a word is the number of bits in the word. It is denoted by $|w|$, where w is some word. For instance, $|0110110| = 7$.

A subset H of G is called a group code, if $< H, \oplus >$ is a subgroup of $< G, \oplus >$, where '$\oplus$' is a binary operation such that

$$z = x \oplus y$$

is a word that has 1s in those positions where x and y differ, and has 0s in those positions where x and y are same.

For instance, let $x = 0110101$ and $y = 1001111$.

Then
$$\begin{aligned} z &= x \oplus y \\ &= 0110101 \oplus 1001111 \\ &= 1111010. \end{aligned}$$

Further, we define **weight of a word** x, denoted as $w(x)$, to be number of 1s in x. Therefore, $w(1111010) = 5$. We also define

$$\text{Distance } d(x, y) = w(x \oplus y).$$

Consider a coding function defined as.

$$f : X^3 \rightarrow X^6,$$

where $X^3 = \{x_1x_2x_3 \mid x_1, x_2, x_3 \in \{0, 1\}\}$. Let the list of codewords is given in Table 5.4. Here X^3 is a set of words to be transmitted. Before transmission, words in X^3 are coded to get a set of codewords C.

Table 5.4 Codeword

S. No.	X^3 = *word*	*Codeword* $C = f(word)$
1	000	000000
2	001	001100
3	010	010011
4	011	011111
5	100	100101
6	101	101001
7	110	110110
8	111	111010

The rationale behind selecting this list of codewords is that C = {000000, 001100, 010011, 011111, 100101, 101001, 110110, 111010} forms a subgroup of $< X^6, \oplus >$, where X^6 is the set containing words of length 6 over the alphabet {0, 1}. Further, we can see that the first three digits are from the set X^3.

Now, let us find the cosets of C. We can find cosets by operating each element of X^6 with the elements of C. Since, the cosets of a subgroup are disjoint and form the partition of the group, therefore, in this case, there will be eight cosets in all, including C itself. Other cosets will be $S_1, S_2, S_3, \ldots, S_7$. These cosets are listed in Table 5.5.

Table 5.5 Cosets of C

S. No.	*C*	*Cosets*						
		S_1	S_2	S_3	S_4	S_5	S_6	S_7
1	000000	000001	000010	000100	010000	100000	000110	010100
2	001100	001101	001110	001000	011100	101100	001010	011000
3	010011	010010	010010	010111	000011	110011	010101	000111
4	011111	011110	011101	011011	011111	111111	011001	001011
5	100101	100100	100111	100001	110101	000101	100011	110001
6	101001	101000	101011	101101	111001	001001	101111	111101
7	110110	110111	110100	110010	**100110**	010110	110000	100010
8	111010	111011	111000	111110	101010	011010	111100	101110

Now, question is how to decode the received word. The mechanism is very simple. Suppose we receive the word 100110, which is shown in bold and underlined in row 7, and under the column S_4. It is decoded as 110110, which is the first word of row 7. From this word, we can separate the first three digits, i.e. 110, to get the actual word. Again the rationale is that 100110 is closest to 110110, since $d(110110, 100110) = w(110110 \oplus 100110) = w(010000) = 1$.

The distance of other elements of C from 100110 is as follows:

$$w(000000 \oplus 100110) = 3$$
$$w(001100 \oplus 100110) = 3$$
$$w(010011 \oplus 100110) = 4$$
$$w(011111 \oplus 100110) = 4$$
$$w(100101 \oplus 100110) = 2$$
$$w(101001 \oplus 100110) = 4$$
$$w(111010 \oplus 100110) = 3$$

Similarly, we can decode other words. Any received word is decoded as the first word of the row in which the received word belong, and then from this, actual word can be separated. Group code mechanism can further be improved by combining it with parity check scheme.

The motive to discuss coding theory is to show an important application of group theory.

5.14 Solved Problems

PROBLEM 5.1 Prove that the set $S = \{a + b\sqrt{3} \mid a, b \in \mathbb{Q}$, where $\mathbb{Q}$ is the set of rational numbers$\}$, forms a group with respect to addition.

Solution

Closure Let $x = a_1 + b_1\sqrt{3}$ and $y = a_2 + b_2\sqrt{3}$.

Then, $c = x + y = (a_1 + a_2) + (b_1 + b_2)\sqrt{3}$.
Therefore, S is closed.

Associative Since $a + b\sqrt{3}$ is a real number and addition of real numbers is associative, therefore, associativity is satisfied.

Identity $0 + 0\sqrt{3}$ will be the identity of the group.

Inverse $(-a) + (-b)\sqrt{3}$ will be the inverse of $a + b\sqrt{3}$, i.e., inverse of each element exists.

Thus, S forms a group with respect to addition. Further, it is an abelian group, since addition of real numbers is commutative.

PROBLEM 5.2 Prove that the group $G = \langle \{1, 2, 3, 4, 5, 6\}, \times_7 \rangle$ is cyclic.

Solution

If G is cyclic then we can find the generators. By observation, we can see that 1 and 2 are not generators. However, 3 is a generator, since all elements may be generated by it. This is obvious from the following steps.

$$3^1 = 3$$
$$3^2 = 3 \times_7 3 = 2$$
$$3^3 = 3 \times_7 3 \times_7 3 = 6$$

$$3^4 = 3 \times_7 3 \times_7 3 \times_7 3 = 6 \times_7 3 = 4$$

$$3^5 = 3 \times_7 3 \times_7 3 \times_7 3 \times_7 3 = 4 \times_7 3 = 5$$

$$3^6 = 3 \times_7 3 \times_7 3 \times_7 3 \times_7 3 \times_7 3 = 5 \times_7 3 = 1$$

Since, all elements are generated by one element 3, therefore, G is cyclic. Further, $3^6 = 1$, therefore, order of 3 is 6.

Thus, $G = < \{3^1, 3^2, 3^3, 3^4, 3^5, 3^6\}, \times_7 >$.

Also, 5 is relatively prime to 6, therefore, $3^5 = 5$ is also a generator.

{refer Section 5.7, theorem 4}

PROBLEM 5.3 Prove that $G_1 = <\{0, 1, 2, 3\}, +_4>$ is isomorphic to $G_2 = <\{1, 2, 3, 4\}, \times_5>$.

Solution

Since, identity of G_1 is 0 and identity of G_2 is 1, therefore, we can take $f(0) = 1$.

Further, in group G_1 : $o(1) = 4$

$o(2) = 2$

$o(3) = 4$.

Similarly, in group G_2 : $o(2) = 4$

$o(3) = 4$

$o(4) = 2$.

Here, we observe that $f(2) = 4$. In addition, we can take $f(3) = 3$ and $f(1) = 2$. Thus, the mapping is completely defined.

Tables 5.6 and 5.7 of two groups are given.

Table 5.6 Cayley Table for Addition Modulo

$+_4$	0	1	2	3
0	0	1	2	3
1	1	2	3	0
2	2	3	0	1
3	3	0	1	2

Table 5.7 Cayley Table for Multiplication Modulo

$\times_5$	1	2	4	3
1	1	2	4	3
2	2	4	3	1
4	4	3	1	2
3	3	1	2	4

Observe that the two tables are identical except the changes in the names of the elements. Therefore, there is one-one onto mapping between two groups; i.e., two groups are isomorphic.

PROBLEM 5.4 Let $< Z_2, +_2 >$ and $< Z_3, +_3 >$ be two groups. Find the external direct product.

Solution

Here, $Z_2 = \{0, 1\}$ and $Z_3 = \{0, 1, 2\}$.

Therefore, $Z_2 \times Z_3 = \{(0, 0), (0, 1), (0, 2), (1, 0), (1, 1), (1, 2)\}$.

It is a group of order 6.

Exercises

5.1 Prove that the identity in a group is unique.

5.2 Prove that left and right cancellation laws hold in a group G.

5.3 Which of the following groups are cyclic? Also, find the generators in each case.

(i) $G = < \mathbb{Z}, + >$

(ii) $G = < \mathbb{Q}, + >$

(iii) $G = < 6\mathbb{Z}, + >$

(iv) $G = \{a + b\sqrt{2} \mid a, b \in \mathbb{Z}\}$

5.4 Prove that a subgroup of a cyclic group is cyclic.

5.5 Let $\mathbb{Z}$ be the set of integers and $\mathbb{E}$ be the set of all even integers. Prove that the groups $< \mathbb{Z}, + >$ and $< \mathbb{E}, + >$ are isomorphic with respect to mapping $f : \mathbb{Z} \to \mathbb{E}$ given by $f(a) = 2a$.

5.6 Let S be the set of ordered pairs of integers. An operation '$*$' is defined on the elements of S by

$$< a_1, b_1 > * < a_2, b_2 > = < a_1 + a_2, b_1 + b_2 >.$$

Show that $< S, * >$ is a commutative group.

5.7 Consider the set $\mathbb{Q}$ of rational numbers, and let '$*$' be the operation on $\mathbb{Q}$ defined by

$$a * b = a + b - a \cdot b.$$

Prove that $< \mathbb{Q}, * >$ is a group.

5.8 Prove that an infinite cyclic group is isomorphic to the group $\mathbb{Z}$ of integers under addition.

5.9 Prove that every group is isomorphic to a group of permutation (Cayley's theorem).

Ring Theory

6.1 Introduction

In the previous chapters, we discussed only those algebraic structures that involve one binary operation. In this chapter, we shall study an algebraic structure which is defined on two binary operations, usually taken as addition and multiplication. It is called ring. The ring has many important applications in the field of science.

6.2 Definition of Ring

A ring $< R, +, \cdot >$ is a set, R, together with two binary operations usually '+' and '·' which are called addition and multiplication respectively, that satisfies the following axioms:

1. **Closure** For all $a, b \in R$
   (a) $a + b \in R$.
   (b) $a \cdot b \in R$.
2. **Associative** For all $a, b, c \in R$.
   (a) $(a + b) + c = a + (b + c)$.
   (b) $(a \cdot b) \cdot c = a \cdot (b \cdot c)$.
3. **Identity** There exists an element e (additive identity), such that
   $$\forall a \in R, e + a = a + e = a.$$
4. For each $a \in R$, there exists a^{-1}, such that
   $$a + a^{-1} = a^{-1} + a = e.$$

5. For all $a, b \in R$

$$a + b = b + a.$$

6. For all $a, b, c \in R$
   (a) $a \cdot (b + c) = a \cdot b + a \cdot c.$
   (b) $(a + b) \cdot c = a \cdot c + b \cdot c.$

In other words, $< R, +, \cdot >$ is a ring if the following axioms are satisfied.

1. $< R, + >$ is an additive abelian group.
2. $< R, \cdot >$ is a multiplicative semigroup.
3. For all $a, b, c \in R$,
   (i) $a \cdot (b + c) = a \cdot b + a \cdot c.$
   (ii) $(a + b) \cdot c = a \cdot c + b \cdot c.$

In addition, if multiplicative identity which is generally 1, also exists, then the ring is called **ring with unity**. Further, if $a \cdot b = b \cdot a$ for all a, b in R, then $< R, +, \cdot >$ is called a **commutative ring**.

Remark: Usually, we write ab in place of $a \cdot b$.

EXAMPLE 6.1 Ring with one element: Let R be a set which contains only one element, 0. Then, the set R together with two operations '+' and '·' is a commutative ring. It is generally called Zero ring.

EXAMPLE 6.2 Ring of integers: The set of integers, with operation addition and multiplication, i.e., $< \mathbb{Z}, +, \cdot >$ is a ring with 0 as additive identity. It is a commutative ring.

EXAMPLE 6.3 Ring of rational numbers: The set of rational numbers, with operation addition and multiplication, i.e., $< \mathbb{Q}, +, \cdot >$ is a commutative ring with 0 as additive identity.

Similarly, $< R, +, \cdot >$ is a commutative ring with 0 as additive identity. It is a ring of rational numbers.

EXAMPLE 6.4 Let $R = \{0, 1, 2, 3, 4, 5\}$. Then $< R, +_6, \times_6 >$ is a commutative ring with unity. Here 0 and 1 are additive and multiplicative identities, respectively.

EXAMPLE 6.5 Ring of complex numbers: The set of complex numbers, with operation addition and multiplication, i.e., $< \mathbb{C}, +, \cdot >$ is a commutative ring with 0 as additive identity.

EXAMPLE 6.6 Ring of matrices: Let R be the set of all square matrices of order n. Then it will form a ring with respect to matrix addition and matrix multiplication. Identity, in this case will be the null matrix.

6.3 Ring of Polynomials

Suppose R be a set of polynomials. Then $< R, +, \cdot >$ will be a ring with respect to the operations: polynomial addition and polynomial multiplication.

Let

$$P_1 = a_0 + a_1x + a_2x^2 + \cdots + a_nx^n$$

and

$$P_2 = b_0 + b_1x + b_2x^2 + \cdots + b_mx^m$$

be two polynomials then, the operations addition and multiplication are defined as:

$$P_1 + P_2 = (a_0 + b_0) + (a_1 + b_1)x + (a_2 + b_2)x^2 + \cdots$$

$$P_1P_2 = c_0 + c_1x + c_2x^2 + \cdots + c_{m+n}x^{m+n}$$

where $$c_i = \Sigma_{j+k=i}\, a_j\, b_k,\ 0 \le i \le m + n.$$

6.4 Subring and Ideal

Let $< R, +, \cdot >$ be a ring, and S be a nonempty subset of R. Then $< S, +, \cdot >$ will be called a subring, if it is a ring itself.

Formally,

Let $< R, +, \cdot >$ be a ring and $S \subseteq R$, then S will be called a subring, if

1. $< S, + >$ is an additive abelian group.
2. $< S, \cdot >$ is a multiplicative semigroup.
3. For all $a, b, c \in S$,
   (i) $a \cdot (b + c) = a \cdot b + a \cdot c$.
   (ii) $(a + b) \cdot c = a \cdot c + b \cdot c$.

Every ring R has two **trivial subrings**: $< \{0\}, +, \cdot >$ and R itself.

The concept of **ideal** in ring theory is analogous to the normal subgroup in group theory.

Let $< R, +, \cdot >$ be a ring and $S \subseteq R$. Then S is called

1. **a left ideal** of $< R, +, \cdot >$, if
   (i) $a, b \in S \Rightarrow a - b \in S$,
   (ii) $a \in S$ and $r \in R \Rightarrow ra \in S$.
2. **a right ideal** of $< R, +, \cdot >$ if
   (i) $a, b \in S \Rightarrow a - b \in S$,
   (ii) $a \in S$ and $r \in R \Rightarrow ar \in S$.

If an ideal is both, a left ideal and a right ideal, then it is called an **ideal** (or a **two-sided ideal**).

EXAMPLE 6.7 In a ring of integers $< \mathbb{Z}, +, \cdot >$, every subring is an ideal. For instance, let S be a subring of $\mathbb{Z}$. Also, let $a \in S$ and $r \in \mathbb{Z}$, then

$$\begin{aligned} ar &= a + a + a + \cdots + a\ \{r \text{ times}\}, && \text{if } r > 0 \\ &= 0, && \text{if } r = 0 \\ &= -a - a - a - \cdots - a\ \{r \text{ times}\}, && \text{if } r < 0 \end{aligned}$$

$$\Rightarrow \qquad ar \in S.$$

Thus, S is an ideal of $\mathbb{Z}$.

6.5 Maximal Ideal

An ideal $I \neq R$ in a ring $< R, +, \cdot >$ is said to be a maximal ideal of R, if whenever S is an ideal of R and $I \subseteq S \subseteq R$, then either $R = S$ or $I = S$.

6.6 Prime Ideal

Let I be a proper ideal of a commutative ring $< R, +, \cdot >$. Then I is said to be a prime ideal of R if for all $a, b \in R$, $ab \in I \Rightarrow a \in I$ or $b \in I$.

6.7 Principal Ideal

Let $< R, +, \cdot >$ be a ring. An ideal I of R is said to be principal ideal if there exists an element $a \in I$, such that any other ideal I' of R containing a also contains I. If every ideal in a ring $(R, +, \cdot)$ is a principal ideal then $(R, +, \cdot)$ is said to be a principal ideal ring.

6.8 Characteristic of a Ring

Let $< R, +, \cdot >$ be a ring. If there exists a positive integer n such that $na = 0$ for each $a \in R$, then the smallest such positive integer is called the characteristic of R. However, R is said to be of characteristic zero if no such positive integer exists. For instance, the ring of integers has characteristic zero.

6.9 Ring Homomorphism

Let $< R_1, +, \cdot >$ and $< R_2, +, \cdot >$ be two rings. A mapping $f: R_1 \rightarrow R_2$, is called a ring homomorphism if, for all $a, b \in R$,

1. $f(a + b) = f(a) + f(b)$ and
2. $f(ab) = f(a) f(b)$.

In addition, if mapping f is one-to-one and onto, then it is called isomorphism. In other words, two rings R_1 and R_2 are said to be isomorphic, if $f: R_1 \rightarrow R_2$ is a one-to-one and onto. Symbolically, it is written as $R_1 \approx R_2$. An isomorphism of a commutative ring from R to R is called an **automorphism** of R.

EXAMPLE 6.8 The function that maps each integer x to its remainder modulo 4 (a number in $\{0, 1, 2, 3\}$) is a homomorphism from the ring Z to the ring Z_4.

6.10 Field and Integral Domain

A ring $< R, +, \cdot >$ whose nonzero elements form a group under multiplication is called a **division ring**. If in addition, R is also commutative then it is called a **field**. Further, a ring $< R, +, \cdot >$ is called **integral domain** if for $a, b \in R$,

$$ab = 0,$$

$$\Rightarrow \quad a = 0 \text{ or } b = 0.$$

EXAMPLE 6.9

1. The set of integers form a commutative ring without zero divisors. It is also an **integral domain**.
2. Every field is an integral domain and every integral domain is a commutative ring. Furthermore, any finite integral domain is a field.
3. The ring of all polynomials with real coefficients is also an integral domain, but the ring of all real valued functions is not an integral domain.

6.11 Kernel of Homomorphism

Let f: $R_1 \to R_2$ be a ring homomorphism. The set I of those elements of R_1 which are mapped on the additive identity of R_2 is called the kernel of the homomorphism f, i.e., $I = \{x \in R_1 \mid f(x) = 0_2$, where 0_2 is the additive identity of $R_2\}$. Further, it can be seen that I is an ideal of the ring R_1.

6.12 Quotient Ring

Let $< I, +, \cdot >$ be an ideal of a ring $< R, +, \cdot >$. Then R/I is the set of all equivalence classes or additive cosets given by, $R/I = \{I + a \mid a \in R\}$. The algebraic structure $< R/I, +, \cdot >$ is called the **quotient ring** under the operations defined as

$$(I + a) + (I + b) = I + (a + b)$$

and

$$(I + a) \cdot (I + b) = I + (a \cdot b).$$

6.13 Direct Product of Rings

Let $< R, +, \cdot >$ and $< S, +, \cdot >$ be two rings. Then the direct product of R and S forms a ring $< R \times S, \oplus, \otimes >$ with operations $\oplus$ and $\otimes$ defined as

$$(a, b) \oplus (c, d) = (a + c, b + d)$$

and

$$(a, b) \otimes (c, d) = a \cdot c, b \cdot d).$$

In general, if $R_1, R_2, R_3, \ldots, R_n$ are n rings, then the direct product of these rings will consist of a set of n-tuples, which will also form a ring under the operation, addition and multiplication of n-tuples, similar to $\oplus$ and $\otimes$, respectively.

6.14 Solved Problems

PROBLEM 6.1 Let $< R, +, \cdot >$ be a ring, such that $x^2 = x$, $\forall x \in R$. Prove that

(i) $a + a = 0$
(ii) if $a + b = 0$ then $a = b$
(iii) R is a commutative ring.

Solution

(i) Since R is a ring, therefore,

$$a \in R \Rightarrow a + a \in R.$$

Also, $(a + a)^2 = a + a$ {given}

$\Rightarrow$ $(a + a)(a + a) = a + a$

$\Rightarrow$ $a(a + a) + a(a + a) = a + a$ {using distributive law}

$\Rightarrow$ $(a^2 + a^2) + (a^2 + a^2) = a + a$

$\Rightarrow$ $(a + a) + (a + a) = a + a$

$\Rightarrow$ $(a + a) + (a + a) = (a + a) + 0$ {0 is additive identity}

$\Rightarrow$ $(a + a) = 0.$

(ii) $a + b = 0$

$\Rightarrow$ $a + b = a + a$

$\Rightarrow$ $a = b.$ {by left cancellation}

(iii) Since $x^2 = x,$ {given}

$\Rightarrow$ $(a + b)^2 = a + b$ {taking $x = a + b$}

$\Rightarrow$ $(a + b)(a + b) = a + b$

$\Rightarrow$ $a(a + b) + b(a + b) = a + b$ {using distributive law}

$\Rightarrow$ $(a^2 + ab) + (ba + b^2) = a + b$

$\Rightarrow$ $(a + ab) + (ba + b) = a + b$

$\Rightarrow$ $(a + b) + (ab + ba) = (a + b) + 0$ {0 is additive identity}

$\Rightarrow$ $ab + ba = 0$ {by left cancellation}

$\Rightarrow$ $ab = ba.$ {using part (ii)}

PROBLEM 6.2 Let $< X, +, \cdot >$ be a ring, where $X = \{a + b\sqrt{2} \mid a, b \in \mathbb{R}\}$. Prove that, it is an integral domain.

Solution

(i) Let us show that X is abelian group under addition.

Closure

Let $a_1 + b_1\sqrt{2},\ a_2 + b_2\sqrt{2} \in X.$

Then $a_1 + b_1\sqrt{2} + a_2 + b_2\sqrt{2} = (a_1 + a_2) + (b_1 + b_2)\sqrt{2}.$

Also $(a_1 + a_2),\ (b_1 + b_2) \in \mathbb{R},$

$\Rightarrow$ $(a_1 + a_2) + (b_1 + b_2)\sqrt{2} \in X.$

Therefore, X is closed.

Associative

Since $a + b\sqrt{2}$ is a real number and addition of real numbers is associative, therefore, associativity is satisfied.

Identity

$0 + 0\sqrt{2}$ will be the identity of the group.

Inverse

$(-a) + (-b)\sqrt{2}$ will be the inverse of $a + b\sqrt{2}$ i.e., inverse of each element exists. Thus, S forms a group with respect to addition. Further, it is abelian group, since addition of real numbers is commutative.

(ii) Let us show that X is semigroup under multiplication.

Closure

$$(a_1 + b_1\sqrt{2})(a_2 + b_2\sqrt{2}) = a_1a_2 + a_1b_2\sqrt{2} + a_2b_1\sqrt{2} + 2b_1b_2$$
$$= (a_1a_2 + 2b_1b_2) + (a_1b_2 + a_2b_1)\sqrt{2}$$

Here $\quad (a_1a_2 + 2b_1b_2),\ (a_1b_2 + a_2b_1)\sqrt{2} \in \mathbb{R}.$

$\Rightarrow \quad (a_1a_2 + 2b_1b_2) + (a_1b_2 + a_2b_1)\sqrt{2} \in X.$

Therefore, X is closed.

Associative

Since $a + b\sqrt{2}$ is a real number and multiplication of real numbers is associative, therefore, associativity is satisfied.

(iii) The operation in X are usual addition and multiplication, hence distributive axiom will also be satisfied.

Thus, $< X, +, \cdot >$ is a ring.

In addition, $1 + 0\sqrt{2}$ is multiplicative identity.

Further, $\dfrac{1}{a+\sqrt{2}}$ will be the multiplicative inverse of $a + b\sqrt{2}$.

Hence, $< X, +, \cdot >$ is a field.

PROBLEM 6.3 Consider two rings $< R_1, +, \cdot >$ and $< R_2, +, * >$, where operations + and $\cdot$ are usual addition and multiplication. However, operation $*$ is defined as

$$a * b = \frac{ab}{2}.$$

Prove that R_1 is isomorphic to R_2.

Solution Consider a mapping $f: R_1 \to R_2$, such that

$$f(x) = 2x,\ \forall x \in R_1.$$

Obviously, f is one-one onto, since

$$f(x_1) = f(x_2)$$

$\Rightarrow \quad 2x_1 = 2x_2$

$\Rightarrow \quad x_1 = x_2;\ \forall x_1, x_2 \in R_1.$

Further,
$$f(x_1 + x_2) = 2(x_1 + x_2)$$
$$= 2x_1 + 2x_2$$
$$= f(x_1) + f(x_2).$$

Also,
$$f(x_1 \cdot x_2) = 2(x_1 \cdot x_2)$$
$$= \frac{2x_1 \cdot 2x_2}{2}$$

$$= 2x_1 * 2x_2$$
$$= f(x_1) * f(x_2).$$

It proves that R_1 is isomorphic to R_2.

PROBLEM 6.4 Prove that every homomorphic image of a commutative ring is commutative.

Solution Let $< R_1, +, \cdot >$ and $< R_2, +, * >$ be two rings and $f: R_1 \to R_2$ be a homomorphism. In addition, let R_1 be commutative.

Let $x_1, y_1 \in R_1$ such that

$$f(x_1) = x_2$$

and

$$f(y_1) = y_2.$$

Therefore,

$$\begin{aligned} x_2 y_2 &= f(x_1) f(y_1) \\ &= f(x_1 y_1) && \{f \text{ is homomorphism}\} \\ &= f(y_1 x_1) && \{\text{since } R_1, \text{ is commutative}\} \\ &= f(y_1) f(x_1) \\ &= y_2 x_2. \end{aligned}$$

Hence, R_2 is commutative.

Exercises

6.1 If $< R, +, \cdot >$ is a ring with additive identity 0 and $a, b \in R$, then prove that

(i) $0a = a0 = 0$,
(ii) $a(-b) = (-a)b - (ab)$,
(iii) $(-a)(-b) = ab$.

6.2 Find the units in each of the following rings.

(i) $\mathbb{Z}$
(ii) $\mathbb{Z} \times \mathbb{Z}$
(iii) $\mathbb{Z}_4$
(iv) $\mathbb{Z} \times \mathbb{Q} \times \mathbb{Z}$

(An element u in R is called a **unit** of R if it has a multiplicative inverse in R)

6.3 Show that the rings $2\mathbb{Z}$ and $3\mathbb{Z}$ are not isomorphic.

6.4 Show that the fields $\mathbb{R}$ and $\mathbb{C}$ are not isomorphic.

6.5 A ring R is a Boolean ring if $a^2 = a$ for all $a \in R$. Show that every Boolean ring is commutative.

6.6 Prove that, in a ring $\mathbb{Z}_n$, the divisors of zero are those elements which are not relatively prime to n.

6.7 Prove that every field F is an integral domain.

6.8 Prove that, if p is prime then $\mathbb{Z}_p$, has no divisors of zero.

6.9 Find the characteristics of each of the following rings:

(i) $2\mathbb{Z}$
(ii) $\mathbb{Z}_3 \times 3\mathbb{Z}$
(iii) $\mathbb{Z}_3 \times \mathbb{Z}_4$
(iv) $\mathbb{Z}_6 \times \mathbb{Z}_{15}$

Part B
MATHEMATICAL LOGIC

7

Statement Calculus

7.1 Introduction

In this chapter, the aim is to symbolize the language sentences and transform them into a form that is independent of the context. The rules for reasoning and drawing conclusions are also developed.

7.2 Propositional Logic

Propositional logic is also called symbolic logic. The symbols for statements and their connectives (logical operators) are used to form a language which is suitable for computer programming. Here, natural language English is used as the basis or the first language. The language we obtain after symbolization is the **target** or **object** language. The syntax of the object language is well defined. The natural language (English in this case) is called the 'meta language'.

7.3 Statement

A statement is a declarative sentence of the meta language. It must have one and only one of the two possible truth-values: true or false. A statement may be simple or compound. A simple statement is the primitive declarative sentence which cannot be broken down further. It is also called atomic.

A compound statement is formed by joining two or more simple statements using connectives. A statement is also called a proposition. It must always be true or false but not both, i.e., there must not be any ambiguity in the truth-value of the statement.

EXAMPLE 7.1 Consider the following propositions:

Statements	Truth-value
1. Canada is a country.	(True)
2. The sun sets in the north.	(False)
3. New Delhi is the capital of India.	(True)
4. Paris is in France.	(True)
5. Denmark is the capital of U.K.	(False)
6. 2 + 2 = 4.	(True)

All the statements given above are simple (or atomic). Their truth-values are shown against them.

The compound statements are as follows:

1. Canada is a country and London is a city.
2. If Paris is in France then 2 + 2 makes four.
3. I am either reading or watching the television.

Truth-values of these statements will depend on the truth-values of their constituent simple statements.

The following sentences cannot be taken as statements, and hence cannot be considered for translation into the **target language**.

1. Close the door.
2. Who are you?
3. There are hundred millions of stars in the sky.
4. She is beautiful.
5. Water level of the sea will rise by five metres in the next hundred years.

Reasons: As per the definition of statement, sentences 1 and 2 cannot be considered as statements, because these sentences are the commands, and hence do not have any truth-value. The truth-value of sentence 3 is uncertain and is subject to physical verification. Further, 'beauty' is person dependent. Some may consider a person beautiful while others may not. Therefore, the truth-value of the sentence 4 is ambiguous. Truth-value of sentence 5 cannot be determined at present. It is time dependent and may take even hundred years for its determination.

7.4 Symbols

Simple statements are represented by capital letters $A, B, C, \ldots, Z$ etc. Generally, the first letter of the prominent word in the statement is chosen to represent the statement, although there is no such rule. In fact, any letter can be selected arbitrarily to represent the atomic statement.

EXAMPLE 7.2

Statements	Symbols
1. Canada is a country.	C
2. Man is mortal.	M
3. 2 + 2 = 4.	P (taken arbitrarily)
4. The Earth is revolving around the Sun.	R (revolving)

In statement 1, Canada and country both are prominent words, therefore, C is taken as symbol. Similar is the case in statement 2. P is chosen arbitrarily for statement 3. Letter R is chosen to represent statement 4, because it is the first letter of the word 'revolving', which can be considered as prominent.

7.5 Formula

A mathematical model of a statement into a form that contains symbols (variables), logic operators and parentheses '( )' is called a formula. Parentheses are used to remove the ambiguity. Logic operators will be discussed in Section 7.7.

EXAMPLE 7.3

1. I am eating. : E
2. I am reading. : R

Take representative variables (or symbols) as E and R. Table 7.1 illustrates the statements and their formulas.

Table 7.1 Statements and Formulas

Statements	*Formulas*
I am not eating.	$\sim E$
I am eating and reading.	$E \wedge R$
I am eating or reading.	$E \vee R$
If there are clouds then it may rain.	$C \rightarrow R$

7.6 Some More Formulas

Some examples of formulas are:

$C \wedge R$, $E \uparrow R$, $(P \oplus (\sim Q)) \uparrow (P \vee Q)$, $C \leftrightarrow R$, $E \rightarrow R$, $(P \vee (\sim Q)) \wedge (P \vee Q)$ and so on.

The well-formed formulas will be discussed in Section 7.8, after introducing the logic operators.

7.7 Logic Operators or Connectives

There are some words which connect atomic statements to form compound statements. These are called logic operators or connectives.

Most of the operators join two statements and hence are called binary operators. However, negation operates on a single statement, and therefore, is known as unary operator. The simple statements do not have any connectives. Compound statements are formed by combining simple statements using connectives. The truth-value associated with each connective is discussed with the help of truth table. The **Truth Table** is a representation used to determine the truth-values of a compound statement. Rows of the truth table represent all possible combinations of the truth-values of the constituent atomic statements. In fact, the truth table is a summary of truth-values of the resulting statements for all possible assignments of the values to the symbols appearing in the statement. Table 7.2 gives brief details of important logic operators. The detailed definitions of these operators will be given in the succeeding sections. In these sections, we will describe the English word used, the symbol, format of the formula, and pronunciation of each connective.

Table 7.2 List of Logic Operators

Words (connectives)	*Symbol*	*Format*	*Pronunciation*
NOT	$\sim$	$\sim A$	not A
AND	$\wedge$	$A \wedge B$	A and B
OR	$\vee$	$A \vee B$	A or B
If...then	$\rightarrow$	$A \rightarrow B$	A implies B
If and only if	$\leftrightarrow$	$A \leftrightarrow B$	A if and only if B
Either...or	$\oplus$	$A \oplus B$	A exor B
..not...and	$\uparrow$	$A \uparrow B$	A nand B
..not...or	$\downarrow$	$A \downarrow B$	A nor B

7.7.1 Conjunction

A conjunction, known as 'AND', is a logical connective used to combine two simple (atomic) statements. The English word used, its symbol, format of the formula and pronunciation are given below.

English word : and
Symbol : $\wedge$
Formula : $A \wedge B$
Pronounced as : A and B

Table 7.3 gives the truth-values associated with the conjunction. The formula will be true, only if all the symbols in it have truth-value true.

Table 7.3 Conjunction

A	B	$A \wedge B$
T	T	T
T	F	F
F	T	F
F	F	F

Here, A and B are symbols of constituent atomic statements.

This connective is symmetric, i.e.,

$$A \wedge B = B \wedge A.$$

EXAMPLE 7.4

1. It is raining today.
2. Two is less than three.
3. It is raining today and two is less than three.

Statement 3 is the conjunction of simple statements 1 and 2. Although statements 1 and 2 make the statement 3 almost funny, yet statement 3 is a valid construction in logic as per our definitions. To symbolize statement 3, we first assign symbols to each of its constituent statements.

Statements	**Symbols**
1. It is raining today.	R
2. Two is less than three.	L

Then, the statement 3 above becomes '$R \wedge L$' in symbolic form and is pronounced as 'R and L'. '$R \wedge L$' is in target language. Further, since there are two symbols, therefore, the truth table of this connective will contain four rows (i.e., $2^n = 4$ for $n = 2$). The last column in the table gives the truth-value of '$R \wedge L$'. The value of '$R \wedge L$' is true only in the first row, i.e., when R and L both, are true. The compound statement '$R \wedge L$' is false in all other cases (i.e., when any of the two symbols is false). Truth table given in Table 7.4 shows the logical values of $R \wedge L$.

Table 7.4 *R L*, Conjunction

R	L	$R \wedge L$
T	T	T
T	F	F
F	T	F
F	F	F

EXAMPLE 7.5

1. Jack and Jill went up the hill.

The two constituent atomic statements are:

(a) Jack went up the hill.
(b) Jill went up the hill.

These two, statements can be combined to form a compound statement using connective '$\wedge$' as given below:

2. Jack went up the hill and Jill went up the hill.

Notice that statements 1 and 2 carry the same meaning, but translating 2 is easier. Therefore, for transforming natural language (English) sentence into target language, sometimes, it is better to reorganize the sentence into a suitable form.

In these examples, the word 'and' has been translated to connective 'AND' ($\wedge$). However, it may not be always wise to do so. Consider following examples (Example 7.6 and 7.7).

EXAMPLE 7.6

1. He opened the book and started to read.

Here, the word 'and' cannot be taken as connective. If so, then we can find two simple statements (a) He opened the book, and (b) He started to read, joined together to form the following statement.

2. He started to read and he opened the book.

Since connective '$\wedge$' is symmetric, therefore, the compound statement 2 must have the same meaning as statement 1. However, statement 2 seems absurd because the sequence of actions is wrong. Therefore, statement 1 cannot be treated as compound statement. In fact, it is an atomic statement.

EXAMPLE 7.7

Jack and John are brothers.

Here, in this statement, the word 'and' cannot be treated as connective 'AND' (∧), because that will make the statement 'Jack is brother and John is brother', meaningless. Therefore, in our target language, it is atomic statement and must be assigned only one letter symbol.

Consequently, it is suggested that, before translating a natural language sentence into object language formula, proper analysis of its semantic should be done.

7.7.2 Disjunction

Two simple (atomic) statements can be combined by connective 'OR' to form a compound statement. It is called disjunction. Formally, it is described as:

English word : or
Symbol : $\vee$
Formula : $A \vee B$
Pronounced as : A or B

Table 7.5, which is the truth table of disjunction, defines the truth-values associated with it.

Table 7.5 Disjunction

A	B	$A \vee B$
T	T	T
T	F	T
F	T	T
F	F	F

It is also known as 'inclusive or' and is symmetric. The truth-value of disjunction is true whenever any of its constituent symbol is true as is the case in row 1, 2 and 3. It will only be false when all the symbols involved in the formula are false.

EXAMPLE 7.8

1. There is some problem with the tube-light or with the switch.

Its constituent statements are:

(a) There is some problem with the tube-light. T
(b) There is some problem with the switch. S

Here word 'or' in the statement 1 is in 'inclusive' sense, i.e., tube-light may be defective, switch may be defective or both may be defective. Therefore, it can be transformed into object language formula as '$T \vee S$'. Sometimes, the word 'or' cannot be considered as a connective 'or'. Examples are given below.

EXAMPLE 7.9

1. I shall watch the television or play the game.

Let us symbolize the statement

(a) I shall watch the television W
(b) I shall play the game P

The word 'or' in statement 1 is not in 'inclusive' sense, therefore, it is not connective 'OR'. However, it is 'exclusive or', i.e., both events cannot happen together. We will assign a separate symbol to 'exclusive or', and discuss it in Section 7.7.6.

EXAMPLE 7.10

Fifty or sixty people were injured in a road accident.

In this statement, 'or' is used for approximation and not as connective.

7.7.3 Negation

The negation, popularly known as NOT, is an unary operator. It reverses the logical value of the operand. Formally, it is described as:

English word : not, it is not the case, etc.
Symbol : ~
Formula : $\sim A$
Pronounced as : not A

Table 7.6, which is the truth table of negation, defines the truth-values associated with it. The formula $\sim A$ is true when A is false, and it is false when A is true.

Table 7.6 Negation

A	$\sim A$
T	F
F	T

EXAMPLE 7.11

It is raining.

Its negation is, 'It is not raining.'

Therefore, if symbol 'R' is taken for 'It is raining' then '$\sim R$' is for 'It is not raining'. It can also be written as 'It is not the case that it is raining'. There may be statements, which may not contain the word 'not' but still it may negate other statements.

EXAMPLE 7.12

He is poor.

This statement is the negation of 'He is rich.' The truth-value of negation of a statement is opposite to that of the statement.

7.7.4 Conditional

Two simple (atomic) statements can be combined by the connective **'conditional'** to form a compound statement. The formal description is as follows:

English word : If…then
Symbol : $\rightarrow$
Formula : $A \rightarrow B$
Pronounced as : A implies B, A then B

Table 7.7, which is the truth table of conditional, defines the truth-values associated with it. The formula $A \rightarrow B$ is false when A is true and B is false. In all other cases, it is true.

Table 7.7 Conditional

A	B	$A \rightarrow B$
T	T	T
T	F	F
F	T	T
F	F	T

Here A is called antecedent and B, consequent.

EXAMPLE 7.13

1. If there are clouds then it will rain.

The constituent statements are:

(a) There are clouds. C
(b) It will rain. R

Target formula will be '$C \rightarrow R$'.

Let us analyse the truth-value of the statement 1 with respect to truth table given in Table 7.7.

If the statements (a) and (b) are true, then intuitively the English sentence 1 is true, which justifies the entries in row 1. Further, if statement (a) is true but (b) is false, then statement 1 will be false. Therefore, entries in row 2 are justified. In this case, statement will take the form:

If there are clouds, it will not rain.

Let us consider entries in row 4: 'There are no clouds', 'It will not rain'. Intuitively statement 1 will be true. It may be a little difficult to justify the entries in row 3: 'There are no clouds', 'It will rain'. The entry '$C \rightarrow R$' in this case is true which looks absurd. However, there are English sentences, which may justify these entries. Consider another example.

EXAMPLE 7.14

1. If I work hard, I will pass.

Its constituent statements are:

(a) I work hard. H
(b) I will pass. P

The formula will be '$H \rightarrow P$'. Its truth table is given in Table 7.8.

Table 7.8 Conditional Truth Table

H	P	$H \rightarrow P$
T	T	T
T	F	F
F	T	T
F	F	T

Entries in the third row in Table 7.8, correspond to the following compound statement.

2. 'I do not work hard and still I pass.

Logically, it may be true. In fact, statement 1 says that 'I will pass if I work hard'. It does not say that 'I will not pass if I do not work hard'. Well! I may pass even if I do not work hard. Therefore, statement 2 is true and hence row 3 is justified.

Note: Many students may be unwilling to accept the entries of row 3. However, note that the above truth table is the definition of 'conditional' and, therefore, one should not show over enthusiasm in justifying the entries. In fact, best possible options for the values have been taken. Other options will give rise to the definitions of some other connectives.

7.7.5 Biconditional

Two simple (atomic) statements can be combined by connective **'biconditional'** to form a compound statement. Formally, it is described as:

English word : *if and only if*
Symbol : $\leftrightarrow$
Formula : $A \leftrightarrow B$
Pronounced as : *A if and only if B*

Table 7.9, which is the truth table of **biconditional**, defines the truth-values associated with it. The formula $A \leftrightarrow B$ is true when A and B have same truth-values, otherwise, it is false.

Table 7.9 Biconditional

A	B	$A \leftrightarrow B$
T	T	T
T	F	F
F	T	F
F	F	T

Biconditional is also symmetric.

EXAMPLE 7.15

1. It will rain if and only if there are clouds.

The constituent statements are:

(a) There are clouds. C
(b) It will rain. R

Target formula will be '$C \leftrightarrow R$'. Other equivalent forms of statement 1 are:

(i) If and only if there are clouds, it will rain.
(ii) It will rain only if there are clouds.

7.7.6 Exclusive Disjunction

Two simple (atomic) statements can be combined by connective 'EXOR' to form a compound statement. It is also known as Exclusive conjunction. Its formal description is:

English word : either … or, or
Symbol : $\oplus$
Formula : $A \oplus B$
Pronounced as : A exor B

Table 7.10, which is the truth table of Exclusive or, defines the truth-values associated with it. The formula $A \oplus B$ is true when A and B have opposite truth values, otherwise, it is false.

Table 7.10 Exclusive Disjunction

A	B	$A \oplus B$
T	T	F
T	F	T
F	T	T
F	F	F

In this formula, either A or B, not both will be true. It is symmetric.

EXAMPLE 7.16

1. I shall either watch the television or play the game.

The constituent statements are:

(a) I shall watch the television. W
(b) I shall play the game. P

Formula will be '$W \oplus P$'

Statement 1 can also be written as: I shall watch the television or play the game. However, the semantic of the statement is enough to declare word 'or' as 'exor' as the two events (watching and playing) cannot happen together.

7.7.7 Negation Conjunction

Two simple (atomic) statements can be combined by connective 'NAND' to form a compound statement. It is also known as negation conjunction. Its formal description is:

English word : … not…and…, …not the case…and…etc.
Symbol : $\uparrow$
Formula : $A \uparrow B$
Pronounced as : A nand B

Table 7.11, which is the truth table of nand, defines the truth-values associated with it. The formula $A \uparrow B$ is false when both A and B are true, otherwise, it is true. It is symmetric.

Table 7.11 Nand Operation

A	B	$A \uparrow B$
T	T	F
T	F	T
F	T	T
F	F	T

EXAMPLE 7.17

1. It is not the case that you and I are watching the television.

The constituent statements are:

(a) I am watching the television. I

(b) You are watching the television. Y

Formula will be $Y \uparrow I$.

One should be careful while creating the formula for a statement looking similar to 1. In fact, this operator negates the meaning of the conjunction of two statements. Therefore, it should be ascertained whether the conjunction is correct, otherwise, the negation conjunction will also be wrong. Truth table for formula $Y \uparrow I$ is given in Table 7.12.

Table 7.12 Negation Conjunction (NAND Operation)

I	Y	$Y \uparrow I$
T	T	F
T	F	T
F	T	T
F	F	T

7.7.8 Negation Disjunction

Two simple (atomic) statements can be combined by connective 'NOR' to form a compound statement. It is also known as negation disjunction. Formally, it is described as:

English word : ... not...or..., ...not the case...or...etc.
Symbol : $\downarrow$
Formula : $A \downarrow B$
Pronounced as : A nor B

'NOR' operation is negation of the disjunction. Its truth-values are defined by Table 7.13. Formula $A \downarrow B$ is true when both A and B are false, otherwise, it is false. 'NOR' is also symmetric.

Table 7.13 NOR Operation

A	B	$A \downarrow B$
T	T	F
T	F	F
F	T	F
F	F	T

EXAMPLE 7.18

1. It is not the case that you or I are watching the television.

The constituent statements are:

(a) I am watching the television. I
(b) You are watching the television. Y

Formula will be '$I \downarrow Y$'.
Truth table for formula '$I \downarrow Y$' is given by Table 7.14.

Table 7.14 Truth Table for Formula $I \downarrow Y$ (Nor Operation)

I	Y	$I \downarrow Y$
T	T	F
T	F	T
F	T	T
F	F	T

7.8 Well-formed Formula (wff)

We have already defined the formula. Well-formed formula (or statement formula) is itself the formula with clarity in its construction. It is an expression consisting of variables (capital letters), logic operators (connective symbols) and parentheses. It is generated by the following rules.

1. A statement variable (capital letter symbol) assigned to an atomic statement is a well-formed formula.
2. If A is a well-formed formula, then $\sim A$, and $(\sim A)$ are also well-formed formulas.
3. If A and B are well-formed formulas, then $A \wedge B$, $A \vee B$, $A \rightarrow B$, $A \rightleftharpoons B$, $A \oplus B$, $A \uparrow B$, $A \downarrow B$ are also well-formed formulas.
4. Any formula obtained by using rules 1 to 3 any number of times, is a well-formed formula. The outermost parentheses in a formula are, however, considered as optional and may be omitted.

EXAMPLE 7.19 Following are the examples of well-formed formulas:

1. P, $\sim P$, and $(\sim P)$ etc.
2. $(P \wedge Q)$, $(P \rightarrow (Q \wedge R))$, $P \vee (Q \wedge R)$, $P \rightarrow (Q \rightarrow R)$, and $P \uparrow (Q \uparrow R)$ etc.

EXAMPLE 7.20 Following formulas are not well-formed formulas due to the reasons given against them.

1. $\sim P \vee Q$: It is ambiguous and must be either $(\sim P) \vee Q$ or $\sim(P \vee Q)$, depending on the scope of negation.
2. $\sim P \vee \sim Q$: Two operators are placed together. It must be $(\sim P) \vee (\sim Q)$.
3. $P \vee Q \wedge R$: Must be either $(P \vee Q) \wedge R$ or $P \vee (Q \wedge R)$.
4. $((P \vee Q) \vee R$: Rightmost parenthesis is missing.
5. $P \rightarrow Q \rightarrow R$: It is ambiguous. Use of parenthesis is required. Must be $(P \rightarrow Q) \rightarrow R$ or $P \rightarrow (Q \rightarrow R)$.

7.9 Well-formed Sub Formula

A well-formed sub formula is a part of the well-formed formula which is itself a formula and satisfy the rules given in Section 7.8 for well-formed formula.

EXAMPLE 7.21 Consider the well formed-formula $(P \to Q) \to R$. Here, $P \to Q$ is a well-formed sub formula of $(P \to Q) \to R$.

7.10 Construction of Truth Table for a Formula

The truth table for a formula can be constructed by the following procedure.

1. Identify the variables. The number of rows will be 2^n, where n is the number of variables.
2. Assign possible truth-values (true or false) to all variables individually, and write their all-possible row combinations.
3. Divide the formula into sub formulas and find their truth-values in the table.
4. Combine the sub formula to complete the truth table.

EXAMPLE 7.22 Consider a well-formed formula $((\sim A) \vee B) \to C$. There are three variables A, B and C, therefore, there will be 8 rows ($2^3 = 8$) in total. Columns 1, 2 and 3 contains truth-values for the variables involved. Truth-values of two sub formulas are obtained in columns 4 and 5, which are used in obtaining the truth-values of the required formula. The complete solution is given below in Table 7.15.

Table 7.15 Table for $((\sim A) \vee B) \to C$

A	B	C	$\sim A$	$((\sim A) \vee B)$	$((\sim A) \vee B) \to C$
T	T	T	F	T	T
T	T	F	F	T	F
T	F	T	F	F	T
T	F	F	F	F	T
F	T	T	T	T	T
F	T	F	T	T	F
F	F	T	T	T	T
F	F	F	T	T	F

Entries in the truth table are obtained by using the truth-values of the tables of the logical operators already discussed in Section 7.7.

7.11 Equivalence of Formulas

If two formulas have identical truth-values for all possible assignments of truth-values in the rows, then these are called equivalent. It is written as $P \equiv Q$, i.e., P and Q are equivalent. These have identical truth-values in the final columns of truth tables.

EXAMPLE 7.23 Formulas $(\sim A) \vee B$ and $A \rightarrow B$ are equivalent which is shown in Table 7.16.

Table 7.16 Table for $(\sim A) \vee B$ and $A \rightarrow B$

A	B	$\sim A$	$(\sim A) \vee B$	$A \rightarrow B$
T	T	F	T	T
T	F	F	F	F
F	T	T	T	T
F	F	T	T	T

Note that, columns 4 and 5 are identical. Therefore, these formulas are equivalent.

That is $$((\sim A) \vee B) \equiv (A \rightarrow B).$$

7.12 Duality Law

Two formulas A and B are called duals of each other if either one can be obtained from the other by replacing operator $\wedge$ by $\vee$ and $\vee$ by $\wedge$. Operators $\wedge$ and $\vee$ are called duals of each other. If the formula A also contains special variables T or F, then in addition to above interchanges, its dual, B will be obtained by replacing T by F and F by T. Table 7.17 gives some formulas with their duals.

Table 7.17 Formulas with Their Duals

Formula	*Dual of the formula*
$(P \wedge Q) \wedge R$	$(P \vee Q) \vee R$
$(P \wedge Q) \wedge T$	$(P \vee Q) \vee R$
$(P \vee Q) \wedge F$	$(P \wedge Q) \vee T$

Note: If two formulas are equivalent then their duals will also be equivalent. Sometimes, it becomes easier to prove the validity of a formula using its dual.

7.13 Some Useful Equivalent Formulas

Some important formulas and their equivalents are given in Table 7.18.

Table 7.18 Equivalent Formulas

Formula	*Dual of formula*	*Name*
$A \vee A \equiv A$	$A \wedge A \equiv A$	Idempotent laws
$(A \vee B) \vee C \equiv A \vee (B \vee C)$	$(A \wedge B) \wedge C \equiv A \wedge (B \wedge C)$	Associativity
$A \vee (B \wedge C) \equiv (A \vee B) \wedge (A \vee C)$	$A \wedge (B \vee C) \equiv (A \wedge B) \vee (A \wedge C)$	Distributive laws
$A \vee F \equiv A$	$A \wedge T \equiv A$	Identity laws

(Contd.)

Table 7.18 Equivalent Formulas (*Contd.*)

Formula	*Dual of formula*	*Name*
$A \vee T \equiv T$ (Tautology) $A \vee (\sim A) \equiv T$ (Tautology), $\sim T \equiv F$	$A \wedge F \equiv F$ (Contradiction) $A \wedge (\sim F) \equiv F$ (Contradiction), $\sim F \equiv T$	Complement laws
$A \vee (A \wedge B) \equiv A$	$A \wedge (A \vee B) \equiv A$	Absorption laws
$\sim (\sim P) \equiv P$	—	Involution law
$\sim (A \vee B) \equiv (\sim A) \wedge (\sim B)$	$\sim (A \wedge B) \equiv (\sim A) \vee (\sim B)$	De Morgan's laws

7.14 Converse, Inverse and Contrapositive

Consider the formula $A \rightarrow B$. Then $B \rightarrow A$ is called converse, $(\sim A) \rightarrow (\sim B)$ is inverse and $(\sim B) \rightarrow (\sim A)$ contrapositive of it. Table 7.19 summarizes the truth-values.

Table 7.19 Converse, Inverse and Contrapositive

A	B	$\sim A$	$\sim B$	$A \rightarrow B$	$B \rightarrow A$ *Converse*	$(\sim A) \rightarrow (\sim B)$ *Inverse*	$(\sim B) \rightarrow (\sim A)$ *Contrapositive*
T	T	F	F	T	T	T	T
T	F	F	T	F	T	T	F
F	T	T	F	T	F	F	T
F	F	T	T	T	T	T	T

7.15 Tautology

The final column in the truth table of a formula contains both T and F. However, if it contains only T, at all places then it is called tautology or universally valid formula. In such case, the formula is always valid irrespective of the truth-values assigned to its variables.

EXAMPLE 7.24 Consider the formula $P \vee (\sim P)$.

It is a tautology since its truth-value is always true and is independent of the truth-value of 'P'. It is evident from Table 7.20 given below.

Table 7.20 Tautology

P	$\sim P$	$P \vee (\sim P)$
T	F	T
F	T	T

7.16 Construction of New Formulas

A formula which is tautology, can be used to form new formulas by using replacement process, i.e., one can replace atomic variables in the formula by new variables, or by new formulas at its all occurrences. The resulting formula will itself remain a tautology. A well-formed sub formula can also be replaced by its equivalent well-formed formula.

EXAMPLE 7.25 Consider the formula $P \vee (\sim P)$ which is a tautology. A number of formulas can be obtained from it, replacing 'P' by any other formula. Every new formula so formed will also be a tautology. These may be:

1. $(P \vee (\sim P)) \vee (\sim(P \vee (\sim P)))$
2. $(P \wedge Q) \vee (\sim(P \wedge Q))$
3. $((P \oplus (\sim Q)) \rightarrow R) \vee (\sim((P \oplus (\sim Q)) \rightarrow R))$
4. $(P \oplus Q) \vee (\sim(P \oplus Q))$
5. $(P \rightarrow Q) \vee (\sim(P \rightarrow Q))$
6. $(P \leftrightarrow Q) \vee (\sim(P \leftrightarrow Q))$

Formula at serial number 5, can give rise to another formula by replacing $P \rightarrow Q$ by its equivalent formula $(\sim P) \vee Q$.

For example, $((\sim P) \vee Q) \vee (\sim((\sim P) \vee Q))$.

7.17 Contradiction

A formula is called a contradiction or a fallacy if its truth-value is always false. Its value does not depend on the truth-values of the variables involved in it. In this case, the final column of the truth table contains only F, which is evident from Table 7.21.

EXAMPLE 7.26 Consider the formula $P \wedge (\sim P)$. It is a contradiction.

Table 7.21 Contradiction

P	$\sim P$	$P \wedge (\sim P)$
T	F	F
F	T	F

7.18 Argument

An argument consists of two parts: a collection of premises (simple or compound statements) and a conclusion. Conclusion is another proposition that yields (results) from the premises. Generally, it is written in the following forms:

(1) $P_1, P_2, P_3, \ldots, P_n$, therefore, C.

(2) $P_1, P_2, P_3, \ldots, P_n \Rightarrow C$.

(3)

1. P_1
2. P_2
3. P_3

...

...

$\therefore C$

Here P_1, P_2, P_3, ..., P_n are premises, and C is conclusion. Generally, the first form is preferred when argument is written in Meta language, while second and third are used when premises and conclusion in the argument are in terms of well formed formulas.

EXAMPLE 7.27 If prices are high then there is inflation. If there is inflation then life will be difficult. The prices are high. Therefore, life is difficult.

Premises

1. If prices are high then there is inflation.
2. If there is inflation then life will be difficult.
3. The prices are high.

Conclusion

Therefore, life is difficult.

EXAMPLE 7.28

1. If John is reading the book then dog is barking. (premise)
2. If dog is barking then cat is running. (premise)
3. John is reading the book. (premise)

Therefore cat is running. (conclusion)

In this argument, the first three statements 1, 2 and 3 are premises, and the last one is called the conclusion.

EXAMPLE 7.29 If there are clouds then it may rain. If it rains, I shall stay at home and play chess. If I play chess, then I may learn more tricks and consequently, I may participate in a championship tournament. Playing in a championship tournament may lead to winning the title. Since I won the championship, therefore, I know more tricks.

Premises

1. If there are clouds then it may rain.
2. If it rains, I shall stay at home and play chess.
3. If I play chess, then I may learn more tricks and consequently, I may participate in a championship tournament.
4. Playing in championship tournament may lead to winning the title.
5. I won the championship.

Conclusion

Therefore, I know more tricks.

EXAMPLE 7.30 If he is suffering with body ache then he may have fever. If there is common cold then there may be fever. Also high body temperature leads to fever. If there is fever then doctors advise rest. John is suffering with common cold. Therefore, John is advised to take rest.

Premises

1. If he is suffering with body ache then he may have fever.
2. If there is common cold then there may be fever.
3. High body temperature leads to fever.
4. If there is fever, then doctors advise rest.
5. John is suffering with common cold.

Conclusion

Therefore, John is advised to take rest.

7.19 Valid Argument

An argument is said to be valid, if conclusion is true whenever all the premises are true. Truth table can be used to find the validity of the argument. Rows of truth table are examined for the purpose. In fact, for the validity of an argument, the rows having truth-values true of all the premises must also have truth-values true for their corresponding conclusions in the truth table. An argument, which is not valid, is called an invalid argument. Validity of an argument can also be proved by forming a well-formed formula using its premises and conclusion. If P_1, P_2, P_3, …, P_n are the premises of an argument and C is its conclusion, then form the formula:

$$P_1 \wedge P_2 \wedge P_3 \wedge \ldots \wedge P_n \rightarrow C.$$

If this formula is a tautology then the argument is said to be valid.

Note: The validity of an argument does not depend on the truth-values (in actual semantic sense) of the statements appearing in the argument but on the particular form of the argument. This will become clear after going through the examples.

7.20 Proving the Validity by Truth Table Methods

There are two ways to find the validity of an argument using truth table. These are discussed below in this section.

7.20.1 Truth Table Method 1

Transform the argument into symbolic form and construct the truth table. If all rows with premises having truth value true (T) also contain truth value true (T) in the column of the conclusion, then the argument is said to be valid. The procedure is discussed with the help of the following examples.

EXAMPLE 7.31 Consider the argument given in Example 7.27.

If prices are high then there is inflation. If there is inflation then life will be difficult. The prices are high. Therefore, life is difficult.

Now, write the premises and conclusion separately and assign symbols to them.

Premises

1. If prices are high then there is inflation.
   (i) prices are high. P
   (ii) there is inflation. I

   wff : $P \to I$
2. If there is inflation then life will be difficult.
   (i) life will be difficult. D
   wff : $I \to D$
3. The prices are high. P

Conclusion

Therefore, life is difficult. D

Since there are three variables, therefore, there will be eight rows in the truth table. Table 7.22 is the complete truth table constructed below for this argument.

Table 7.22 Truth Table For Method 1

P *Premise*	I	D *Conclusion*	$P \to I$ *Premise*	$I \to D$ *Premise*
T	**T**	**T**	**T**	**T**
T	T	F	T	F
T	F	T	F	T
T	F	F	F	T
F	T	T	T	T
F	T	F	T	F
F	F	T	T	T
F	F	F	T	T

Observe that the first row is the only row in Table 7.22, which contains truth value true (T) for all—the three premises. Further, this row also contains truth value true (T) for the conclusion, therefore, this argument is valid.

7.20.2 Truth Table Method 2

Transform the argument into symbolic form and construct a well-formed formula using its premises and conclusion. All the premises are connected with 'AND' to form a well-formed formula, which is further connected with conclusion using implication (→).

For instance, $P_1 \wedge P_2 \wedge P_3 \wedge \ldots \wedge P_n \to C$

where $P_1, P_2, P_3, \ldots, P_n$ are all premises of the argument and C is its conclusion. Now, construct the truth table and examine the truth-values.

If this formula is a tautology then the argument is said to be valid.

EXAMPLE 7.32 Again, consider the argument given in Example 7.27. Let us discuss its validity.

Premises

1. If prices are high then there is inflation.
   (i) prices are high. P
   (ii) there is inflation. I
   wff : $P \to I$
2. If there is inflation then life will be difficult.
   (i) life will be difficult. D
   wff : $I \to D$
3. The prices are high P

Conclusion

Therefore, life is difficult. D

Now, the well-formed formula using its premises and conclusion will be:

$$((P \to I) \wedge (I \to D) \wedge P) \to D.$$

The truth table for this is given below.

Table 7.23 Truth Table Method 2

P *Premise*	I	D *Conclusion*	$P \to I$ *Premise*	$I \to D$ *Premise*	$((P \to I) \wedge (I \to D) \wedge P) \to D$
T	T	T	T	T	T
T	T	F	T	F	T
T	F	T	F	T	T
T	F	F	F	T	T
F	T	T	T	T	T
F	T	F	T	F	T
F	F	T	T	T	T
F	F	F	T	T	T

Observe that the truth-values in the last column of Table 7.23 are all true, therefore, it is a tautology. Hence, the argument is valid.

7.21 Inference Theory of Statement Calculus

Theory for providing rules of inference for inferring conclusions is known as inference theory. Inference theory has been developed for constructing proofs of theorems and proving the validity of the arguments.

7.21.1 Proving Validity by Deduction Method

The procedure used for proving the validity of an argument by using deductions is called deduction method. In this method, conclusions are drawn from certain premises. Inference is made using well-established rules. New premises are deduced from the premises given in the argument. While constructing the proof, all the premises given in the argument are listed. The aim is then to deduce the conclusion. The success depends on one's capability in making the proper selections. A wrong choice in deduction may lead to a complex situation and even to failure. In truth table method, it is always possible to find the validity of an argument, i.e., to declare it valid or invalid. But, if a formula is invalid, then it is difficult to prove it by deduction method. However, in some cases, this method is short and simple. The logical implications and equivalences used in this method are listed in Table 7.24, and Table 7.25, respectively. The whole system is called deduction system.

Logical implications

A statement A is said to tautologically imply another statement B, if and only if formula $A \to B$ is a tautology. Generally, it is written as $A \Rightarrow B$, and is read as A tautologically implies B, or in short, A implies B. Some well established logical implications are tabulated in Table 7.24.

Table 7.24 Logical Implications

S. No	*Logical implications*	*Special names*
1	$P \Rightarrow P \vee Q$	Addition
2	$Q \Rightarrow P \vee Q$	Addition
3	$P \wedge Q \Rightarrow P$	Simplification
4	$P \wedge Q \Rightarrow Q$	Simplification
5	$\sim P, P \vee Q \Rightarrow Q$	Disjunctive syllogism
6	$\sim Q, P \to Q \Rightarrow \sim P$	Modus tollens
7	$P, P \to Q \Rightarrow Q$	Modus ponens
8	$P \to Q, Q \to R \Rightarrow P \to R$	Hypothetical syllogism
9	$P \vee Q, P \to R, Q \to R \Rightarrow R$	Dilemma
10	$\sim P \Rightarrow P \to Q$	
11	$Q \Rightarrow P \to Q$	
12	$\sim (P \to Q) \Rightarrow P$	
13	$\sim (P \to Q) \Rightarrow \sim Q$	
14	$P, Q \Rightarrow P \wedge Q$	
15	$(\sim P) \vee Q, P \vee Q \Rightarrow Q$	

Equivalences

While deducing conclusions from the given premises, the formulas equivalent to these premises can also be used. A partial list of equivalences was given in Table 7.18. However, it is reproduced here as Table 7.25 with more additions.

Table 7.25 Equivalences

S. No.	*Formula*	*Dual of formula*	*Special names*
1	$A \vee A \equiv A$	$A \wedge A \equiv A$	Idempotent laws
2	$(A \vee B) \vee C \equiv A \vee (B \vee C)$	$(A \wedge B) \wedge C \equiv A \wedge (B \wedge C)$	Associative laws
3	$A \vee (B \wedge C) \equiv (A \vee B) \wedge (A \vee C)$	$A \wedge (B \vee C) \equiv (A \wedge B) \vee (A \wedge C)$	Distributive laws
4	$A \vee F \equiv A$	$A \wedge T \equiv A$	Identity laws
5	$A \vee T \equiv$ T (Tautology)	$A \wedge F \equiv F$ (Contradiction)	

(Contd.)

Table 7.25 Equivalences

S. No.	*Formula*	*Dual of formula*	*Special names*
6	$A \vee (\sim A) \equiv T$ (Tautology) $\sim T \equiv F$	$A \wedge (\sim A) \equiv F$ (Contradiction) $\sim F \equiv T$	Complement laws
7	$A \vee (A \wedge B) \equiv A$	$A \wedge (A \vee B) \equiv A$	Absorption laws
8	$\sim (\sim A) \equiv A$	–	Involution law/ Double negation
9	$\sim (A \vee B) \equiv (\sim A) \wedge (\sim B)$	$\sim (A \wedge B) \equiv (\sim A) \vee (\sim B)$	De Morgan's laws
10	$A \oplus B \equiv ((\sim A) \wedge B) \vee (A \wedge (\sim B))$		
11	$A \vee B \equiv B \vee A$	$A \wedge B \equiv B \wedge A$	Commutative laws
12	$A \vee (B \wedge (\sim B)) \equiv A$	$A \wedge (B \vee (\sim B)) \equiv A$	
13	$A \vee (B \vee (\sim B)) \equiv T$	$A \wedge (B \wedge (\sim B)) \equiv F$	
14	$A \rightarrow B \equiv (\sim A) \vee B$		
15	$\sim (A \rightarrow B) \equiv A \wedge (\sim B)$		
16	$A \rightarrow B \equiv (\sim B) \rightarrow (\sim A)$		
17	$A \rightarrow (B \rightarrow C) \equiv (A \wedge B) \rightarrow C$		
18	$(A \leftrightarrow B) \equiv (A \rightarrow B) \wedge (B \rightarrow A)$		
19	$\sim (A \leftrightarrow B) \equiv A \leftrightarrow (\sim B)$		
20	$(A \leftrightarrow B) \equiv (A \wedge B) \vee ((\sim A) \wedge (\sim B))$		

Inference rules

The following inference rules are used for proving the validity of arguments.

Rule 1: A premise may be introduced at any place in the derivation.

Rule 2: A formula may be introduced in a derivation if it is logically implied by one or more of the preceding formulas or premises.

Rule 3: A formula may be replaced by its equivalent formula.

Using these rules, if it is possible to deduce the conclusion then the argument is said to be valid. However, if conclusion is not deduced in some steps, then one should not immediately declare that the argument is invalid. Sometimes, one may fail to infer the conclusion due to wrong inference at some step in the derivation. In fact, constructing proof by deduction is a matter of intelligent thinking. In case of doubt, other methods may be tried before declaring the validity.

A derivation consists of a sequence of formulas. Each formula in the sequence is either a premise, or a logical implication of the formulas, appeared before. We are using a tabular representation for showing the derivation steps of each proof system. Column 1 of the table gives the serial number of the derivation step. Column 2 contains formulas, which are either premises of the argument, or the deductions obtained using previous steps. Column 3 gives the reference numbers which are the serial numbers given in column 1. These reference numbers are of those formulas that have been used in deducing the current formula. Finally, column 4 of the table gives the complete explanation.

EXAMPLE 7.33 Consider the following argument.

If Mars mission fails, then Venus mission will be delayed. Delay in Venus mission will influence technological advancement in the astronomy. However, Mars mission failed. Therefore, technological advancement in astronomy is influenced.

Let us assign symbols to the premises and conclusion.

Premises

1. If Mars mission fails, then Venus mission will be delayed.
   (i) The Mars mission fails. F
   (ii) The Venus mission will be delayed. D
   wff: $F \rightarrow D$
2. Delay in Mars mission will influence technological advancement in astronomy.
   (i) There is delay in Venus mission. D
   (ii) Technological advancements in astronomy are influenced. A
   wff: $D \rightarrow A$
3. Mars mission failed. F
   wff: F

Conclusion

Technological advancement in astronomy is influenced. A

wff: A

Now, to prove the validity, the argument is written in its symbolic form as given below.

1. $F \rightarrow D$
2. $D \rightarrow A$
3. F

$\therefore A$

All derivation steps are listed in Table 7.26.

Table 7.26 Proving Validity

S. No.	*Formula*	*Reference*	*Explanation*
1	$F \rightarrow D$		Premise introduced
2	$D \rightarrow A$		Premise introduced
3	F		Premise introduced
4	$F \rightarrow A$	1, 2	Hypothetical syllogism
5	A	3, 4	Modus ponens

Observe that at step 5, conclusion is deduced, therefore, the argument is valid.

7.21.2 Proving the Validity by Method of Contradiction

This method is similar to the deduction method and uses Tables 7.24 and 7.25 for derivations. In this method, an additional premise, which is negation of the conclusion, is introduced and then, the derivations are made to obtain a contradiction. If a contradiction is obtained then the given argument is valid. In fact, contradiction shows that the given set of premises is inconsistent, i.e., their conjunction is always false. It implies that the negation of the conclusion, which was introduced as an additional premise, is wrong. Hence, the conclusion follows from the given premises.

EXAMPLE 7.34 Consider the following argument:

1. $A \to B$
2. $C \to D$
3. $A \vee C$

$\therefore\ B \vee D$

The complete derivation is given in Table 7.27.

Table 7.27 Proving Validity

S. No.	*Formula*	*Reference*	*Explanation*
1	$A \to B$		Given Premise
2	$C \to D$		Given Premise
3	$A \vee C$		Given Premise
4	$\sim (B \vee D)$		Negation of conclusion is introduced as an additional premise
5	$(\sim B) \wedge (\sim D)$	4	De Morgan's law
6	$\sim B$	5	Simplification
7	$\sim D$	5	Simplification
8	$(\sim A) \to C$	3	Equivalence $A \to B \equiv (\sim A) \vee B$
9	$(\sim C) \to A$	8	Equivalence $A \to B \equiv (\sim B) \to (\sim A)$
10	$(\sim C) \to B$	1, 9	Hypothetical syllogism
11	$(\sim B) \to C$	10	Equivalence $A \to B \equiv (\sim B) \to (\sim A)$
12	$B \vee C$	11	Equivalence $A \to B \equiv (\sim A) \vee B$
13	C	6, 12	Disjunctive syllogism
14	$(\sim C) \vee D$	2	Equivalence $A \to B \equiv (\sim A) \vee B$
15	D	13, 14	Disjunctive syllogism
16	$(\sim D) \wedge D$	7, 15	Since $P, Q \Rightarrow P \wedge Q$

A contradiction is obtained at step 16 of the table, therefore, the given argument is valid.

7.21.3 Proving Validity by Tree Tableau Method

In this method, a tree (for tree, refer chapter 14) is constructed by taking negation of the argument at the root of the tree. The tree is expanded by taking sub formulas as descendent nodes. If two sub formulas are connected with the conjunction then they will lie on the same path. However, if the sub formulas are connected with disjunction then they will lie on different paths. A path starts from root and ends at a leaf. A path is said to be closed if it contains a variable and its negation. An argument is said to be valid if all its paths are closed. It is also called a closed formula.

For example, consider an argument

$$((\sim (A \vee B)) \wedge ((\sim C) \vee B)) \to (\sim C).$$

Now, take negation of this argument,

i.e., $\sim (((\sim (A \vee B)) \wedge ((\sim C) \vee B)) \rightarrow (\sim C))$.

This formula will be root of the tree tableau. Then apply logical equivalences and laws to expand the tree.

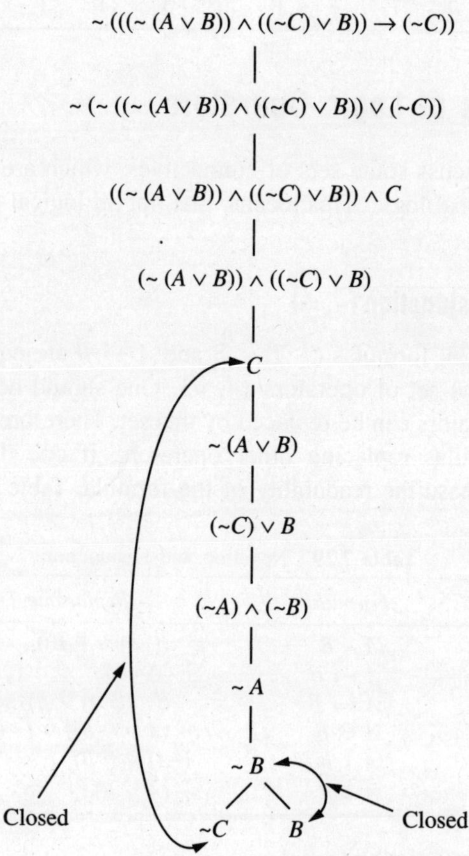

Figure 7.1 Tree tableau.

In this tableau, there are two paths ending at ~C and B. Further, since ~C and C occur on the same path, therefore, it is a closed path. Similarly, ~B and B also occur on same path, therefore, it is also a closed path. Thus, both the paths are closed. Hence, argument is valid.

7.22 Complete List of Formulas

If the number of variables is n, then the number of formulas will be 2^{2^n}. Let the value of n be 2, then there will be 16 numbers of formulas or truth tables, which are given below in Table 7.28.

Table 7.28 Complete List of Formulas for $n = 2$

A	B	1	2	3	4	5	6	7	8	9	10	11	12	13	14	15	16
T	T	F	F	F	F	F	F	F	F	T	T	T	T	T	T	T	T
T	F	F	F	F	F	T	T	T	T	F	F	F	F	T	T	T	T
F	T	F	F	T	T	F	F	T	T	F	F	T	T	F	F	T	T
F	F	F	T	F	T	F	T	F	T	F	T	F	T	F	T	F	T

7.23 Minimal Sets of Logic Operators

In this section, we will discuss some sets of connectives, which are sufficient, rather complete for the study of mathematical logic. This means, that not all logical connectives are required to be used in a formula.

7.23.1 Negation and Disjunction {~, ∨}

As given in truth Table 7.25, formulas $(\sim A) \vee B$ and $A \rightarrow B$ are equivalent. It means operator '→' can be replaced by the set of operators {~, ∨}. One should note that this is not the only case; in fact, all other operators can be replaced by this set. Therefore, the set {~, ∨} is sufficient to construct all the formulas replacing other operators, if one desires so. However, other operators are used to increase the readability of the formula. Table 7.29 gives the equivalence of operators.

Table 7.29 Negation and Disjunction

Operator	*Formula*	*Replacement by* {~, ∨}
∧	$A \wedge B$	$\sim((\sim A) \vee (\sim B))$
→	$A \rightarrow B$	$(\sim A) \vee B$
↔	$A \leftrightarrow B$	$\sim((\sim((\sim A) \vee B)) \vee (\sim(A \vee (\sim B))))$
⊕	$A \oplus B$	$\sim(A \vee (\sim B)) \vee (\sim((\sim A) \vee B))$
↑	$A \uparrow B$	$(\sim A) \vee (\sim B)$
↓	$A \downarrow B$	$\sim(A \vee B)$

7.23.2 Negation and Conjunction {~, ∧}

A combination of negation and conjunction can also be used to replace all other operators, exactly in the similar way as negation and disjunction given in Section 7.23.1. Details are given in Table 7.30.

Table 7.30 Negation and Conjunction

Operator	*Formula*	*Replacement by* {~, ∧}
∨	$A \vee B$	$\sim((\sim A) \wedge (\sim B))$
→	$A \rightarrow B$	$\sim(A \wedge (\sim B))$
↔	$A \leftrightarrow B$	$\sim(A \wedge (\sim B)) \wedge (\sim((\sim A) \wedge B))$
⊕	$A \oplus B$	$\sim((\sim((\sim A) \wedge B)) \wedge (\sim(A \wedge (\sim B))))$
↑	$A \uparrow B$	$\sim(A \wedge B)$
↓	$A \downarrow B$	$(\sim A) \wedge (\sim B)$

7.23.3 NAND { ↑ } and NOR { ↓ }

Any logic formula can be constructed using only NAND or NOR operators. Therefore, NAND {↑} is a functionally complete set of operators. Similarly, NOR {↓} is also a functionally complete set of operators. Some details showing the replacement are given in Table 7.31. Others are left for the readers to workout.

Table 7.31 Functionally Complete Operators {↑} and {↓}

Operator	*Formula*	*Replacement by* {↑}	*Replacement by* {↓}
$\vee$	$A \vee B$	$(A \uparrow A) \uparrow (B \uparrow B)$	$(A \downarrow B) \downarrow (A \downarrow B)$
$\wedge$	$A \wedge B$	$(A \uparrow B) \uparrow (A \uparrow B)$	$(A \downarrow A) \downarrow (B \downarrow B)$
$\sim$	$\sim A$	$A \uparrow A$	$A \downarrow A$
$\to$	$A \to B$	$((A \uparrow A) \uparrow (A \uparrow A)) \uparrow (B \uparrow B)$	$((A \downarrow A) \downarrow B) \downarrow ((A \downarrow A) \downarrow B)$

7.24 Solved Problems

PROBLEM 7.1 Construct the well-formed formulas for the following statements:

1. John is reading the book and dog is barking.
2. Cat or rat is running.
3. Jack is rich but unhappy.
4. It is not the case, that the houses are cold or haunted and it is false that the cottages are warm or houses are ugly.

Solution

1. Atomic statements are:
   (i) John is reading the book. R
   (ii) Dog is barking. B
   $\therefore$ wff is $R \wedge B$
2. Atomic statements are:
   (i) Cat is running. C
   (ii) Rat is running. R
   $\therefore$ wff is $C \vee R$
3. Atomic statements are:
   (i) Jack is rich. R
   (ii) Jack is unhappy. U
   $\therefore$ wff is $R \wedge U$
4. Atomic statements are:
   (i) Houses are cold. C
   (ii) Houses are haunted. H
   (iii) Cottages are warm. W
   (iv) Houses are ugly. U
   $\therefore$ wff is $(\sim (C \vee H)) \wedge (\sim (W \vee U))$

PROBLEM 7.2 Construct the truth table for the following formulas:

1. $(\sim A) \vee B$
2. $(A \to B) \to C$
3. $(A \wedge B) \vee C$
4. $((\sim A) \vee (\sim B)) \wedge C$

Solution

1. Truth table for the formula $(\sim A) \vee B$ is given in Table 7.32.

Table 7.32 Table for $(\sim A) \vee B$

A	B	$\sim A$	$(\sim A) \vee B$
T	T	F	T
T	F	F	F
F	T	T	T
F	F	T	T

From this example, you can see that $(\sim A) \vee B$ is equivalent to $A \to B$ because both have identical entries in the last column of the truth table.

2. Truth table for $(A \to B) \to C$ is given in Table 7.33.

Table 7.33 Table for $(A \to B) \to C$

A	B	C	$A \to B$	$(A \to B) \to C$
T	T	T	T	T
T	T	F	T	F
T	F	T	F	T
T	F	F	F	T
F	T	T	T	T
F	T	F	T	F
F	F	T	T	T
F	F	F	T	F

3. Truth-values of $(A \wedge B) \vee C$ are given in Table 7.34.

Table 7.34 Table for $(A \wedge B) \vee C$

A	B	C	$A \wedge B$	$(A \wedge B) \vee C$
T	T	T	T	T
T	T	F	T	T
T	F	T	F	T
T	F	F	F	F
F	T	T	F	T
F	T	F	F	F
F	F	T	F	T
F	F	F	F	F

4. Truth-values of $((\sim A) \vee (\sim B)) \wedge C$ are given in Table 7.35.

Table 7.35 Table for $((\sim A) \vee (\sim B)) \wedge C$

A	B	C	$\sim A$	$\sim B$	$((\sim A) \vee (\sim B))$	$((\sim A) \vee (\sim B)) \wedge C$
T	T	T	F	F	F	F
T	T	F	F	F	F	F
T	F	T	F	T	T	T
T	F	F	F	T	T	F
F	T	T	T	F	T	T
F	T	F	T	F	T	F
F	F	T	T	T	T	T
F	F	F	T	T	T	F

PROBLEM 7.3 Prove the validity of the following argument:

If dark clouds are dangerous then everybody will stay at home. If dark clouds are there, then I will not stay at home. However, I stay at home. Therefore, dark clouds are not dangerous.

Solution

Premises

1. If dark clouds are dangerous, then everybody will stay at home.
   (i) Dark clouds are dangerous. D
   (ii) Stay at home. S
   wff : $D \to S$
2. If dark clouds are there, then I will not stay at home.
   (i) Dark clouds are there. C
   (ii) I will not stay at home. $\sim S$
   wff : $C \to (\sim S)$
3. I stay at home. S
   wff : S

Conclusion

Therefore, dark clouds are not dangerous.
wff : $\sim D$

Truth table is given in Table 7.36.

Table 7.36 Truth Table of Problem 7.3

D	S Premise	C	$D \to S$ Premise	$C \to (\sim S)$ Premise	$\sim D$ Conclusion
T	T	T	T	F	F
T	T	F	T	T	F
T	F	T	F	T	F
T	F	F	F	T	F
F	T	T	T	F	T
F	T	F	T	T	T
F	F	T	T	T	T
F	F	F	T	T	T

On analyzing the values in row 2 of Table 7.36, we observe that, all the premises are true, but, the conclusion is not true. Therefore, the argument is not valid.

PROBLEM 7.4 Prove the validity of the argument given in Problem 7.3 by truth table method 2 discussed in Section 7.20.2.

Solution

The well formed formula of the given argument will be:

$$((D \to S) \wedge (C \to (\sim S)) \wedge S) \to (\sim D)$$

The truth table is shown below in Table 7.37.

Table 7.37 Truth Table of Problem 7.4

D	$\sim D$	S Premise	C	$D \to S$ Premise	$C \to (\sim S)$ Premise	$((D \to S) \wedge (C \to (\sim S)) \wedge S) \to (\sim D)$
T	F	T	T	T	F	T
T	**F**	**T**	**F**	**T**	**T**	**F**
T	F	F	T	F	T	T
T	F	F	F	F	T	T
F	T	T	T	T	F	T
F	T	T	F	T	T	T
F	T	F	T	T	T	T
F	T	F	F	T	T	T

From last column of Table 7.37, it is obvious that the formula is not a tautology, therefore, the given argument is not valid.

PROBLEM 7.5 Prove the validity of the following argument by deduction method.

1. $A \to B$
2. $C \to D$
3. $A \vee C$

$\therefore\ B \vee D$

Solution

Derivation steps are shown in Table 7.38.

Table 7.38 Proving Validity of Problem 7.5

S. No.	*Formula*	*Reference*	*Explanation*
1	$A \to B$		Given premise
2	$C \to D$		Given premise
3	$A \vee C$		Given premise
4	$(\sim A) \to C$	3	Equivalence $A \to B \equiv (\sim A) \vee B$
5	$(\sim C) \to A$	4	Equivalence $\mathrm{A} \to B \equiv (\sim B) \to (\sim A)$
6	$(\sim C) \to B$	1, 5	Hypothetical syllogism
7	$(\sim B) \to C$	6	Equivalence $A \to B \equiv (\sim B) \to (\sim A)$
8	$(\sim B) \to D$	2, 7	Hypothetical syllogism
9	$B \vee D$	8	Equivalence $A \to B \equiv (\sim A) \vee B$

Since the conclusion is deduced at step 9, therefore, the argument is valid.

PROBLEM 7.6 Prove that the following argument is valid. Use deduction method.

1. $A \to B$
2. $A \to C$
3. $\sim (B \wedge C)$
4. $D \vee A$

$\therefore\ D$

Solution

Derivation steps are shown in Table 7.39, below.

Table 7.39 Proving Validity of Problem 7.6

S. No.	*Formula*	*Reference*	*Explanation*
1	$A \to B$		Given premise
2	$A \to C$		Given premise
3	$\sim (B \wedge C)$		Given premise
4	$D \vee A$		Given premise
5	$(\sim B) \vee (\sim C)$	3	De Morgan's law
6	$B \to (\sim C)$	5	Equivalence $A \to \text{B} \equiv (\sim) \vee B$
7	$A \to (\sim C)$	1, 6	Hypothetical syllogism
8	$(\sim A) \vee C$	2	Equivalence $A \to B \equiv (\sim A) \vee B$
9	$(\sim A) \vee (\sim C)$	7	Equivalence $A \to B \equiv (\sim A) \vee B$
10	$\sim A$	8, 9	$P \vee Q, P \vee (\sim Q) \Rightarrow P$
11	D	4, 10	$A \vee B, (\sim B) \Rightarrow A$

Since we obtain the conclusion at step 11, therefore, the given argument is valid.

PROBLEM 7.7 Prove the validity of the argument given in Problem 7.6, by the method of contradiction.

Solution

Derivation steps are shown in Table 7.40.

Table 7.40 Proving Validity of Problem 7.7

S. No.	*Formula*	*Reference*	*Explanation*
1	$A \to B$		Given premise
2	$A \to C$		Given premise
3	$\sim (B \wedge C)$		Given premise
4	$D \vee A$		Given premise
5	$\sim D$		Negation of conclusion introduced
6	$(\sim B) \vee (\sim C)$	3	De Morgan's law
7	A	4, 5	$A \vee B, (\sim A) \Rightarrow B$
8	$(\sim A) \vee C$	2	$A \to C \equiv (\sim A) \vee C$
9	C	7, 8	$A \vee B, (\sim A) \Rightarrow B$
10	$(\sim A) \vee B$	1	$A \to B \equiv (\sim A) \vee B$
11	B	7, 10	$(\sim A) \vee B, A \Rightarrow B$
12	$\sim C$	6, 11	$(\sim B) \vee (\sim C), B \Rightarrow \sim C$
13	$C \wedge (\sim C)$	9, 12	Conjunction

A contradiction is obtained at step number 13 of Table 7.40, therefore, the given argument is valid.

Exercises

7.1 Using the symbols given against the statements, construct the formulas for the following sentences:

John is rich. R
John is happy. H

(i) John is poor but happy.
(ii) John is neither rich nor happy.
(iii) John is rich or he is both poor and happy.
(iv) If John is poor then he is unhappy.
(v) John is poor if and only if he is unhappy.
(vi) It is not the case that John is both rich and unhappy.

7.2 Construct the truth tables for the following:

(i) $((P \to (\sim Q)) \wedge R) \downarrow P$
(ii) $(((\sim P) \downarrow (\sim Q)) \wedge R) \uparrow P$
(iii) $(((\sim P) \to Q) \vee R) \downarrow Q$
(iv) $(P \oplus (\sim Q)) \to R$
(v) $((P \to Q) \wedge R) \oplus P$
(vi) $(((\sim P) \uparrow (\sim Q)) \wedge R) \wedge S$

7.3 Using the truth tables, prove that following are equivalent:

(i) $(P \leftrightarrow Q)$ and $((\sim P) \vee Q) \wedge (P \vee (\sim Q))$
(ii) $(P \oplus Q)$ and $((\sim P) \wedge Q) \vee (P \wedge (\sim Q))$
(iii) $(\sim P) \vee Q$ and $P \to Q$
(iv) $(P \wedge Q) \wedge R$ and $P \wedge (Q \wedge R)$

7.4 Discuss the validity of the following arguments by Truth table method:

(i) 1. $J \to M$
2. $(M \vee G) \to (\sim J)$
3. $J \vee G$
$\therefore\ G$

(ii) 1. $A \vee H$
2. $(A \vee B) \to (\sim H)$
3. $(\sim H) \to (A \wedge (\sim B))$
$\therefore\ B \vee H$

7.5 Prove the validity of the following arguments by deduction method:

(i) 1. $(\sim J) \to (M \vee V)$
2. $(H \vee G) \to (\sim J)$
3. $H \vee G$
$\therefore\ M \vee V$

(ii) 1. $C \vee D$
2. $(C \vee D) \to (\sim H)$
3. $(\sim H) \to (A \wedge (\sim B))$
4. $(A \wedge (\sim B)) \to (R \vee S)$
$\therefore\ R \vee S$

7.6 Prove the validity of the following arguments by method of contradiction:

(i) 1. $((\sim J) \wedge V) \rightarrow (K \vee V)$
   2. $(H \wedge G) \rightarrow (\sim J)$
   3. $H \vee G$
   $\therefore\ J$

(ii) 1. $\sim(A \vee B)$
   2. $(A \vee B) \rightarrow (\sim B)$
   3. $(\sim C) \rightarrow (A \wedge (\sim B))$
   4. $(A \wedge (\sim B)) \rightarrow C \vee D$
   $\therefore\ B$

7.7 Prove the validity of the following argument by deduction and contradiction methods.
"If there was a traffic jam then travelling was difficult. If they arrived in time then travelling was not difficult. They arrived in time. Therefore, there was no traffic jam."

7.8 Discuss the validity of the following argument using contradiction and tree tableau methods. Also, if the argument is invalid then suggest the suitable changes in the premises to make it valid.
"If two sides of a triangle are equal, then the opposite angles are equal. Two sides of a triangle are not equal. Therefore, the opposite angles are not equal."

7.9 Prove the validity of the following argument by tree tableau method:

1. $((\sim A) \wedge V) \rightarrow ((\sim B) \vee V)$
2. $(H \vee V) \rightarrow (\sim A)$
3. $H \wedge B$

$\therefore\ A$

8

Predicate Calculus

8.1 Introduction

In Chapter 7, we discussed theory of statement calculus. In this chapter, the aim is to introduce another class of arguments that contain predicates and quantifiers. The study of arguments, which contain predicates is called **predicate logic** or **predicate calculus**. Predicate calculus is a generalized study of logic theory. The validity of arguments in predicate calculus cannot be proved with the help of statement calculus, alone. However, the procedures used to prove the validity of arguments in predicate calculus, are almost similar to those used for statement calculus. Further, the statement calculus is an indispensable part of predicate calculus.

In statement calculus, a separate symbol is assigned to each atomic statement ignoring the common property, if any, among the statements. Such common property must also be dealt with proper consideration, otherwise, the validity of argument, thus proved, will be incorrect. This common property is called predicate.

8.2 Predicate

A part of a statement that tells something about the subject is called a predicate. Let us understand this with the help of examples.

EXAMPLE 8.1

All human beings are mortal.
Joy is a human being.
Therefore, Joy is mortal.

Here the words 'mortal', 'human beings', etc. are called predicates.

EXAMPLE 8.2

All bachelors die young.
Joy is a bachelor.
Therefore, Joy dies young.

Here also, the words 'bachelor' and 'young' are predicates.

Now, if we symbolize arguments given in Examples 8.1 and 8.2, simply as we did in Chapter 7, and prove their validity, then we are sure to get wrong conclusions.

Premises	**Statements**	**Symbols**
	1. All human beings are mortal.	H
	2. Joy is a human being.	J
Conclusion		
	Therefore, Joy is mortal.	M

All possible truth-values of the argument are given in Table 8.1. It is clear from the last column of this table that the formula $H \wedge J \to M$ is not a tautology. Therefore, the given argument is invalid. However, this inference is incorrect, because the argument is very much valid. To deal with such arguments, theory of statement calculus is slightly improved. We call it predicate calculus. In the following sections, we will discuss new terms, symbols and methods related with predicate calculus.

Table 8.1 Table for $H \wedge J \to M$

H	J	M	$H \wedge J$	$H \wedge J \to M$
T	T	T	T	T
T	T	F	T	F
T	F	T	F	T
T	F	F	F	T
F	T	T	F	T
F	T	F	F	T
F	F	T	F	T
F	F	F	F	T

8.3 Simple Statement Function and Statement

A simple statement function is defined as an expression consisting of a predicate symbol and a variable, written inside the parentheses. A statement function becomes a statement when some object replaces the variable. Like statement in statement calculus, the statement function may be simple or compound. A single statement function is called simple or atomic statement function. A compound statement function is formed by joining two or more simple statement functions with logical connectives. A statement assumes truth-value true or false when corresponding objects replace the variables within the parentheses.

8.4 Predicate Symbols

As discussed in Chapter 7, simple statements are represented by capital letters A, B, C, ..., Z etc. For statement functions, capital letters, along with their subjects or variable symbols inside the parentheses, are used. Generally, the first letter of the predicate is chosen to represent the statement function and small letters are used for variable symbols. All the symbols of logical operators discussed in Chapter 7 will be used in this chapter in a similar way.

EXAMPLE 8.3

Statements	Symbols
1. Canada is a country.	$C(x)$
2. Man is mortal.	$M(x)$
3. Joy is a human being.	$H(x)$

Here, symbol C represents the country and x inside the parentheses is a variable symbol. The variable symbol may assume names of the countries, for example, Canada, USA, UK, Italy, India, China and so on.

The symbol M represents the predicate 'mortal', and the variable x represents 'man'. Similarly, symbol H is taken for the predicate 'human being'.

8.5 One or More Variable Predicate Functions

If one variable is associated with the predicate, then it is called 1-variable predicate function. Similarly, if two variables are associated with the predicate, then it is called 2-variable predicate function. In general, there may be an n-variable predicate function. The statement calculus is, in fact, a 0-variable predicate calculus.

For example,

1-variable predicate function	$M(x)$
2-variable predicate function	$M(x_1, x_2)$
3-variable predicate function	$M(x_1, x_2, x_3)$
⋮ ⋮	⋮
n-variable predicate function	$M(x_1, x_2, x_3, \ldots, x_n)$

Let us elaborate it with the help of English sentences and symbols:

Statements	Symbols
1. P is mother of Q.	$M(x, y)$
2. P is taller than Q.	$T(x, y)$
3. P is greater than Q but less than R.	$G(x, y, z)$

Here x, y, and z are the variables representing P, Q, and R respectively.

8.6 Quantifiers

The quantifiers are used for quantification of variables. Two types of quantifiers: 'for all' (∀) and 'there exists' (∃), are used.

Quantifiers	Symbols
Universal quantifier	∀
Existential quantifier	∃

For example,

1. For all x, x is a man. $\forall x\, M(x)$
2. There exists a man. $\exists x\, M(x)$

8.7 Predicate Formula

An n-variable predicate function $P(x_1, x_2, x_3, \ldots, x_n)$ is called a simple predicate formula. It is also called atomic predicate formula.

8.7.1 Well-formed Formula

In Chapter 7, the definition of a well-formed formula was given for statement calculus. Similar definition for a well-formed formula of predicate calculus is also given here.

1. An atomic formula is a well-formed formula.
2. If A is a well-formed formula then $\sim A$, and $(\sim A)$ are also well-formed formulas.
3. If A and B are well-formed formulas then $A \wedge B$, $A \vee B$, $A \rightarrow B$, $A \leftrightarrow B$, $A \oplus B$, $A \uparrow B$, $A \downarrow B$ are also well-formed formulas.
4. If A is a well-formed formula, and x is a variable, then $\forall x(A)$, and $\exists x(A)$ are also well-formed formulas.
5. A formula obtained by using rule number 1 to 4, any number of times, is a well-formed formula. The outermost parentheses in a formula are however, considered as optional, and may be omitted.

EXAMPLE 8.4

Statements	Formulas
1. All human beings are mortal.	$\forall x(H(x) \rightarrow M(x))$
2. Some human beings are clever.	$\exists x(H(x) \wedge C(x))$

The well-formed formulas, given for statement 1 and statement 2, are read as: 'for all x, if x is a human being then x is mortal', and 'there exists x, such that x is a human being and x is clever', respectively.

8.7.2 Well-formed Sub Formula

A well-formed sub formula is a part of a well-formed formula, and is itself a well-formed formula satisfying the rules given in Section 8.7.1.

EXAMPLE 8.5 Consider the well-formed formula, $\exists x(H(x) \wedge C(x))$. Here $C(x)$ is a well-formed sub formula.

8.8 Scope of a Quantifier

The scope of a quantifier is the formula, associated with the quantifier.

EXAMPLE 8.6

1. Consider a formula
$$\exists x(H(x) \wedge C(x)).$$
In this formula, '$H(x) \wedge C(x)$' is the scope of the existential quantifier, '$\exists$'.
2. In formula
$$\forall x(H(x) \rightarrow M(x)) \wedge C(x),$$
the scope of universal quantifier '$\forall$' is
$$H(x) \rightarrow M(x).$$
3. The scope of universal quantifier '$\forall$' in formula
$$\forall x((H(x) \rightarrow M(x)) \wedge C(x, y))$$
is formula
$$(H(x) \rightarrow M(x)) \wedge C(x, y).$$

8.9 Free and Bound Variables

A variable, which is not quantified in the scope of a quantifier, is called a free variable. Further, if a variable is quantified in the scope of a quantifier, then all its occurrences are bound in that scope, and therefore it is called bound variable.

EXAMPLE 8.7
1. $\forall x(H(x) \rightarrow M(x, y))$
2. $\exists x H(x) \wedge C(y)$

In above examples, x is bound and y is free. If a bound variable is replaced by another variable at all its occurrences, then the meaning (or the truth-value) of the formula will not change. Sometimes, it is useful to do so for proving the validity of an argument.

8.10 Universe of Discourse

The variables, which are quantified, may take only those objects which are members of a particular set or class. Such set or class is called universe of discourse. For example, if the discussion refers to human beings, then the universe of discourse is the set or class of 'human beings'. The universe of discourse must be stated explicitly, because the truth-value of the statement depends on it.

EXAMPLE 8.8 Consider formula $L(x)$: 'x is less than 6'. Then the formula $\forall x\ L(x)$' is true for the set of values {..., –2, –1, 0, 1, 2, 3, 4, 5}. Therefore, this may be the universe of discourse. However, if the set of real numbers, as a whole, is taken as the universe of discourse, then the above formula will be invalid, because it may contain numbers, which may not be less than six.

8.11 Proving the Validity by Truth Table Methods

The truth table methods, discussed in Chapter 7, are also applicable in predicate calculus. Before using these methods, the formulas must be made free from quantifiers. It can be done by replacing the quantified variables by the actual names or values of objects from the universe of discourse, and then treat the predicate symbol along with the object as one letter symbol in the truth table. The validity, then, is discussed by examining the truth table. The complete procedure will become clear from the following example.

EXAMPLE 8.9 Consider the following argument:

All religions are sacred. *H* is a religion. Therefore, *H* is sacred.

Let the universe of discourse be: Religion = $\{B, C, H, I\}$.

Symbolize the argument and replace the variable occurrences by the names of the religions from the universe of discourse. Once the objects replace the variables, the quantifiers are dropped from the statements. The argument is written in the following form:

Premises

1. All religions are sacred.
   wff : $\forall x(R(x) \rightarrow S(x))$ {wff means well-formed formula}
2. *H* is a relation.
   wff : $R(H)$

Conclusion

Therefore, *H* is sacred.
wff : $\therefore\ S(H)$

Now, replace the variable x in formula

$$\forall x(R(x) \rightarrow S(x))$$

by the name of the religion and drop the quantifier. The argument will become:

1. $R(H) \rightarrow S(H)$
2. $R(H)$

   $\therefore\ S(H)$

Finally, replace the terms in the argument by single letter symbols to get the following form of the argument:

1. $R \rightarrow S$
2. R

   $\therefore\ S$

Now, it has become the argument of statement calculus, therefore, truth table methods discussed in Chapter 7 are applicable here. The truth table for this is given in Table 8.2.

Table 8.2 Table for $((R \to S) \wedge R) \to S$

R *Premise*	S	$R \to S$ *Premise*	$(R \to S) \wedge R$ *Premise*	$((R \to S) \wedge R) \to S$
T	T	T	T	T
T	F	F	F	T
F	T	T	F	T
F	F	T	F	T

Since, formula in the last column of Table 8.2 is a tautology, therefore, the argument is valid.

8.12 Inference Theory of Predicate Calculus

Inference theory of predicate calculus provides a complete system of rules consisting of logical implication and equivalences for constructing proofs of theorems and arguments. It is an extension of inference theory of statement calculus.

8.12.1 Proving the Validity by Deduction Method

Since the truth table method is feasible only for small number of variables with small universe of discourse, hence, deduction method is used for proving the validity of an argument in predicate calculus. The deduction system consists of all implications and equivalences used in statement calculus along with some additions in it. The enhanced lists are given in Table 8.3 and Table 8.4, respectively.

Logical Implications

Some important logical implications are given in Table 8.3.

Table 8.3 Logical Implications

S. No	*Logical implications*	*Special names*
1	$P \Rightarrow P \vee Q$	Addition
2	$Q \Rightarrow P \vee Q$	Addition
3	$P \wedge Q \Rightarrow P$	Simplification
4	$P \wedge Q \Rightarrow Q$	Simplification
5	$\sim P, P \vee Q \Rightarrow Q$	Disjunctive syllogism
6	$\sim Q, P \to Q \Rightarrow \sim P$	Modus tollens
7	$P, P \to Q \Rightarrow Q$	Modus ponens
8	$P \to Q, Q \to R \Rightarrow P \to R$	Hypothetical syllogism
9	$P \vee Q, P \to R, Q \to R \Rightarrow R$	Dilemma
10	$\sim P \Rightarrow P \to Q$	
11	$Q \Rightarrow P \to Q$	
12	$\sim (P \to Q) \Rightarrow P$	

(*Contd.*)

Table 8.3 Logical Implications (*Contd.*)

S. No	*Logical implications*	*Special names*
13	$\sim (P \to Q) \Rightarrow \sim Q$	
14	$P, Q \Rightarrow P \wedge Q$	
15	$(\sim P) \vee Q, P \vee Q \Rightarrow Q$	
16	$\forall x A(x) \vee \forall x B(x) \Rightarrow \forall x(A(x) \vee B(x))$	
17	$\exists x(A(x) \wedge B(x)) \Rightarrow \exists x A(x) \wedge \exists x B(x)$	

Equivalences

A list of equivalences is given in Table 8.4.

Table 8.4 Equivalences

S. No.	*Formula*	*Dual of formula*	*Special names*
1	$A \vee A \equiv A$	$A \wedge A \equiv A$	Idempotent laws
2	$(A \vee B) \vee C \equiv A \vee (B \vee C)$	$(A \wedge B) \wedge C \equiv A \wedge (B \wedge C)$	Associative laws
3	$A \vee (B \wedge C) \equiv (A \vee B) \wedge (A \vee C)$	$A \wedge (B \vee C) \equiv (A \wedge B) \vee (A \wedge C)$	Distributive laws
4	$A \vee F \equiv A$	$A \wedge T \equiv A$	Identity laws
5	$A \vee T \equiv T$ (Tautology)	$A \wedge F \equiv F$ (Contradiction)	
6	$A \vee (\sim A) \equiv T$ (Tautology) $\sim T \equiv F$	$A \wedge (\sim A) \equiv F$ (Contradiction) $\sim F \equiv T$	Complement laws
7	$A \vee (A \wedge B) \equiv A$	$A \wedge (A \vee B) \equiv A$	Absorption laws
8	$\sim (\sim A) \equiv A$		Involution law/ Double negation
9	$\sim (A \vee B) \equiv (\sim A) \wedge (\sim B)$	$\sim (A \wedge B) \equiv (\sim A) \vee (\sim B)$	De Morgan's laws
10	$A \oplus B \equiv ((\sim A) \wedge B) \vee (A \wedge (\sim B))$		
11	$A \vee B \equiv B \vee A$	$A \wedge B \equiv B \wedge A$	Commutative laws
12	$A \vee (B \wedge (\sim B)) \equiv A$	$A \wedge (B \vee (\sim B)) \equiv A$	
13	$A \vee (B \vee (\sim B)) \equiv T$	$A \wedge (B \wedge (\sim B)) \equiv F$	
14	$A \to B \equiv (\sim A) \vee B$		
15	$\sim (A \to B) \equiv A \wedge (\sim B)$		
16	$A \to B \equiv (\sim B) \to (\sim A)$		
17	$A \to (B \to C) \equiv (A \wedge B) \to C$		
18	$(A \leftrightarrow B) \equiv (A \to B) \wedge (B \to A)$		
19	$\sim (A \leftrightarrow B) \equiv A \leftrightarrow (\sim B))$		
20	$(A \leftrightarrow B) \equiv (A \wedge B) \vee ((\sim A) \vee (\sim B))$		
21	$\exists x(A(x) \vee B(x)) \equiv \exists x A(x) \vee \exists x B(x)$	$\exists x(A(x) \wedge B(x)) \equiv \exists x A(x) \wedge \exists x B(x)$	
22	$\forall x(A(x) \wedge B(x)) \equiv \forall x A(x) \wedge \forall x B(x)$	$\forall x(A(x) \vee B(x)) \equiv \forall x A(x) \vee x B(x)$	
23	$\sim (\exists x A(x)) \equiv \forall x(\sim A(x))$		
24	$\sim (\forall x A(x)) \equiv \exists x(\sim A(x))$		
25	$\forall x(A \vee B(x)) \equiv A \vee \forall x B(x)$	$\forall x(A \wedge B(x)) \equiv A \wedge \forall x B(x)$	
26	$\exists x(A \wedge B(x)) \equiv A \wedge \exists x B(x)$	$\exists x(A \vee B(x)) \equiv A \vee \exists x B(x)$	
27	$\exists x A(x) \to B \equiv \exists x(A(x) \to B))$		
28	$\forall x A(x) \to B \equiv \forall x(A(x) \to B))$		
29	$A \to \forall x B(x) \equiv \forall x(A \to B(x))$		
30	$A \to \exists x B(x) \equiv \exists x(A \to B(x))$		

Inference Rules

Inference rules are used for deducing the conclusion. These rules are as follows:

Rule 1: A premise may be introduced at any place in the derivation.

Rule 2: A formula may be introduced in a derivation, if it is logically implied by one or more of the preceding formulas, or premises.

Rule 3: A formula can be replaced by its equivalent formula.

Rule 4: $\forall x A(x) \Rightarrow A(y)$. It is known as universal specification (U.S.)

Rule 5: $\exists x A(x) \Rightarrow A(y)$, provided that y is not free in any given premise and also not free in any prior step of the derivation. It is known as existential specification (E.S.). Use new variable each time, E.S. is used.

Rule 6: $A(x) \Rightarrow \exists y A(y)$. It is called existential generalization (E.G.).

Rule 7: $A(x) \Rightarrow \forall y A(y)$, provided that x is not free in any of the given premises, and further, if x is free in a prior step which resulted from the use of E.S., then no variable introduced by that use of E.S. appear free in $A(x)$. It is called universal generalization (U.G.).

By using these rules, if it is possible to deduce the conclusion, then the argument is said to be valid. A tabular representation, which was used in Chapter 7 for listing derivation steps of an argument, is again our choice for showing proof steps. Column 1 of each table, showing proof steps, contains serial numbers, and column 2 contains formulas, which are either premises of the argument, or the deductions obtained using previous steps. Further, column 3 gives the reference numbers, which are the serial numbers given in column 1 of the table. These reference numbers are of those formulas, which have been used in deducing the current formula. Finally, column 4 of the table gives the complete explanation.

EXAMPLE 8.10 Consider the following argument.

"All human beings are mortal. John is a human being. Therefore, John is mortal."

This argument can be symbolized as:

Premises

1. All human beings are mortal.
   wff: $\forall x(H(x) \rightarrow M(x))$.
2. John is a human being.
   wff: $H(\text{John})$.

Conclusion

Therefore, John is mortal.

wff: $M(\text{John})$.

Derivation steps are shown in Table 8.5.

Table 8.5 Derivation Steps

S. No.	*Formula*	*Reference*	*Explanation*
1	$\forall x(H(x) \to M(x))$		Given premise
2	$H(\text{John})$		Given premise
3	$H(\text{John}) \to M(\text{John})$	1	U.S.
4	$(\sim H(\text{John})) \vee M(\text{John})$	3	Equivalence $A \to B \equiv (\sim A) \vee B$
5	$M(\text{John})$	2, 4	Disjunctive syllogism

Since the conclusion is deduced at step 5, therefore, the argument is valid.

8.12.2 Proving the Validity by Method of Contradiction

The method of contradiction, discussed earlier in Chapter 7, is also applicable here. As discussed, in this method, an additional premise which is the negation of the conclusion, is introduced, and then the derivations are made to deduce a contradiction. If a contradiction is deduced, then the given argument is valid. It shows that the given set of premises is inconsistent, i.e., their conjunction is always false. It implies that the negation of conclusion, which was introduced as an additional premise, is wrong. Hence, the conclusion follows from the given premises.

EXAMPLE 8.11 Consider the argument

$$\forall x(A(x) \vee B(x)) \Rightarrow \forall x A(x) \vee \exists x B(x).$$

In this argument, formula $\forall x(A(x) \vee B(x))$ is premise, while formula $\forall x A(x) \vee \exists x B(x)$ is the conclusion. Introduce negation of conclusion as an additional premise. Then, using standard equivalences and implications, we deduce a contradiction, which shows that the given argument is valid. The derivation steps are shown in Table 8.6.

Table 8.6 Derivation Steps

S. No.	*Formula*	*Reference*	*Explanation*
1	$\forall x(A(x) \vee B(x))$		Given premise
2	$\sim (\forall x A(x) \vee \exists x\, B(x))$	1	Negation of conjunction introduced
3	$\sim (\forall x A(x)) \wedge (\sim \exists x B(x))$	2	De Morgan's law
4	$\sim (\forall x A(x))$	3	$P \wedge Q \Rightarrow P$
5	$\sim \exists x B(x)$	3	$P \wedge Q \Rightarrow Q$
6	$\exists x(\sim A(x))$	4	$\sim (\forall x A(x)) \equiv \exists x(\sim A(x))$
7	$\forall x(\sim B(x))$	5	$\sim (\exists x A(x)) \equiv \forall x(\sim A(x))$
8	$\sim A(y)$	6	E.S.
9	$\sim B(y)$	7	U.S.
10	$(\sim A(y)) \wedge (\sim B(y))$	8, 9	$A, B \Rightarrow A \wedge B$
11	$\sim (A(y) \vee B(y))$	10	De Morgan's law
12	$A(y) \vee B(y)$	1	U.S.
13	$(\sim (A(y) \vee B(y))) \wedge (A(y) \vee B(y))$	11, 12	$A, B \Rightarrow A \wedge B$

Since at S. No. 13, a contradiction is obtained, therefore, the argument is valid.

8.13 Solved Problems

PROBLEM 8.1 Prove that the following argument is valid:

1. $\forall x(A(x) \rightarrow B(x))$
2. $\forall x(B(x) \rightarrow C(x))$
3. $A(y) \vee C(y)$

$\therefore C(y)$

Solution

The derivation steps are shown in Table 8.7.

Table 8.7 Derivation Steps of Problem 8.1

S. No.	*Formula*	*Reference*	*Explanation*
1	$\forall x(A(x) \rightarrow B(x))$		Given premise
2	$\forall x(B(x) \rightarrow C(x))$		Given premise
3	$A(y) \vee C(y)$		Given premise
4	$A(y) \rightarrow B(y)$	1	U.S.
5	$B(y) \rightarrow C(y)$	2	U.S.
6	$A(y) \rightarrow C(y)$	4, 5	Hypothetical syllogism
7	$(\sim A(y)) \vee C(y)$	6	Equivalence $A \rightarrow B \equiv (\sim A)\ B$
8	$C(y)$	3, 7	$(\sim A) \vee B,\ A \vee B \Rightarrow B$

Since we obtain the conclusion at step 8, therefore, the argument is valid.

PROBLEM 8.2 Prove that the following argument is valid.

$$\exists x(A(x) \wedge B(x)) \quad \text{(premise)}$$

$$\therefore \quad \exists xB(x) \wedge \exists xC(x) \quad \text{(conclusion)}$$

Solution

The complete derivation is given in Table 8.8.

Table 8.8 Derivation Steps of Problem 8.2

S. No.	*Formula*	*Reference*	*Explanation*
1	$\exists x(A(x) \wedge B(x))$		Premise
2	$A(y) \wedge B(y)$	1	E.S.
3	$A(y)$	2	$(A \wedge B) \rightarrow A$
4	$B(y)$	2	$(A \wedge B) \rightarrow B$
5	$\exists xA(x)$	3	E.G.
6	$\exists xB(x)$	4	E.G.
7	$\exists xA(x) \wedge \exists xB(x)$	5, 6	$A, B \Rightarrow A \wedge B$

Note that the conclusion is obtained at step 7 in the table, therefore, the given argument is valid.

PROBLEM 8.3 Prove that the following argument is valid by the method of contradiction.

All human beings are mortal. If John is mortal then John is clever. But, it is not the case that John is a human being and John is clever. However, John is a human being or John is a robot. Therefore, John is a robot.

Further verify your inference by truth table method.

Solution

The symbolic form of the argument is given below. Symbol J is taken for the object 'John'. Symbol H represents 'human being', M represents 'mortal', C represents 'clever', and R represents 'robot'.

1. $\forall x(H(x) \rightarrow M(x))$
2. $M(J) \rightarrow C(J)$
3. $\sim (H(J) \wedge C(J))$
4. $H(J) \vee R(J)$

$\therefore R(J)$

The derivation steps are shown in table 8.9.

Table 8.9 Derivation Steps of Problem 8.3

S. No.	*Formula*	*Reference*	*Explanation*
1	$\forall x(H(x) \rightarrow M(x))$		Given premise
2	$M(J) \rightarrow C(J)$		Given premise
3	$\sim (H(J) \wedge C(J))$		Given premise
4	$H(J) \vee R(J)$.		Given premise
5	$\sim R(J)$		Negation of conclusion introduced
6	$(\sim (H(J)) \vee (\sim C(J))$	3	De Morgan's law
7	$H(J) \rightarrow M(J)$	1	U.S.
8	$H(J) \rightarrow C(J)$	2, 7	Hypothetical syllogism
9	$(\sim H(J)) \vee C(J)$	8	$P \rightarrow Q \equiv (\sim P) \vee Q$
10	$\sim H(J)$	6, 9	$P \vee Q, (\sim P) \vee Q \Rightarrow Q$
11	$R(J)$	4, 10	$\sim P, P \vee Q \Rightarrow Q$
12	$\sim R(J) \wedge R(J)$	5, 11	

A contradiction is found at step 12 in Table 8.9, therefore, the given argument is valid. Also, let us try to prove the validity of this argument by truth table method. While constructing the truth table, we can use one letter symbol after dropping attached quantifiers and words used for each object. Next, construct well-formed formula for the given argument. The formula thus obtained for the argument under discussion, is as follows:

$$((H \rightarrow M) \wedge (M \rightarrow C) \wedge (\sim (H \wedge C)) \wedge (H \vee R)) \rightarrow R$$

It is represented by symbol X in Table 8.10.

Table 8.10 Truth Table for Problem 8.3

H	M	C	R Conclusion	$H \to M$ Premise	$M \to C$ Premise	$H \wedge C$	$\sim (H \wedge C)$ Premise	$H \vee R$ Premise	X
T	T	T	T	T	T	T	F	T	T
T	T	T	F	T	T	T	F	T	T
T	T	F	T	T	F	F	T	T	T
T	T	F	F	T	F	F	T	T	T
T	F	T	T	F	T	T	F	T	T
T	F	T	F	F	T	T	F	T	T
T	F	F	T	F	T	F	T	T	T
T	F	F	F	F	T	F	T	T	T
F	T	T	T	T	T	F	T	T	T
F	T	T	F	T	T	F	T	F	T
F	T	F	T	T	F	F	T	T	T
F	T	F	F	T	F	F	T	F	T
F	F	T	T	T	T	F	T	T	T
F	F	T	F	T	T	F	T	F	T
F	F	F	T	T	T	F	T	T	T
F	F	F	F	T	T	F	T	F	T

The formula X, in the last column is a tautology, therefore, the given argument is valid.

PROBLEM 8.4 Show that the following argument is valid:

1. $\exists x(F(x) \wedge S(x)) \to \forall y(M(y) \to W(y))$
2. $\exists y(M(y) \wedge (\sim W(y))$

   $\therefore\ \forall x(F(x) \to (\sim S(x))$

Solution

The derivation steps are shown in Table 8.11.

Table 8.11 Derivation Steps of Problem 8.4

S. No.	*Formula*	*Reference*	*Explanation*
1	$\exists y(M(y) \wedge (\sim W(y)))$		Given premise
2	$M(z) \wedge (\sim W(z))$	1	E.S.
3	$\sim (\sim (M(z) \wedge (\sim W(z))))$	2	Double negation
4	$\sim (\sim (M(z) \vee W(z)))$	3	De Morgan's law
5	$\sim (M(z) \to W(z))$	4	Equivalence $A \to B \equiv (\sim A) \vee B$
6	$\exists y(\sim(M(y) \to W(y)))$	5	E.G.
7	$\sim \forall y\ (M(y) \to W(y))$	6	$\sim\forall x(A(x)) \equiv \exists x \sim A(x)$
8	$\exists x(F(x) \wedge S(x)) \to \forall y(M(y) \to W(y))$		Given premise
9	$\sim (\exists x(F(x) \wedge S(x)))$	7, 8	$\sim B,\ A \to B \Rightarrow (\sim A)$
10	$\forall x(\sim(F(x) \wedge S(x)))$	9	$\forall x(\sim A(x)) \equiv \sim (\exists x A(x))$
11	$\forall x\ ((\sim F(x)) \vee (\sim S(x)))$	10	De Morgan's law
12	$\forall x\ (F(x) \to (\sim S(x)))$	11	Equivalence $A \to B \equiv (\sim A) \vee B$

Conclusion is derived at step 12 in Table 8.11, therefore, the argument is valid.

Exercises

8.1 Indicate the free and bound variables in each of the following formulas. Also, find the scope of each variable.

1. $\forall x(H(x) \rightarrow M(x)) \wedge H(y)$
2. $\exists x(H(x) \rightarrow M(x)) \wedge (H(y) \rightarrow M(z))$
3. $\forall x F(x) \rightarrow (\sim S(x))$
4. $\forall x(H(x) \rightarrow M(x)) \wedge (\sim (\exists y A(y)))$

8.2 Let the universe of discourse for x and y be $\{a, b, c\}$. Obtain the formulas that are free from quantifiers. Construct their equivalent formulas of statement calculus.

1. $\forall x(H(x) \rightarrow M(x))$
2. $\exists y(A(y) \wedge R(x))$
3. $\forall x(F(x) \rightarrow (\sim S(x)))$
4. $\forall x(F(x) \vee (\sim S(x)))$

8.3 Prove the validity of the following argument by deduction and contradiction methods.

1. $\forall x(\sim J(x)) \rightarrow (H(y) \vee J(y))$
2. $\exists x((H(x) \vee G(x)) \rightarrow (\sim J(x)))$
3. $H(y) \wedge G(y)$

$\therefore \; J(y)$

8.4 Prove the validity of the following argument by deduction and contradiction methods.

1. $\forall x(C(x) \vee D(x))$
2. $\forall x((C(x) \vee D(x)) \rightarrow (\sim H(x)))$
3. $(\sim H(y)) \rightarrow (A(y) \wedge (\sim B(y)))$
4. $\forall x((A(x) \wedge (\sim B(x))) \rightarrow (R(x) \vee S(y)))$

$\therefore \; R(y) \vee S(y)$

8.5 Prove the validity of the following argument by the method of contradiction. Verify your result with truth table method.

1. $\forall x((\sim J(x) \wedge V(x)) \rightarrow C(x))$
2. $\exists x((\sim J(x)) \vee G(x))$
3. $\forall x((\sim C(x)) \wedge V(x))$

$\therefore \; G(x)$

8.6 Symbolize the following argument using suitable symbols and prove its validity.

All cats are animals. Some animals are not cats. If X is an apple then X is not a cat. However, X is an apple. Therefore, X is not an animal.

Part C

LATTICE AND BOOLEAN STRUCTURE

9

Lattice Theory

9.1 Introduction

Lattice theory is very helpful in developing the theory of switching logic. It is the foundation of Boolean algebra, which is an integral part of circuit design. In this chapter, we will discuss partial order relation, Hasse diagramms and lattices.

9.2 Partial Order Relation

A relation R on a set S is called a partial order relation if it has the following three properties:

1. R is reflexive, i.e., xRx, for every x in S.
2. R is antisymmetric, i.e., whenever xRy and yRx, then $x = y$.
3. R is transitive, i.e., if xRy and yRz, then xRz.

9.3 Partially Ordered Set

The set S with partial order relation R, defined above is called partially ordered set or poset and is denoted by $< P, R >$. Let $x, y \in S$, then we say that x and y are comparable if either $x \leq y$ or $y \leq x$. Symbol $\leq$ is used for partial order relation R. The word 'partial' is used in defining partially ordered set because some of the elements of S may not be comparable (i.e., neither $x \leq y$ nor $y \leq x$).

9.4 Totally Ordered Set

If every pair of elements of S is comparable with respect to relation R, then S is said to be totally ordered set, or a chain. The relation R is called a total order relation or total ordering.

EXAMPLE 9.1 Consider the set $\mathbb{Z}$ of integers and $x, y \in \mathbb{Z}$. Let the relation xRy be defined as

$$x \geq y.$$

Then R is a partial order relation. It is

(i) reflexive: $(x, x) \in R, \forall x \in \mathbb{Z}$.
(ii) antisymmetric: $(x, y) \in R$ and $(y, x) \in R \Rightarrow x = y$, and
(iii) transitive: $(x, y) \in R$ and $(y, z) \in R \Rightarrow (x, z) \in R$.

EXAMPLE 9.2 Let S be a collection of all subsets of a set (i.e. power set). Consider a relation 'A is a subset of B' for $A, B \in S$. It will be partial order relation, since

$A \subseteq A$: reflexive,
$A \subseteq B$ and $B \subseteq A \Rightarrow A = B$: antisymmetric
and $A \subseteq B, B \subseteq C \Rightarrow A \subseteq C$ transitive.

It is not total ordering. For instance, consider the power set of $\{a, b\} = \{\{\ \}, \{a\}, \{b\}, \{a, b\}\}$. Observe that, in this case, $\{a\}, \{b\} \in S$ but neither $\{a\} \subseteq \{b\}$ nor $\{b\} \subseteq \{a\}$.

EXAMPLE 9.3 Suppose $\mathbb{I}^+$ is the set of positive integers and $x, y \in \mathbb{I}^+$. Let xRy be defined as 'x divides y', then R is partial order relation. Note that, R is not total order relation. For instance, neither 3 divides 5 nor 5 divides 3, therefore, 3 and 5 are not comparable. Further, if $x, y \in \mathbb{Z}$ (set of integers), then R is not partial order relation. Since x divides $-x$, and $-x$ divides x, but $x \neq -x$, therefore, it is not antisymmetric.

EXAMPLE 9.4 The set $\mathbb{Z}$ of positive integers with usual order '$\leq$' (less than or equal to) is total order relation, since every pair of elements is comparable.

9.5 Hasse Diagrams

Hasse diagram is a directed graph (see Chapter 13) that represents partially ordered set. An edge connects two elements of the set, which are comparable. Elements, which are not comparable, are shown isolated in the Hasse diagram. Hasse diagram is read bottom up and elements at higher level are said to succeed the elements at lower level. For example, if $a \leq b$ (i.e. aRb), then a will be at lower end of the edge.

EXAMPLE 9.5 Let

$$A = \{a, b, c, d, e, f, g, h\}.$$

Consider relation, 'usual ordering', i.e., '$\leq$', then it is totally ordered set. It is represented by the Hasse diagram given by Figure 9.1.

EXAMPLE 9.6 Let

$$A = \{1, 2, 3, 4, 6, 8, 9, 12, 18, 24\}.$$

Consider relation, '*x* divides *y*', then it is partially ordered set and the relation is partial order relation. It is represented by the Hasse diagram given in Figure 9.2.

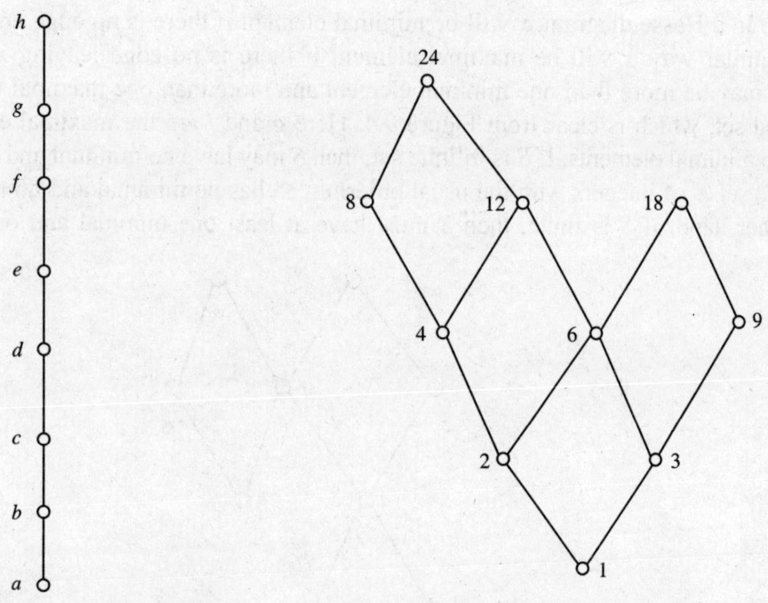

Figure 9.1 Hasse diagram (totally ordered set).

Figure 9.2 Hasse diagram (partially ordered set).

EXAMPLE 9.7 Let

$$A = \{2, 3, 6, 12, 24, 36\}.$$

Consider relation, '*x* divides *y*'. Then it is a partially ordered set and the relation is partial order relation. It is represented by the Hasse diagram given in Figure 9.3.

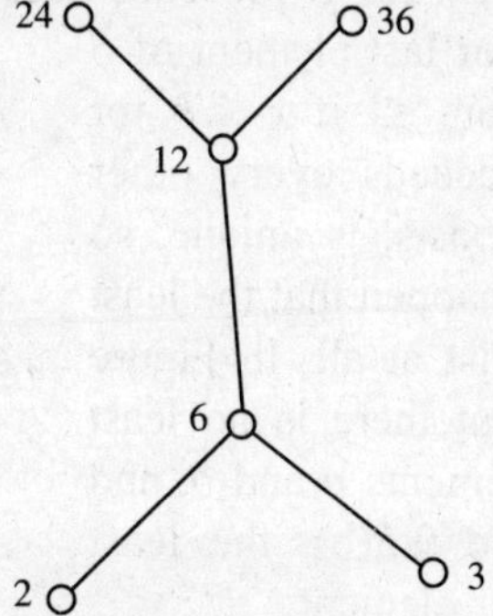

Figure 9.3 Hasse diagram for set $A\{2, 3, 6, 12, 24, 36\}$.

9.6 Minimal and Maximal Elements; Least and Greatest Elements

Consider a partially ordered set S, an element $a \in S$ is called minimal element if no other element of S precedes a (i.e., there is no $b \in S$, such that $b \leq a$). Similarly, an element $a \in S$ is called maximal element if no other element of S succeeds a (i.e., there is no $b \in S$, such that $a \leq b$). In a Hasse diagram, x will be minimal element if there is no edge entering x from below. In a similar way, x will be maximal element if there is no edge leaving x and going upward. There may be more than one minimal element and more than one maximal element in a partially ordered set, which is clear from Figure 9.4. Here g and j are the maximal elements and a and h are the minimal elements. If S is infinite set, then S may have no minimal and no maximal element, e.g. the set $\mathbb{Z}$ of integers with the usual ordering '$\leq$' has no minimal and no maximal element. On the other hand, if S is finite, then S must have at least one minimal and one maximal element.

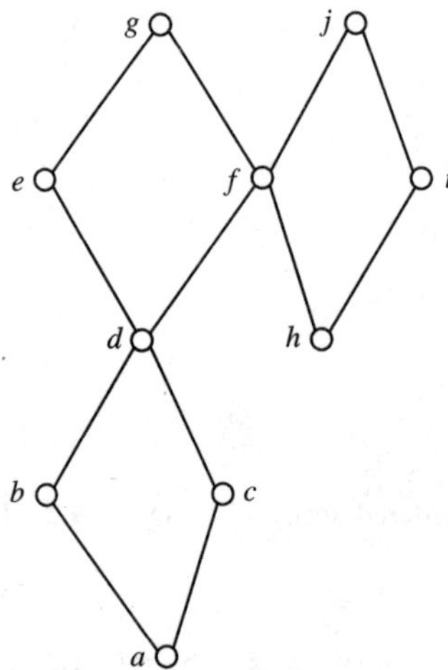

Figure 9.4 Hasse diagram (minimal and maximal elements).

An element $a \in S$ is called a least or first element of S with respect to partial order relation $\leq$, if $a \leq x$ for every element x in S, i.e., a precedes every other element in S. Similarly, an element $b \in S$ is called a greatest or last element of S with respect to partial order relation '$\leq$' if $x \leq b$ for every element x in S, i.e., b succeeds every other element in S. A least element, if exists, is unique, so also is the greatest element. It may happen that the least or the greatest elements do not exist at all. In Figure 9.5(a), the greatest element is c but there is no least element. There are two minimal elements a and b, and one maximal element c. In Figure 9.5(b), the least element is f but there is no greatest element.

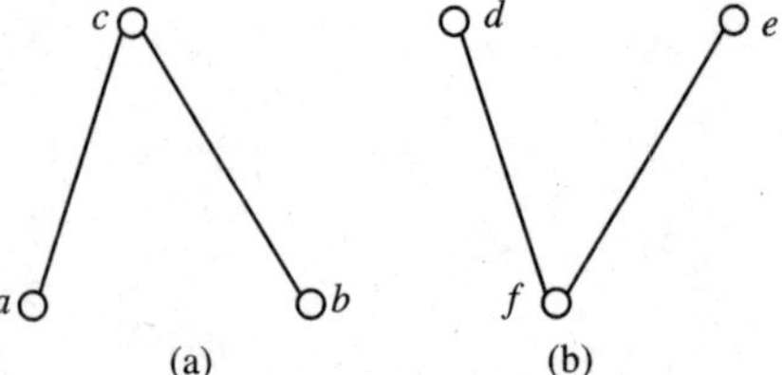

Figure 9.5 Hasse diagram (least and greatest elements).

9.7 Supremum and Infimum

Consider a partially ordered set S, and let A be the subset of S. An element U in S is called an upper bound of A if U succeeds every element of A, i.e., $x \le U$ for every x in A. If U precedes every other upper bound of A, then it is called the supremum (least upper bound, lub) of A, sup(A). Analogously, an element L in a partially ordered set S is called a lower bound of a subset A of S, if L precedes every element of A, i.e., $L \le x$ for every x in A. If L succeeds every other lower bound of A, then it is called the infimum (greatest lower bound, glb) of A, inf (A).

If $x \le y$ for $x, y \in S$, then

$$\sup (x, y) = y$$

and

$$\inf (x, y) = x.$$

Supremum is denoted by $x + y$ and infimum by $x \times y$.

If set A has an upper bound, it is said to be bounded above, and if it has a lower bound, we say A is bounded below. If it is both, bounded above and bounded below, we say A is bounded. Also, note that the partially ordered set $< P, \ge >$ is the dual of $< P, \le >$ which can be obtained by rotating Hasse diagram by 180°. In this case, the least and the greatest elements are interchanged.

EXAMPLE 9.8 Consider the Hasse diagram given in Figure 9.6.

Here

$$S = (p, q, r, s, t, u)$$

Let $A = \{q, r, s\}$, then the upper bounds of A are t and u, since t and u succeed every element of A. The lower bounds of A are p and q, since p and q precedes every element of A. Sup (A) does not exist since t and u are not comparable. Inf (A) = q, since q succeeds p.

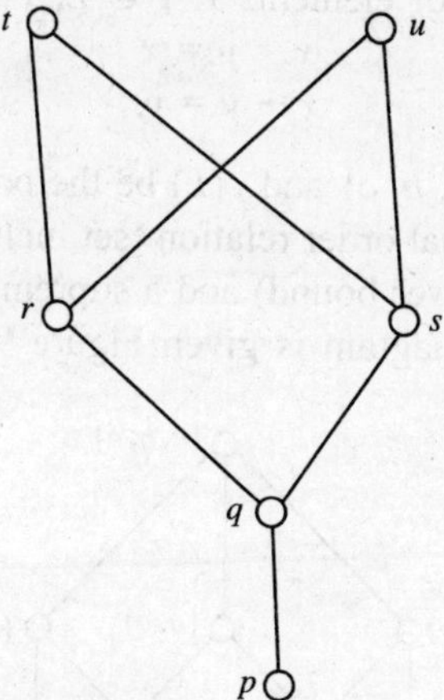

Figure 9.6 Hasse diagram for finding upper and lower bounds.

EXAMPLE 9.9 Consider the Hasse diagram given in Figure 9.3. Let $A = \{2, 3, 6\}$, then the upper bounds of A are 6, 12, 24 and 36. Sup (A) is 6. Inf(A) does not exist. For the subset $\{6, 12\}$, supremum is 12 and infimum is 6.

9.8 Lattice

A partially ordered set $< L, \leq >$ in which every pair of elements $x, y \in L$ has a infimum (greatest lower bound) and a supremum (least upper bound) is called a lattice. A totally ordered set whose first element is greatest lower bound and the last element is least upper bound is a trivial lattice. The number of elements in L is called the **order of the lattice**.

9.9 Lattice as an Algebraic System

Lattice is also defined as an algebraic system $< L, *, + >$, with nonempty set L which is closed under two binary operations $*$ and $+$ termed meet and join respectively, satisfying the following axioms:

1. (a) $x * y = y * x$; (b) $x + y = y + x$: Commutative law
2. (a) $(x * y) * z = x * (y * z)$; (b) $(x + y) + z = x + (y + z)$: Associative law
3. (a) $x * (x + y) = x$; (b) $x + (x * y) = x$: Absorption law

Axiom (3) implies another important identity known as idempotent law, which is given below.

4. (a) $x * x = x * [x + (x * y)]$ [using axiom 3(b)]
$= x$ [using axiom 3(a)]
Therefore, $x * x = x.$: Idempotent law
(b) $x + x = x + (x * (x + y)$ [using axiom 3(a)]
$= x$ [using axiom 3(b)]
Therefore, $x + x = x.$: Idempotent law

Axioms 1 to 4 are also called properties of lattices.
In a lattice $< L, \leq >$, if for a pair of elements $x, y \in L$, $x \leq y$, then

$$x * y = x$$

and

$$x + y = y.$$

EXAMPLE 9.10 Suppose $L = \{a, b, c\}$ and $\rho(L)$ be the power set of L. Then $< \rho(L), \cap, \cup >$ will be a lattice with respect to partial order relation 'set inclusion', since every pair of elements of $\rho(L)$ has a infimum (greatest lower bound) and a supremum (least upper bound). It is called the lattice of subsets. The Hasse diagram is given Figure 9.7.

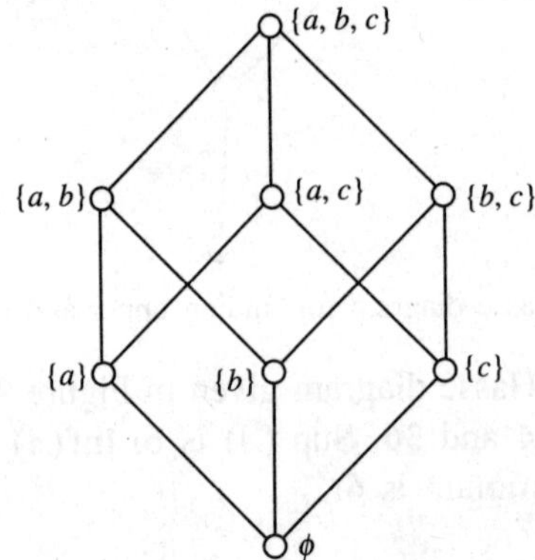

Figure 9.7 Hasse diagram for set inclusion.

EXAMPLE 9.11 Consider the set of divisors of 42 denoted by D_{42}. Then $< D_{42}, *, + >$ will be a lattice with respect to partial order relation 'divides'; since every pair of elements of D_{42} has an infimum (greatest lower bound) and a supremum (least upper bound), where operation '*' corresponds to GCD (greatest common divisor) and '+' to LCM (least common multiple). The Hasse diagram is given in Figure 9.8.

$$D_{42} = \{1, 2, 3, 6, 7, 14, 21, 42\}$$

The first and the last elements of the lattice are 1 and 42 respectively.

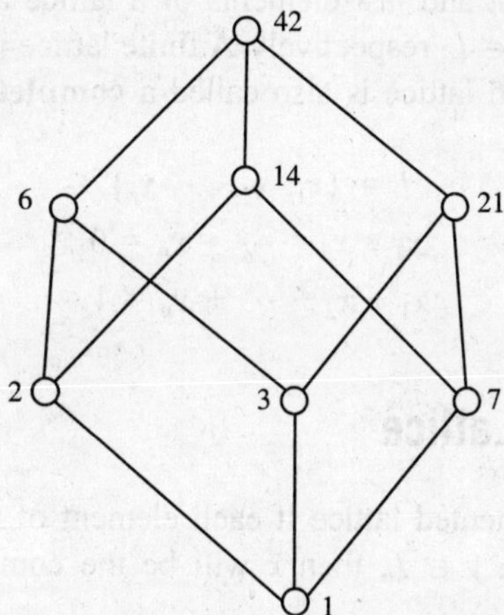

Figure 9.8 Hasse diagram for D_{42}.

9.10 Duality

The dual of a statement is obtained by interchanging binary operators * and +.

For instance,

$$(x * y) * z = x * (y * z)$$

is the dual of

$$(x + y) + z = x + (y + z).$$

It is clear from the axioms 1 to 4, given in Section 9.9 that the dual of each axiom is also an axiom. Accordingly, the principle of duality can be stated as, 'The dual of a theorem, in a lattice, is also a theorem'.

9.11 Sub Lattice

Consider a lattice $< L, *, + >$ and let S be the subset of L. Then, the algebraic structure $< S, *, + >$ will also be a lattice and will be called a sub lattice if S is closed under both the binary operations * and +, and satisfy axioms 1 to 4 given in Section 9.9.

EXAMPLE 9.12 Consider the lattice given in Example 9.11.

Here $$D_{42} = \{1, 2, 3, 6, 7, 14, 21, 42\}.$$

Now, let $$S = \{1, 2, 7, 14\}.$$

Then $< S, *, + >$ will be the sub lattice of $< D_{42}, *, + >$. However, another subset of it, {1, 3, 7, 14} is not a lattice, since it is not closed under $*$ and $+$. Therefore, it is clear from the example that all subsets are not lattices.

9.12 Bounded Lattice

A lattice L is called bounded if it is both, bounded above and bounded below, i.e., it has the first and last elements. The first and last elements of a lattice are generally denoted by 0 and 1, such that $0 \leq x$ and $x \leq 1$, $\forall x \in L$, respectively. A finite lattice is always bounded. It is denoted by $< L, *, +, 0, 1 >$. A bounded lattice is also called a **complete lattice**. Let $< L, *, +, 0, 1 >$ be a finite lattice where

$$L = \{x_1, x_2, \ldots, x_n\}.$$

Then $$x_1 * x_2 * \cdots * x_n = 0$$

and $$x_1 + x_2 + \cdots + x_n = 1.$$

9.13 Complemented Lattice

A lattice L is called a complemented lattice if each element of it possesses a complement. Let $< L, *, + >$ be a lattice and $x, y \in L$, then x will be the complement of y or y, will be the complement of x if

$$x * y = 0, \text{ the first element}$$

and $$x + y = 1, \text{ the last element.}$$

It is trivial to show that

$$0 * 1 = 0$$

and $$0 + 1 = 1.$$

That is, 0 and 1 are complements of each other. Complemented lattice is denoted by

$$< L, *, +, 0, 1, ' >.$$

Note that in a complemented lattice, the complements are not unique.

EXAMPLE 9.13 Consider the lattices given in Figures 9.9(a), (b) and (c).

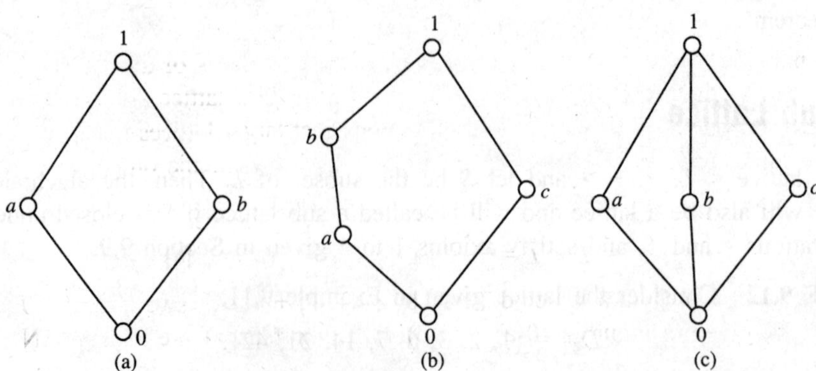

Figure 9.9 Hasse diagrams of lattices.

In Figure 9.9(a), a and b are the complements of each other, while in Figure 9.9(b), a has two complements b and c. In lattice of Figure 9.9(c), the complements of a are b and c, the complements of b are a and c, and that of c are a and b.

9.14 Distributive Lattice

A lattice $< L, *, + >$ is called a distributive lattice if it satisfies the distributive law, i.e., for x, y and z in L,

$$x * (x + y) = (x * y) + (x * z)$$

and
$$x + (x * y) = (x + y) * (x + z).$$

EXAMPLE 9.14 Lattice given in Figure 9.9(a) is distributive, while lattices given in Figures 9.9(b) and (c) are nondistributive.

For instance, in Figure 9.9(b)

$$b * (c + a) = b * 1 = b$$

but
$$(b * c) + (b * a) = 0 + a = a.$$

Therefore,
$$b * (c + a) \neq (b * c) + (b * a).$$

Similarly, in Figure 9.9(c)

$$a * (b + c) = a * 1 = a$$

but
$$(a * b) + (a * c) = 0 + 0 = 0.$$

Hence,
$$a * (b + c) \neq (a * b) + (a * c).$$

9.15 Direct Product

Consider two lattices $< L, *_1, +_1 >$ and $< S, *_2, +_2 >$. The direct product of L and S is given by the algebraic system $< L \times S, \theta, \phi >$, where $L \times S$ is the collection of ordered pairs (x, y), such that $x \in L$ and $y \in S$. The operators θ and ϕ are defined as:

$$(x_1, y_1)\ \theta\ (x_2, y_2) = (x_1 *_1 x_2, y_1 *_2 y_2)$$

and
$$(x_1, y_1)\ \phi\ (x_2, y_2) = (x_1 +_1 x_2, y_1 +_2 y_2)$$

where
$$x_1, x_2 \in L, y_1, y_2 \in S,$$

and
$$(x_1, y_1), (x_2, y_2) \in L \times S.$$

Since the operators θ and ϕ are defined in terms of the operators of the lattices $< L, *_1, +_1 >$ and $< S, *_2, +_2 >$, therefore, the direct product will also be a lattice.

The direct product of the lattices is used to construct larger lattices from smaller ones. The order of the direct product of lattices is equal to the product of the orders of the lattices.

EXAMPLE 9.15 Let $< L, *, + >$ be a lattice where $L = \{0, 1\}$.

Then
$$L^2 = L \times L = \{(0, 0), (0, 1), (1, 0), (1, 1)\}$$

This cross product will form a lattice denoted by $< L^2, *, + >$. The two lattices $< L, *, + >$ and $< L^2, *, + >$ are given in Figure 9.10(a) and (b). Operator * and + are logical 'AND' and 'OR' respectively,

Such that $0 * 0 = 0, 1 * 1 = 1, 0 * 1 = 0, 1 * 0 = 0,$

and $0 + 0 = 0, 1 + 1 = 1, 0 + 1 = 1, 1 + 0 = 1.$

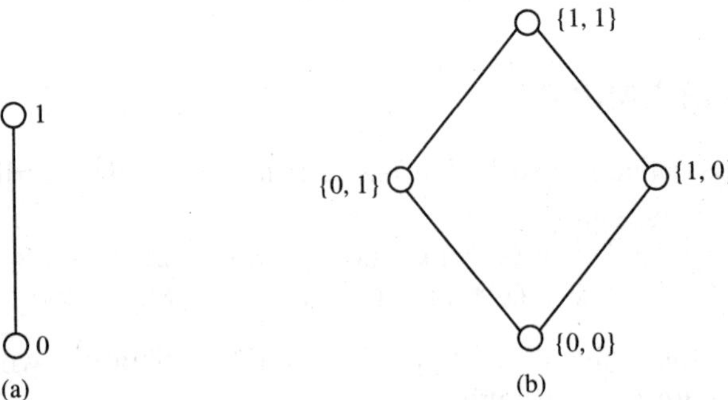

Figure 9.10 Lattices L and L^2.

9.16 Lattice Homomorphism

Let L and S be two lattices denoted by $< L, *_1, +_1 >$ and $< S, *_2, +_2 >$. A mapping $f : L \to S$ is called a lattice homomorphism from L to S if for any $x, y \in L$,

$$f(x *_1 y) = f(x) *_2 f(y)$$

and

$$f(x +_1 y) = f(x) +_2 f(y).$$

Notice that, in lattice homomorphism, both the operations *meet* and *join* are preserved. If mapping f is one-to-one and onto, then it is called **lattice isomorphism**. Further, if the homomorphism of a lattice is defined on to itself, then it is called **lattice endomorphism** and if the endomorphism is one-to-one and onto then it is called **lattice automorphism**.

EXAMPLE 9.16 Consider the lattices given in Examples 9.10 and 9.11.

That is, $< \rho(L), \cap, \cup >$ and $< D_{42}$, gcd, lcm $>$,

Then mapping f: $< \rho(L), \cap, \cup >$ to $< D_{42}, *, + >$ will be an isomorphism.

Here $\rho(L) = \{\phi, \{a\}, \{b\}, \{c\}, \{a, b\}, \{a, c\}, \{b, c\}, \{a, b, c\}\}$

and $D_{42} = \{1, 2, 3, 6, 7, 14, 21, 42\}$.

The correspondence among the elements will be:

$$f(\phi) = 1, f(\{a\}) = 2, f(\{b\}) = 3, f(\{c\}) = 7, f(\{a, b\} = 6$$

$$f(\{a, c\}) = 14, f(\{b, c\}) = 21, f(\{a, b, c\}) = 42.$$

For instance, $f(\{a\}) \cap \{b\}) = f(\phi) \equiv \text{gcd}(2, 3) = 1$

and $f(\{a\} \cup \{b\}) = f(\{a, b\} \equiv \text{lcm}(2, 3) = 6.$

9.17 Join Irreducible

Let $x \in L$ be an element of a Lattice, $< L, *, + >$. Then x is called join irreducible, if it can not be expressed as the join of two distinct elements of L.

That is, $x \in L$ is join irreducible if

$$x \neq x_1 + x_2 \text{ for } x_1, x_2 \in L.$$

In other words $\quad x = x_1 + x_2 \;\Rightarrow\; (x = x_1) \text{ or } (x = x_2).$

In a chain, all elements are join irreducible. Futher, in relation 'divides', the prime numbers are join irreducible.

9.18 Theorems on Lattice

THEOREM 9.1 Let $< L, *, + >$ be a lattice and $a, b \in L$.

Then $$a \leq b \Leftrightarrow a * b = a \Leftrightarrow a + b = b.$$

Proof: Since, $a \leq b$

$\Rightarrow$ $\quad$ glb $(a, b) = a$

$\Rightarrow$ $\quad a * b = a.$ $\quad$ {by definition of operator $*$}

Similarly, $\quad a \leq b$

$\Rightarrow$ $\quad$ lub $(a, b) = b$

$\Rightarrow$ $\quad a + b = b.$ $\quad$ {by definition of operator $+$}

Therefore, $\quad a \leq b \Leftrightarrow a * b = a \Leftrightarrow a + b = b.$

THEOREM 9.2 Let $< L, *, + >$ be a lattice and a, b and c belong to L.

Then $$b \leq c \Rightarrow \begin{cases} a * b \leq a * c \\ a + b \leq a + c \end{cases}$$

This property is known as isotonicity.

Proof:

Part 1 $\quad b \leq c \Rightarrow a * b \leq a * c$

There are four possibilities:

(i) $a * b = a$ and $a * c = a$, i.e., $a \leq b$ and $a \leq c$

(ii) $a * b = b$ and $a * c = c$, i.e., $b \leq a$ and $c \leq a$

(iii) $a * b = a$ and $a * c = c$, i.e., $a \leq b$ and $c \leq a$

(iv) $a * b = b$ and $a * c = a$, i.e., $b \leq a$ and $a \leq c$

From (i), $\quad a * b = a * c.$

$\Rightarrow$ $\quad a * b \leq a * c$ or $a * c \leq a * b.$ $\quad$ (9.1)

From (ii), $\quad a * b = b$ and $a * c = c$

$\Rightarrow$ $\quad a * b \leq a * c$, if $b \leq c.$

Therefore, $\quad b \leq c \Rightarrow a * b \leq a * c.$ $\quad$ (9.2)

From (iii), $c \le a$ and $a \le b$

$\Rightarrow$ $c \le b$, which is not possible since $b \le c$. (9.3)

From (iv), $b \le a$ and $a \le c$

$\Rightarrow$ $b \le c$, i.e., $a * b \le a * c$. (9.4)

Combining Eqs. (9.1) to (9.4) above, it can be deduced that

$$b \le c \Rightarrow a * b \le a * c.$$

Part 2 $b \le c \Rightarrow a + b \le a + c$

There are four possibilities:

(i) $a + b = a$ and $a + c = a$, i.e., $b \le a$ and $c \le a$
(ii) $a + b = b$ and $a + c = c$, i.e., $a \le b$ and $a \le c$
(iii) $a + b = a$ and $a + c = c$, i.e., $b \le a$ and $a \le c$
(iv) $a + b = b$ and $a + c = a$, i.e., $a \le b$ and $c \le a$

From (i), $a + b = a + c$

$\Rightarrow$ $a + b \le a + c$ or $a + c \le a + b$. (9.5)

From (ii), $a + b = b$ and $a + c = c$

$\Rightarrow$ $a + b \le a + c$, if $b \le c$.

Therefore, $b \le c \Rightarrow a + b \le a + c$. (9.6)

From (iii), $b \le a$ and $a \le c$

$\Rightarrow$ $b \le c$, i.e., $b \le c \Rightarrow a + b \le a + c$. (9.7)

From (iv), $a \le b$ and $c \le a$

$\Rightarrow$ $c \le b$, which is not possible since $b \le c$. (9.8)

Combining Eqs. (9.5) to (9.8) above, it can be deduced that

$$b \le c \Rightarrow a + b \le a + c.$$

THEOREM 9.3 Let $< L, *, + >$ be a lattice and a, b and c belong to L. Then, the following distributive inequalities hold:

$$a + (b * c) \le (a + b) * (a + c)$$
$$a * (b + c) \ge (a * b) + (a * c).$$

Proof: Consider a lattice L and let $a, b, c \in L$.
From the definitions of the operators $*$ and $+$, it can easily be concluded that,

$$a \le b \text{ and } a \le c \Rightarrow a \le (b + c) \quad (9.9)$$

and $$a \le b \text{ and } a \le c \Rightarrow a \le (b * c). \quad (9.10)$$

Similarly, by duality, we can write

$$a \ge b \text{ and } a \ge c \Rightarrow a \ge (b * c) \quad (9.11)$$

and $$a \ge b \text{ and } a \ge c \Rightarrow a \ge (b + c). \quad (9.12)$$

Now, from $a \leq (a + b)$, $a \leq (a + c)$ and using principle given in Eq. (9.10), we have

$$a \leq (a + b) * (a + c). \tag{9.13}$$

Also,

$$\left.\begin{aligned} (b * c) \leq b \leq (a + b) \\ (b * c) \leq c \leq (a + c) \end{aligned}\right\} \tag{9.14}$$

From Eq. (9.14), using principle given in Eq. (9.10)

$$(b * c) \leq (a + b) * (a + c). \tag{9.15}$$

Rewriting Eqs. (9.13) and (9.15), we have

$$\left.\begin{aligned} (a + b) * (a + c) \geq a \\ (a + b) * (a + c) \geq (b * c) \end{aligned}\right\}. \tag{9.16}$$

Finally, from Eq. (9.16) and using principle given in Eq. (9.12), we get the first inequality

$$(a + b) * (a + c) \geq a + (b * c)$$

or

$$a + (b * c) \leq (a + b) * (a + c). \tag{9.17}$$

From duality, using Eq. (9.17), we have

$$a * (b + c) \leq (a * b) + (a * c). \tag{9.18}$$

THEOREM 9.4 Let $< L, *, + >$ be a lattice and a, b and c belong to L. Then

$$a \leq c \Rightarrow a + (b * c) \leq (a + b) * c.$$

Proof: Consider a lattice L and let $a, b, c \in L$.

From first distributive inequality given in Theorem 9.3,

$$a + (b * c) \leq (a + b) * (a + c). \tag{9.19}$$

Also

$$a \leq c \Rightarrow a + c = c. \tag{9.20}$$

Replacing $a + c$ by c in Eq. (9.19) we have,

$$a + (b * c) \leq (a + b) * c$$

Therefore,

$$a \leq c \Rightarrow a + (b * c) \leq (a + b) * c.$$

It is also called **modular inequality**. Other modular inequalities are:

$$(a + b) * (a + c) \geq a + (b * (a + c))$$

and

$$(a * b) + (a * c) \leq a * (b + (a * c)), \text{ by duality.}$$

These inequalities can be proved in a similar way.

THEOREM 9.5 The direct product of any two distributive lattices is a distributive lattice.

Proof: Let $< L, *_1, +_1 >$ and $< S, *_2, +_2 >$ be two distributive lattices.

The direct product of L and S is given by the algebraic system $< L \times S, \theta, \phi >$, where $L \times S$ is the collection of ordered pairs, $\{(x, y) \mid x \in L \text{ and } y \in S\}$.

The operators θ and ϕ are defined as:

$$(x_1, y_1)\ \theta\ (x_2, y_2) = (x_1 *_1 x_2, y_1 *_2 y_2)$$

and

$$(x_1, y_1)\ \phi\ (x_2, y_2) = (x_1 +_1 x_2, y_1 +_2 y_2),$$

where $x_1, x_2 \in L, y_1, y_2 \in S$

and $(x_1, y_1), (x_2, y_2) \in L \times S.$

Obviously, in the direct product, operators θ and ϕ are defined in terms of the operators of the given lattices. Therefore, the direct product will also be distributive, since the given lattices are distributive.

THEOREM 9.6 Let < L, $*$, $+$ > be a distributive lattice. Then for any $x, y, z \in L$

$$\left.\begin{array}{l} x * y = x * z \\ x + y = x + z \end{array}\right\} \Rightarrow y = z.$$

Proof: Since, $(x * y) + z = (x * z) + z$ [using $(x * y) = (x * z)$, given above]

$$\left.\begin{array}{l} = x + z \\ = z \end{array}\right\}, \text{if } x \leq z.$$

or $(x * y) + z = (x * z) + z$ [using $(x * y) = (x * z)$, given above]

$$\left.\begin{array}{l} = z + z \\ = z \end{array}\right\}, \text{if } z \leq x.$$

Therefore, $(x * y) + z = z.$ (9.21)

Also, $(x * y) + z = (x + z) * (y + z)$ [using distributive law]

$= (x + y) * (y + z)$ [using $(x + y) = (x + z)$, given above]

$= (x * z) + y$ [using distributive law]

$= (x * y) + y$ [using $x * y = x * z$]

$= y.$ (9.22)

Hence, $(x * y) + z = y$, whether $x \leq y$ or $y \leq x$.

From Eqs. (9.21) and (9.22), we see that $y = z$ follows from

$$(x * y) = (x * z)$$

and

$$(x + y) = (x + z)$$

Hence, the theorem is proved.

THEOREM 9.7 In a distributive lattice < L, $*$, $+$, 0, 1 >, if an element $x \in L$ has a complement, then it will be unique.

Proof: Let $x \in L$ and, y and z be the two complements of it, then

$$(x * y) = 0 \text{ and } (x * z) = 0$$

$\Rightarrow$ $(x * y) = (x * z).$ (9.23)

Also, $(x + y) = 1$ and $(x + z) = 1$

$\Rightarrow$ $(x + y) = (x + z).$ (9.24)

Now, from Theorem 9.6

$$\left.\begin{array}{r} x * y = x * z \\ x + y = x + z \end{array}\right\} \Rightarrow y = z.$$

Therefore, Eqs. (9.23) and (9.24) imply $y = z$.

9.19 Solved Problems

PROBLEM 9.1 Let I^+ be the set of positive integers and the poset defined on I^+ is $< I^+, \leq >$. Also, let E^+ be the set of even positive integers and poset defined is $< E^+, \leq >$. Prove that $f: I^+ \rightarrow E^+$ given by

$f(a) = 2a$ is an isomorphism.

Solution

Suppose $a, b \in I^+$.

(i) $f(a) = f(b) \Rightarrow 2a = 2b$

$\Rightarrow a = b$

Therefore, f is one-to-one.

(ii) For an element $c \in E^+$, there exists an element $\dfrac{c}{2} \in I^+$, such that

$$f\left(\frac{c}{2}\right) = c.$$

Therefore, f is onto.

(iii) $f(a \leq b) \Rightarrow 2a \leq 2b$

$\Rightarrow a \leq b$

$\Rightarrow f$ is homomorphism.

Hence, from (i), (ii) and (iii), it is deduced that f is an isomorphism.

PROBLEM 9.2 Show that the lattice $<\mathbb{Z}^+, *, +>$ under the usual partial order relation, $\leq$ is distributive, where $*$, denoting min (a, b), is minimum and $+$, denoting max (a, b), is maximum of a and b.

Solution

The operators $*$ and $+$ are defined as:

$$* \equiv \min(a, b) = \text{minimum of } a \text{ and } b$$

$$+ \equiv \max(a, b) = \text{maximum of } a \text{ and } b.$$

Now consider,

$$\begin{aligned} a * (b + c) &= \min(a, (b + c)) \\ &= \min(a, \max(b, c)) \\ &= \max(\min(a, b), \min(a, c)) \\ &= (a * b) + (a * c). \end{aligned}$$

Similarly, $a + (b * c) = (a + b) * (a + c)$.

Therefore, $< \mathbb{Z}^+, *, + >$ is distributive

PROBLEM 9.3 Give an example of infinite lattice with finite length.

Solution

Consider the lattice < L, $*$, $+$, 0, 1 > with $L = \{x_1, x_2, x_3, \ldots, x_n\}$. Hasse diagram is given by Figure 9.11. This is an infinite lattice with finite length.

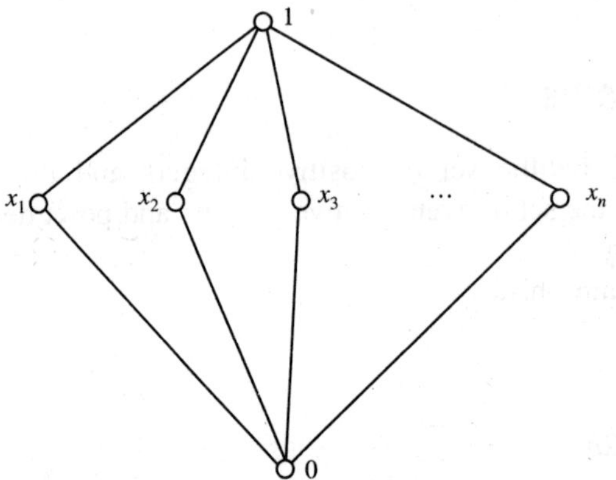

Figure 9.11 Infinite lattice.

PROBLEM 9.4 Let < L, $*$, $+$ > be a lattice where $L = \{0, 1\}$. The operators $*$ and $+$ are logical operators AND and OR respectively. Find < L^3, $*$, $+$ >.

Solution

Given $L = \{0, 1\}$.
Then $L^3 = L \times L \times L = \{(0,0,0), (0,0,1), (0,1,0), (0,1,1), (1,0,0), (1,0,1), (1,1,0), (1,1,1)\}$.

The Hasse diagram of the lattice is given in Figure 9.12. The first element is (0, 0, 0) and last element is (1, 1, 1).

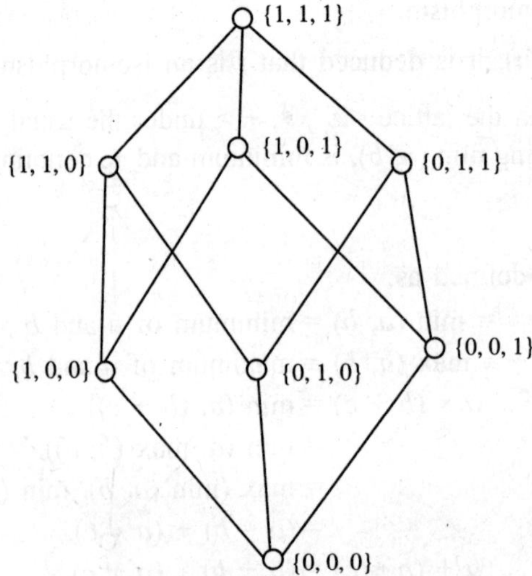

Figure 9.12 Lattice for L^3.

PROBLEM 9.5 Draw Hasse diagrams of lattices of order 4 and 5.

Solution

The Hasse diagrams are given in Figures 9.13(a) and (b).

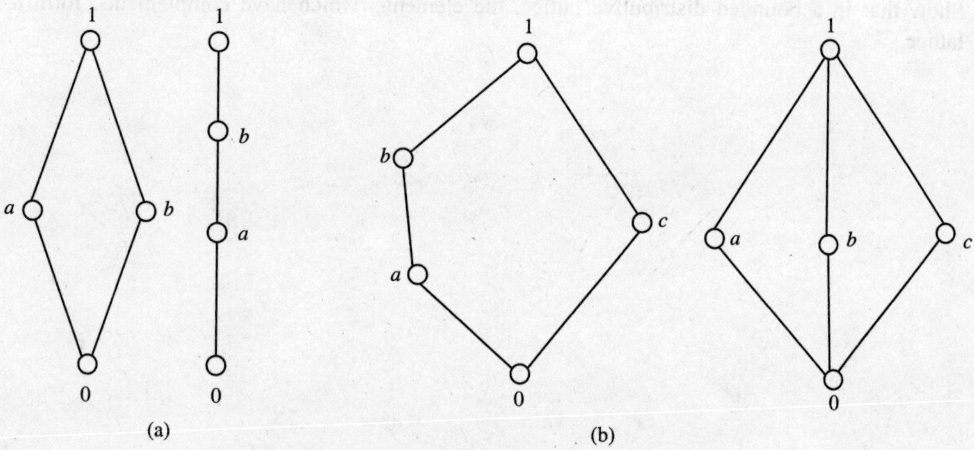

Figure 9.13 Lattices of (a) order 4, and (b) order 5.

Exercises

9.1 Let $S = \{1, 2, 3, 4, 6, 12\}$. Draw a Hasse diagram for partially ordered set $< S, \leq >$, where partial order relation $\leq$ is 'divides'.

9.2 Draw Hasse diagram for $< D_{60}, \leq >$ where partial order relation is 'divides'. Prove that it is a lattice if operators are 'greatest common divisor' and 'least common multiple'.

9.3 Let $< L_1, *, +, 0, 1 >$ and $< L_2, *, +, 0, 1 >$ be two lattices with $L_1 = \{0, 1\}$ and $L_2 = \{0, 1, a, b\}$, where the operators $*$ and $+$ are logical operators AND and OR, respectively. Find $< L_1 \times L_2, *, + >$.

9.4 Prove that the following lattices are nondistributive.

Hint: $a * (b + c) \neq (a * b) + (a * c)$.

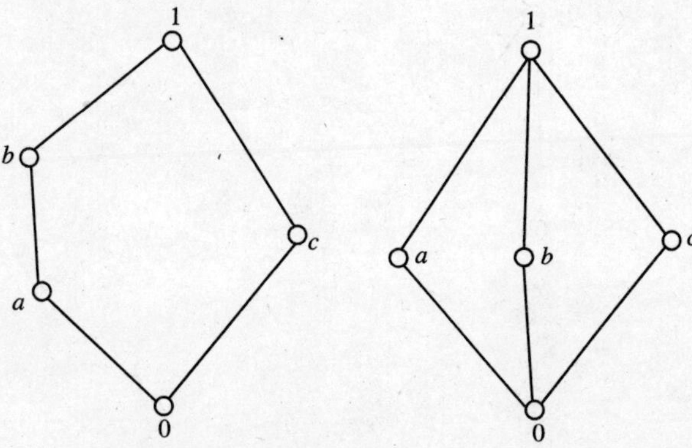

Figure 9.14 Nondistributive lattices.

9.5 Prove that every chain is a bounded lattice.

9.6 Prove that < $\mathbb{Z}^+$, $*$, $+$ >, where operators are 'gcd' and 'lcm', is an infinite and distributive lattice.

9.7 Show that a chain with three or more elements is not a complemented lattice.

9.8 Show that in a bounded distributive lattice, the elements, which have complements, form a sub lattice.

10

Boolean Algebra

10.1 Introduction

George Boole# (1815–1864) introduced the theory of Boolean algebra in 1854, and thereafter, it got the attention of many mathematicians and scientists who extended the work further. In fact, in 1938, C.E. Shannon developed a two-valued Boolean algebra, and named it switching algebra because it exhibited the similarity with bistable electrical circuits. Earlier, in 1904, E.V. Huntington formulated the postulates for defining the Boolean algebra.

10.2 Boolean Algebra

Boolean algebra is an algebraic structure defined on a set of elements B, together with two binary operations + and $*$ provided the following axioms are satisfied:

#**George Boole** (1815–1864), an English **mathematician** and **philosopher** is regarded as one of the founder members of the field of computer science. The crater **Boole** on the Moon is named in his honour. Boole had learned Latin, Greek, German, Italian, and French at an early age. With these languages, he was able to read a wide variety of Christian theology. He had equal interests in mathematics and theology. He was attracted to the Hebrew concept of God as an absolute unity. The personal life and character of Boole inspired many of his friends with the deepest esteem. He was marked by true modesty, and his life was given to the pursuit of truth. Boole was very much influenced by Indian mythology, particularly Hinduism. Boole is also considered as the inventor of **Boolean logic** (the basis of modern digital computer). In 1848, Boole published **The Mathematical Analysis of Logic,** the first of his contributions to symbolic logic. Two systematic treatises were also completed by Boole. The well-known **Treatise on Differential Equations** appeared in 1859, and was followed by a **Treatise on the Calculus of Finite Differences**.

1. **Closure**
   (i) B must be closed with respect to operation $+$.
   (ii) B must be closed with respect to operation $*$.
2. **Identity**
   $\forall\, x \in B$
   (i) There exists an identity element 0, such that
   $$x + 0 = 0 + x = x.$$
   (ii) There exists an identity element 1, such that
   $$x * 1 = 1 * x = x.$$
3. **Commutative**
   $\forall\, x, y \in B$
   (i) $x + y = y + x$.
   (ii) $x * y = y * x$.
4. **Distributive**
   For $x, y, z \in B$
   (i) '$*$' is distributive over '$+$' : $x * (y + z) = (x * y) + (x * z)$.
   (ii) '$+$' is distributive over '$*$' : $x + (y * z) = (x + y) * (x + z)$.
5. **Complements**
   For every element $x \in B$, there exists an element $x' \in B$, such that
   $$x + x' = 1$$
   and
   $$x * x' = 0.$$
   The element x' is called the complement of x.
6. There exists at least two elements $x, y \in B$, such that
   $$x \neq y.$$
   Note that, associative laws are not included as axioms here, although some authors have done so. In fact, associative laws are given as theorems in Boolean algebra.

Note: Readers may write xy in place of $x * y$, as a simple notation.

10.3 Boolean Algebra as Lattice

Boolean algebra is also defined as a complemented distributive lattice and is denoted by 6-tuple, $< B, *, +, ', 0, 1 >$ where the symbols have the meaning as defined in Table 10.1.

Table 10.1 Meaning of Symbols in $< B, *, +, ', 0, 1 >$

S. No.	*Symbol*	*Meaning*
1	B	Set of elements
2	$*$	Meet/product operation
3	$+$	Join/sum operation
4	$'$	Complement
5	0	Zero or least element
6	1	One/unit or greatest element

In lattice theory, $x + 1 = 1$ means $x \leq 1$ and $x * 0 = 0$ implies $0 \leq x$ for any $x \in B$. Therefore, B is bounded. In addition, as given in the definition of Boolean algebra, B is distributive, and the complements of the elements exist, therefore, B is a complemented distributive lattice. If the set B contains binary digits 0 and 1 only, then the operators $*$, $+$ and $'$ are AND, OR and NOT operations of logic respectively. This will be illustrated in Example 10.1, below.

EXAMPLE 10.1 Two-valued Boolean algebra

Consider the Boolean algebra $<B, *, +, ', 0, 1>$. Let $B = \{0, 1\}$, and the operations $*$, $+$ and $'$ be the logical operations AND, OR, and NOT respectively, defined by the truth Tables 10.2, 10.3, and 10.4.

Table 10.2 OR

x	y	$x + y$
0	0	0
0	1	1
1	0	1
1	1	1

Table 10.3 AND

x	y	$x * y$
0	0	0
0	1	0
1	0	0
1	1	1

Table 10.4 NOT

x	x'
0	1
1	0

Now, to verify that B is a Boolean algebra, we construct Cayley Tables 10.5 and 10.6 for the given operations $+$ and $*$, respectively with the help of Tables 10.2, 10.3, and 10.4.

Table 10.5 Cayley Table for +

+	0	1
0	0	1
1	1	1

Table 10.6 Cayley Table for *

*	0	1
0	0	0
1	0	1

From Cayley Tables 10.5 and 10.6, it is clear that operations + and * are closed, and since these tables are symmetric about the main diagonal, therefore, commutative axioms hold true. To prove that distributive axioms too hold, we construct another truth table, Table 10.7, and examine the entries.

Table 10.7 Proof of Distributive Axioms

x (1)	y (2)	z (3)	$x + y$ (4)	$x + z$ (5)	$y + z$ (6)	$x * y$ (7)	$x * z$ (8)	$y * z$ (9)	$x * (y + z)$ (10)	$(x * y) + (x * z)$ (11)	$x + (y * z)$ (12)	$(x + y) * (x + z)$ (13)
0	0	0	0	0	0	0	0	0	0	0	0	0
0	0	1	0	1	1	0	0	0	0	0	0	0
0	1	0	1	0	1	0	0	0	0	0	0	0
0	1	1	1	1	1	0	0	1	0	0	1	1
1	0	0	1	1	0	0	0	0	0	0	1	1
1	0	1	1	1	1	0	1	0	1	1	1	1
1	1	0	1	1	1	1	0	0	1	1	1	1
1	1	1	1	1	1	1	1	1	1	1	1	1

Column 10 in Table 10.7 is identical to column 11, and column 12 is identical to column 13, therefore, distributive axioms hold. Also, as given in Table 10.4, $0' = 1$ and $1' = 0$, therefore, complements exist. Hence, B is a Boolean algebra. It is called **two-valued Boolean algebra**. It is also called **switching algebra** by computer engineers; since its theory is applicable in designing logic circuits.

EXAMPLE 10.2 Let $B = \langle B^n, *, +, ', 0, 1 \rangle$, where

$B^n = B \times B \times B \times \cdots \times B$($n$-times) is a set of n-tuples. The operations $+$, $*$, and $'$ are defined component wise as given in Tables 10.2, 10.3, and 10.4 respectively. Then B^n is a Boolean algebra.
For example, consider $x, y \in B^8$ such that

$$x = < 1, 1, 0, 0, 1, 0, 1, 0 >$$

and $$y = < 1, 0, 0, 0, 1, 0, 0, 0 >.$$

For convenience, write $x = 11001010$, and $y = 10001000$.

Then, $$x + y = 11001010 + 10001000$$
$$= 11001010,$$
$$x * y = 10001000,$$
and $$x' = 00110101.$$

It can easily be verified, that B^n is closed, commutative, and distributive. In addition, $0 = 000 \ldots 0$, is the zero or least element and $1 = 111 \ldots 1$, is the greatest or unit element of B^n. Therefore, B^n is a Boolean algebra.

EXAMPLE 10.3 Consider $B = < \rho(A), *, +, ', 0, 1$, where A is any set, and $\rho(A)$ is the power set of A, then $\rho(A)$ is closed under the set operations union, intersection, and complement. Further, B will satisfy all the axioms, and hence B is a Boolean algebra.

Here, $$0(\text{zero of } B) = \phi$$
and $$1(\text{unity of } B) = \rho(A).$$

EXAMPLE 10.4 Consider the set of divisors of 30, i.e.,

$$D_{30} = \{1, 2, 3, 5, 6, 10, 15, 30\}.$$

Also, define Boolean operations on D_{30} by

$$a + b = \text{lcm}(a, b),$$
$$a * b = \text{gcd}(a, \text{b})$$
and $$a' = \frac{30}{a}.$$

Then $< D_{30}, *, +, ', 0, 1 >$ is a Boolean algebra with 1 as zero element, and 30 as unit element.

10.4 Boolean Expression

A Boolean expression is a finite string of symbols formed by the following rules:

1. 0 and 1 are Boolean expressions.
2. A Boolean element x is a Boolean expression.
3. If x_1 and x_2 are Boolean expressions, then $x_1 + x_2$, $x_1 * x_2$, x_1' and (x_1) are Boolean expressions.
4. Any string formed by using rules 1 to 3 is a Boolean expression.

While writing Boolean expressions, the usual precedence of operators is adopted. That is, if parentheses are not used then complement has precedence over *, and * has precedence over +. For example, $x + y * z$ is treated as $x + (y * z)$. Similarly, $x' * y + z$ means $((x') * y) + z$.

EXAMPLE 10.5 Following are the Boolean expressions.

$$x_1 = x_1 + x_2$$
$$x_2 = x_1 + x_1 + x_2$$
$$x_3 = x_1 + x_1 * x_2$$
$$x_4 = (x_1 + x_2) * (x_3 + x_4)$$
$$x_5 = x_1 * x_2 * (x_1' + x_2)$$
$$x_6 = x_1' + x_2 + 1$$
$$x_7 = (x_1 + x_2) * (0 + x_3 * x_4).$$

10.5 Boolean Functions

Let $< B, *, +, ¢, 0, 1 >$ be a Boolean algebra. A function $f : B^n$ ® B is called a Boolean function, where B^n is a collection of all n-tuples $< x_1, x_2, x_3, \ldots, x_n >$. For instance, Table 10.8 shows a function f from $\{0,1\}^2$ to $\{0,1\}$.

Table 10.8 Boolean Function f

x	y	f
0	0	0
0	1	1
1	0	1
1	1	0

10.6 Principle of Duality

Dual of a Boolean expression can be obtained by interchanging the operations (i.e. * into +, and + into *), and replacing 0 by 1 and 1 by 0 in the original expression.

From the axioms given in Boolean algebra, it can be deduced that principle of duality holds, and therefore, if a theorem is valid in a Boolean algebra, then its dual will also be valid. Some Boolean expressions along with their duals are given below.

Expression	Dual of expression
1. $x + y$	$x * z$
2. $x + y + 1$	$x * y * 0$
3. $x + y * z$	$x * y + z$
4. $(x + y) * (x + z)$	$(x * y) + (x * z)$

10.7 Properties of Boolean Algebra

The important axioms and theorems of Boolean algebra, which are valid for any set *B* are summarized in Tables 10.9 and 10.10.

Table 10.9 Axioms

S. No.	*(a) Axioms*	*(b) Dual of axioms*
1.	$x + 0 = x$	$x * 1 = x$
2.	$x + y = y + x$	$x * y = y * x$
3.	$x + x' = 1$	$x * x' = 0$
4.	$x + (y * z) = (x + y) * (x + z)$	$x * (y + z) = (x * y) + (x * z)$

Table 10.10 Theorems

S. No.	*(a) Theorem*	*(b) Dual of theorem*	*Name*
1	$x + x = x$	$x * x = x$	Idempotent
2	$x + 1 = 1$	$x * 0 = 0$	Identity
3	$(x')' = x$		Double complement
4	$x + x * y = x$	$x * (x + y) = x$	Absorption
5	$x + (y + z) = (x + y) + z$	$x * (y * z) = (x * y) * z$	Associative law
6	$(x + y)' = x' * y'$	$(x * y)' = x' + y'$	De Morgan's theorem or De Morgan's law

Proofs of theorems 1(a), 2(a), 3, 4(a), 5(a) and 6(a), given in Table 10.10 are constructed below, whereas the proofs of 1(b), 2(b), 4(b), 5(b) and 6(b) follow from the principle of duality. Readers may prove these theorems exactly in a similar way.

THEOREM 1(a) $x + x = x$

Proof: Refer axioms of Table 10.8.

$$x + x = (x + x) * 1 \qquad \text{[axiom 1(b)]}$$
$$= (x + x) * (x + x') \qquad \text{[axiom 3(a)]}$$
$$= x + (x * x') \qquad \text{[axiom 4(a)]}$$
$$= x + 0 \qquad \text{[axiom 3(b)]}$$
$$= x \qquad \text{[axiom 1(a)]}$$

THEOREM 2(a) $x + 1 = 1$

Proof:

$$
\begin{aligned}
x + 1 &= (x + 1) * 1 && \text{[axiom 1(b)]}\\
&= (x + 1) * (x + x') && \text{[axiom 3(a)]}\\
&= x + (1 * x') && \text{[axiom 4(a)]}\\
&= x + x' && \text{[axiom 1(b)]}\\
&= 1 && \text{[axiom 3(a)]}
\end{aligned}
$$

THEOREM 3 $(x')' = x$

Proof: The complement of x is x'. Therefore, the complement of x', denoted by $(x')'$ will be x, i.e., $(x')' = x$.

THEOREM 4(a) $x + x * y = x$

Proof:

$$
\begin{aligned}
x + x * y &= x * 1 + x * y && \text{[axiom 1(b)]}\\
&= x * (1 + y) && \text{[axiom 4(b)]}\\
&= x * (y + 1) && \text{[axiom 2(a)]}\\
&= x * 1 && \text{[theorem 2(a)]}\\
&= x && \text{[axiom 1(b)]}
\end{aligned}
$$

THEOREM 5(a) $x + (y + z) = (x + y) + z$

Proof:

$$
\begin{aligned}
x + (y + z) &= (x + (y + z)\,) * 1 && \text{[axiom 1(b)]}\\
&= (x + (y + z)) * (z + z') && \text{[axiom 3(a)]}\\
&= ((x + (y + z)) * z + (x + (y + z)) * z' && \text{[axiom 4(b)]}\\
&= z * ((x + (y + z)) + z' * (x + (y + z)) && \text{[axiom 2(b)]}\\
&= (z * x + z * (y + z)) + (z' * x + z' * (y + z)) && \text{[axiom 4(b)]}\\
&= (z * x + z) + (z' * x + z' * (y + z)) && \text{[theorem 4(b)]}\\
&= (z * x + z * 1) + (z' * x + z' * (y + z)) && \text{[axiom 1(b)]}\\
&= z * (x + 1) + (z' * x + z' * y + z' * z) && \text{[axiom 4(b)]}\\
&= z * (x + 1) + (z' * x + z' * y + 0) && \text{[axiom 3(b)]}\\
&= z * (x + 1) + (z' * x + z' * y) && \text{[axiom 1(a)]}\\
&= z * 1 + (z' * x + z' * y) && \text{[theorem 2(a)]}\\
&= z * 1 + (z' * (x + y)) && \text{[axiom 4(b)]}\\
&= z + (z' * (x + y)) && \text{[axiom 1(b)]}\\
&= (z + z') * (z + (x + y)) && \text{[axiom 4(a)]}\\
&= 1 * (z + (x + y)) && \text{[axiom 3(a)]}\\
&= (z + (x + y)) * 1 && \text{[axiom 2(b)]}\\
&= z + (x + y) && \text{[axiom 1(b)]}\\
&= (x + y) + z && \text{[axiom 2(b)]}
\end{aligned}
$$

THEOREM 6(a) $(x + y)' = x' * y'$

Proof: Theorem states that the complement of $(x + y)$ is $x' * y'$. Therefore, to prove the theorem, it is sufficient to show that $(x + y) + (x' * y') = 1$

Now, $(x + y) + (x' * y') = (x + y + x') * (x + y + y')$ [axiom 4(a)]
$= (x + x' + y) * (x + y + y')$ [axiom 2(a)]
$= (1 + y) * (x + 1)$ [axiom 3(a)]
$= 1 * 1$ [theorem 2(a)]
$= 1$ [theorem 1(b)]

If our discussion is limited to only two-valued Boolean algebra, then above theorems can easily be proved with the help of truth tables. Truth tables for theorems 5 and 6 are constructed as Table 10.11 and Table 10.12. Similar tables can be constructed for other theorems too.

Table 10.11 For Theorem 5

x (1)	y (2)	z (3)	$x + y$ (4)	$y + z$ (5)	$(x + y) + z$ (6)	$x + (y + z)$ (7)	$x * y$ (8)	$y * z$ (9)	$x * (y * z)$ (10)	$(x * y) * z$ (11)
0	0	0	0	0	0	0	0	0	0	0
0	0	1	0	1	1	1	0	0	0	0
0	1	0	1	1	1	1	0	0	0	0
0	1	1	1	1	1	1	0	1	0	0
1	0	0	1	0	1	1	0	0	0	0
1	0	1	1	1	1	1	0	0	0	0
1	1	0	1	1	1	1	1	0	0	0
1	1	1	1	1	1	1	1	1	1	1

Note that the columns 6 and 7 of Table 10.11 are identical. Similarly, columns 10 and 11 are also identical, therefore, theorems 5(a) and 5(b) are valid.

Table 10.12 For Theorem 6

x (1)	y (2)	x' (3)	y' (4)	$x + y$ (5)	$(x + y)'$ (6)	$x' * y'$ (7)	$x * y$ (8)	$(x * y)'$ (9)	$x' + y'$ (10)
0	0	1	1	0	1	1	0	1	1
0	1	1	0	1	0	0	0	1	1
1	0	0	1	1	0	0	0	1	1
1	1	0	0	1	0	0	1	0	0

Notice that the truth-values in column 6 of Table 10.12 are identical to the values in column 7. Similarly, values in column 9 are identical to the values in column 10. Therefore, theorems 6(a) and 6(b) are valid.

10.8 Sub Boolean Algebra

Consider a Boolean algebra $< B, *, +, ', 0, 1 >$ Let $S \subseteq B$ and $\{0, 1\} \in S$, then $< S, *, +, ', 0, 1 >$ will be called a sub Boolean algebra if it satisfies all the axioms given for Boolean algebra.

EXAMPLE 10.6 Consider Example 10.4.

Let $X = < S, *, +, ', 0, 1 >$, where $S = \{1, 2, 15, 30\}$ is a subset of D_{30}. Then X is a sub Boolean algebra of D_{30}.

10.9 Direct Product

Consider the two Boolean algebras B_1 and B_2 given as:

$$B_1 = < X, \theta_1, \varphi_1, ', 0_1, 1_1 >$$

and

$$B_2 = < Y, \theta_2, \varphi_2, '', 0_2, 1_2 >.$$

The direct product of B_1 and B_2 is also a Boolean algebra. It is given by

$$D = < X \times Y, \theta_3, \varphi_3, ''', 0_3, 1_3 >.$$

Here, $X \times Y$ is the set of ordered pairs, and let $< x_1, y_1 >$, $< x_2, y_2 >$, and $< x_3, y_3 >$ belong to $X \times Y$, then the operators are defined as:

$$< x_1, y_1 > \theta_3 < x_2, y_2 > = < x_1\ \theta_1\ x_2, y_1\ \theta_2\ y_2 >$$

$$< x_1, y_1 > \varphi_3 < x_2, y_2 > = < x_1\ \varphi_1\ x_2, y_1\ \varphi_2\ y_2 >$$

$$< x_1, y_1 >''' = < x_1', y_1'' >$$

$$< 0_3 > = < 0_1, 0_2 >$$

$$< 1_3 > = < 1_1, 1_2 >.$$

The direct product can be used to generate new Boolean algebras.

10.10 Boolean Homomorphism

Let $B_1 = < X, \theta_1, \varphi_1, ', 0_1, 1_1 >$ and $B_2 = < Y, \theta_2, \varphi_2, '', 0_2, 1_2 >$ be two Boolean algebras then a mapping $f : X \to Y$, is called a Boolean homomorphism, if all the operations of Boolean algebra are preserved.

That is, $\forall\, x_1, x_2 \in X$

$$f(x_1\ \theta_1\ x_2) = f(x_1)\ \theta_2\ f(x_2)$$

$$f(x_1\ \varphi_1\ x_2) = f(x_1)\ \varphi_2\ f(x_2)$$

$$f(x') = [f(x)]''$$

$$f(0_1) = 0_2$$

$$f(1_1) = 1_2.$$

10.11 Join Irreducible

Let $x \in B$ be an element of Boolean algebra, $< B, *, +, ', 0, 1 >$. Then x is called join irreducible, if it cannot be expressed as the join of two distinct elements of B.

That is, $x \in B$ is join irreducible if

$$x \neq x_1 + x_2 \quad \text{for} \quad x_1, x_2 \in B.$$

In other words, $x = x_1 + x_2 \Rightarrow (x = x_1)$ or $(x = x_2)$.

All elements, which are directly linked with least element 0 are join irreducible. In fact, these are the only elements in a Boolean algebra, which are join irreducible. These are also called atoms of Boolean algebra. Similarly, we can also define meet irreducible elements, i.e., the elements which cannot be expressed as meet of two distinct elements.

10.12 Solved Problems

PROBLEM 10.1 In a Boolean algebra, prove that

$$x * y' = 0 \Leftrightarrow x' + y = 1.$$

Proof:

$$x * y' = 0$$

$\Rightarrow$ $(x * y')' = 0'$

$\Rightarrow$ $x' + y = 1$ [De Morgan's law]

Again $x' + y = 1$

$\Rightarrow$ $(x' + y)' = 1'$

$\Rightarrow$ $x * y' = 0$

Therefore, $x * y' = 0 \Leftrightarrow x' + y = 1.$

PROBLEM 10.2 If $x, y \in B$ then prove that $y \leq x' \Leftrightarrow x * y = 0$.

Proof:

(i) $y \leq x'$

$\Rightarrow$ $y + x' = x'$ $[x \leq y \Rightarrow x + y = y]$

$\Rightarrow$ $y + x' = x' + x'$

$\Rightarrow$ $y = x'$

Now, $x * x' = 0$

$\Rightarrow$ $x * y = 0$

(ii) $x * y = 0$

$\Rightarrow$ $y = x'$

$\Rightarrow$ $y + x' = x' + x'$

$\Rightarrow$ $y + x' = x'$

$\Rightarrow$ $y \leq x'$

Hence (i) and (ii) imply that $y \leq x' \Leftrightarrow x * y = 0$.

PROBLEM 10.3 In a Boolean algebra, prove that

(i) $x * y + z * (x' + y') = (x + z) * (y + z)$.

(ii) $x * y + x' * y' = (x' + y) * (x + y')$.

Proof:

(i) $x * y + z * (x' + y') = x * y + z * (x * y)'$

$= (x * y + z) * (x * y + (x * y)')$ [distributive]

$$= (x * y + z) * 1$$
$$= z + x * y$$
$$= (z + x) * (z + y)$$
$$= (x + z) * (y + z)$$

(ii) $x * y + x' * y' = (x * y + x') * (x * y + y')$ [distributive]
$$= (x' + x * y) * (y' + x * y) \quad \text{[commutative]}$$
$$= (x' + x) * (x' + y) * (y' + x) * (y' + y) \quad \text{[distributive]}$$
$$= (x' + y) * (y' + x) \quad [x + x' = y + y' = 1]$$
$$= (x' + y) * (x + y')$$

PROBLEM 10.4 In a Boolean algebra, prove that

$$x' + y = 1 \Leftrightarrow x + y = y.$$

Proof:

(i) Let $x' + y = 1$

Now, $x + y = x + y$
$$= (x + y) * 1$$
$$= (x + y) * (x' + y) \quad \text{[using } x' + y = 1\text{]}$$
$$= (x + y) * x' + (x + y) * y \quad \text{[distributive]}$$
$$= x * x' + y * x' + x * y + y * y \quad \text{[distributive]}$$
$$= y * x' + x * y + y \quad [x * x' = 1]$$
$$= y + y * x' + x * y \quad \text{[rearranging terms]}$$
$$= y + y * (x + x') \quad \text{[distributive]}$$
$$= y + y$$
$$= y.$$

(ii) $x + y = y$

Consider $1 = 1 + y$
$$= x + x' + y$$
$$= x + y + x'$$
$$= y + x' \quad \text{[using } x + y = y\text{]}$$
$$= x' + y.$$

Therefore, $x' + y = 1.$

Hence (i) and (ii) imply that $x' + y = 1 \Leftrightarrow x + y = y$.

PROBLEM 10.5 Prove that $x + y = x + z$, and $x * y = x * z \Rightarrow y = z$.

Proof: $y = y * 1$
$$= y * (1 + x)$$
$$= y + y * x \quad \text{[distributive]}$$
$$= y * y + y * x$$
$$= y * (y + x) \quad \text{[distributive]}$$

$= y * (x + y)$ [commutative]
$= y * (x + z)$
$= y * x + y * z$ [distributive]
$= x * y + y * z$ [commutative]
$= x * z + y * z$ [using $x * y = x * z$]
$= (x + y) * z$ [distributive]
$= (x + z) * z$ [using $x + y = x + z$]
$= z * (x + z)$ [commutative]
$= z * x + z * z$ [distributive]
$= z * x + z$
$= z * x + z * 1$
$= z * (x + 1)$ [distributive]
$= z$ [$x + 1 = 1$]

Exercises

10.1 Prove that

(i) Complements in a Boolean algebra are unique.
(ii) Identity in a Boolean algebra is unique.

10.2 Show that $< D_{70}, *, +, ', 0, 1 >$ is a Boolean algebra, where D_{70} is the set of divisors of 70. The operators are defined as:

$$x + y = \text{lcm}(x, y),\ x * y = \gcd(x, y) \text{ and } x' = \frac{70}{x} \text{ for } x, y \in D_{70},$$

also, zero (0) = 1
identity (1) = 70.

Also, find a sub Boolean algebra of it.

10.3 Write the dual of each of the following Boolean expression.

(i) $x' * (y + z) = x' * y + x' * z$
(ii) $(x + 0) * (x + 1) = x$
(iii) $x * y + x' * y + 1 = 1$

10.4 Consider set $B = \{0, 1, 2\}$. Operations on B are defined by the following tables:

Table 10.13

+	0	1	2
0	0	0	0
1	0	1	1
2	0	1	2

Table 10.14

*	0	1	2
0	0	1	2
1	1	1	2
2	2	2	2

Prove that $< B, *, +, ', 0, 1 >$ is not a Boolean algebra.

10.5 **Prove the following in a Boolean algebra.**

(i) $(x' * y')' + (x' * y)' = 1$
(ii) $(x + y)' * (x' + y)' = 0$
(iii) $(x + y + z)' = x' * y' * z'$
(iv) $(x * y * z)' = x' + y' + z'$
(v) $x + y + x * y = x + y$
(vi) $(x + y) * (y + z) * (z + x) = x * y + y * z + z * x$
(vii) $(x + y) * (x' + z) = x' * y + x * z$

10.6 **If $x = y$ then prove that $x' * y + x * y' = 0$.**

11

Boolean Function Minimization

11.1 Introduction

In Chapter 10, we discussed the fundamentals of Boolean algebra. The concept of switching algebra was also introduced. In this chapter, we will apply the theory of Boolean algebra for manipulation and minimization of Boolean functions.

11.2 Minterm and Maxterm

Let $< B, *, +, ', 0, 1 >$ be a Boolean algebra. Then a standard product consisting of all its n variables (either in standard form x, or its complement form x') is called a minterm.

For example, $b_1 * b_2 * b_3 * \cdots * b_n, b_1' * b_2 * b_3' * \cdots * b_n$, etc.

Similarly, standard sum consisting of all its n variables (either in standard form x, or its complement form x') is called a maxterm. For example,

$$b_1 + b_2 + b_3 + \cdots + b_n, b_1' + b_2 + b_3' + \cdots + b_n, \text{ etc.}$$

All 2-variable minterms and maxterms are listed in Table 11.1.

Table 11.1 Two-variable Minterms and Maxterms

			Minterms		*Maxterms*	
S. No.	*x*	*y*	*Terms*	*Notations*	*Terms*	*Notations*
1	0	0	$x' * y'$	m_0	$x + y$	M_0
2	0	1	$x' * y$	m_1	$x + y'$	M_1
3	1	0	$x * y'$	m_2	$x' + y$	M_2
4	1	1	$x * y$	m_3	$x' + y'$	M_3

All three-variable minterms and maxterms are listed in Table 11.2.

Table 11.2 Three-variable Minterms and Maxterms

				Minterms		*Maxterms*	
S. No.	*x*	*y*	*z*	*Terms*	*Notations*	*Terms*	*Notations*
1	0	0	0	$x' * y' * z'$	m_0	$x + y + z$	M_0
2	0	0	1	$x' * y' * z$	m_1	$x + y + z'$	M_1
3	0	1	0	$x' * y * z'$	m_2	$x + y' + z$	M_2
4	0	1	1	$x' * y * z$	m_3	$x + y' + z'$	M_3
5	1	0	0	$x * y' * z'$	m_4	$x' + y + z$	M_4
6	1	0	1	$x * y' * z$	m_5	$x' + y + z'$	M_5
7	1	1	0	$x * y * z'$	m_6	$x' + y' + z$	M_6
8	1	1	1	$x * y * z$	m_7	$x' + y' + z'$	M_7

Table 11.3 lists all minterms and maxterms with four variables. Similarly, one can construct tables for listing minterms and maxterms for any number of variables.

Table 11.3 Four-variable Minterms and Maxterms

					Minterms		*Maxterms*	
S. No.	*x*	*y*	*z*	*w*	*Terms*	*Notations*	*Terms*	*Notations*
1	0	0	0	0	$x' * y' * z' * w'$	m_0	$x + y + z + w$	M_0
2	0	0	0	1	$x' * y' * z' * w$	m_1	$x + y + z + w'$	M_1
3	0	0	1	0	$x' * y' * z * w'$	m_2	$x + y + z' + w$	M_2
4	0	0	1	1	$x' * y' * z * w$	m_3	$x + y + z' + w'$	M_3
5	0	1	0	0	$x' * y * z' * w'$	m_4	$x + y' + z + w$	M_4
6	0	1	0	1	$x' * y * z' * w$	m_5	$x + y' + z + w'$	M_5
7	0	1	1	0	$x' * y * z * w'$	m_6	$x + y' + z' + w$	M_6
8	0	1	1	1	$x' * y * z * w$	m_7	$x + y' + z' + w'$	M_7
9	1	0	0	0	$x * y' * z' * w'$	m_8	$x' + y + z + w$	M_8
10	1	0	0	1	$x * y' * z' * w$	m_9	$x' + y + z + w'$	M_9
11	1	0	1	0	$x * y' * z * w'$	m_{10}	$x' + y + z' + w$	M_{10}
12	1	0	1	1	$x * y' * z * w$	m_{11}	$x' + y + z' + w'$	M_{11}
13	1	1	0	0	$x * y * z' * w'$	m_{12}	$x' + y' + z + w$	M_{12}
14	1	1	0	1	$x * y * z' * w$	m_{13}	$x' + y' + z + w'$	M_{13}
15	1	1	1	0	$x * y * z * w'$	m_{14}	$x' + y' + z' + w$	M_{14}
16	1	1	1	1	$x * y * z * w$	m_{15}	$x' + y' + z' + w'$	M_{15}

11.3 Sum of Products and Product of Sums Form

A Boolean expression, which is the sum of the product terms, is called the *sum of products* form. Similarly, the expression, which is product of sum terms, is called *product of sums* form. These are also called **standard sum of products** form, and **standard product of sums** form. These expressions are illustrated in the following examples:

1. Sum of product terms:
   (i) $f_1(x, y, z) = x * y * z + x' * y + y * z'$
   (ii) $f_2(x, y, z) = x * y + x' * y + x * y * z'$
   (iii) $f_3(x, y, z) = x * y * z + x' * y * z + y * z'$
   (iv) $f_4(x, y, z) = x * y' * z + x' * y * z + y * z'$

   Here terms like $x * y * z$, $x' * y$ etc., are called product terms.
2. Product of sum terms:
   (i) $f_1(x, y, z) = (x + y + z) * (x' + y) * (y + z')$
   (ii) $f_2(x, y, z) = (x + y) * (x' + y) * (x + y + z')$
   (iii) $f_3(x, y, z) = (x + y + z) * (x' + y + z) * (y + z')$
   (iv) $f_4(x, y, z) = (x + y' + z) * (x' + y + z) * (y + z')$

   Here $x + y + z$, $x' + y$ etc., are sum terms.

Remark: Now onwards, just for brevity, we will suppress the use of symbol '*' in product terms. That is, in place of $x * y$, we will simply write xy.

11.4 Normal Forms

Sum of products form, and product of sums form defined in Section 11.3 are also called disjunctive normal form and conjunctive normal form, respectively. Further, there are two more normal forms, principal disjunctive normal form and principal conjunctive normal form, which are discussed in the following subsections.

11.4.1 Principal Disjunctive Normal Form (PDNF)

The sum of minterms is called principal disjunctive normal form. It is also called fundamental sum of products. For instance, in a three variable Boolean function, each term of the expression will consist of three literals. A literal is a variable symbol either in its standard form or in its complement form (i.e. x or x'). Some examples of fundamental sum of products of Boolean functions with three variables are as follows:

1. $f_1(x, y, z) = xyz + x'yz + xyz'$
2. $f_2(x, y, z) = xyz' + x'yz + xy'z'$
3. $f_3(x, y, z) = xyz + x'yz + x'yz' + xyz'$
4. $f_4(x, y, z) = xy'z + x'yz + xyz' + x'y'z' + x'yz'$

11.4.2 Principal Conjunctive Normal Form (PCNF)

The product of maxterms is called principal conjunctive normal form or fundamental product of sums. Like PDNF, in a three-variable Boolean function, each term in PCNF will consist of three literals, either in its standard form or in its complement form (i.e. x or x'). Some examples of the fundamental product of sums for a three-variable Boolean function are given as follows.

1. $f_1(x, y, z) = (x + y + z)(x' + y + z)(x' + y + z')$
2. $f_2(x, y, z) = (x + y + z)(x' + y + z)(x + y' + z')$
3. $f_3(x, y, z) = (x + y + z)(x' + y' + z)(x + y + z')(x' + y' + z')$
4. $f_4(x, y, z) = (x + y' + z)(x' + y + z)(x + y + z')(x' + y + z')(x' + y' + z')$

Note: Principal disjunctive normal form and principal conjunctive normal form of an expression are unique. These are also called **canonical forms**. In fact, the terminology has not been standardized, and some authors call these expressions, complete sum of products form and complete product of sums form.

11.5 Conversion of Boolean Expressions into Principal Normal Forms

Sometimes, it is required to obtain principal normal forms of expressions in forms. In the following subsections, we will discuss the procedures for converting Boolean expressions in other forms into the principal normal forms.

11.5.1 Disjunctive Normal Form to Principal Disjunctive Normal Form

Each term in principal disjunctive normal form contains as many variables symbols as the function has. However, an expression in disjunctive normal form may contain terms, which might have lesser number of variables, and therefore, to convert this expression into PDNF, these missing variables need to be included in the respective terms. The process to convert an expression in disjunctive normal form into its principal disjunctive normal form is quite simple. Suppose a term has a variable p missing. Then multiply (means AND operation) $p + p'$ with this term, and use distributive law to get two terms. This process is to be repeated until each term of the expression becomes a minterm. Now remove duplicate terms, if any, to get the desired expression.

EXAMPLE 11.1 Consider the following expression in disjunctive normal form:

$$f_1(x, y, z) = xy + x'yz + yz'$$

In the first term on the right hand side (i.e. xy), symbol z is missing, therefore multiply (AND operation) $z + z'$ with it. Since $z + z'$ is equal to 1, therefore, the logical value of the term will remain unchanged. Similarly, multiply the last term with $x + x'$. Now apply the distributive law in each term and remove duplicate terms from the expression. The complete details of derivation are as follows:

$$\begin{aligned} f_1(x, y, z) &= xy + x'yz + yz' \\ &= xy(z + z') + x'yz + (x + x')yz' \\ &= xyz + xyz' + x'yz \\ &\quad + xyz' + x'yz' && \text{(using distributive law)} \\ &= xyz + xyz' + x'yz + x'yz' && \text{(eliminating duplicates)} \end{aligned}$$

This expression consists of minterms only, therefore, it is in principal disjunctive normal form.

11.5.2 Conjunctive Normal Form to Principal Conjunctive Normal Form

The conversion from conjunctive normal form to principal conjunctive normal form is similar to the procedure, given in Section 11.5.1. Let p be a missing symbol. Then, add (means OR operation) $p*p'$ with the respective term, followed by the use of distributive law and removal of repetition to get desired PCNF.

EXAMPLE 11.2

$$f(x, y, z) = (x + z)\,(x' + y)\,(x + y + z')$$

In the first term (i.e. $x + z$), 'y' is missing, therefore, add yy' .Similarly, add zz' in the second term. The logical values of the terms will remain unaffected by adding these terms. Now, apply distributive law and remove duplicate terms, if any to obtain the desired expression. The detailed derivation is as follows:

$$\begin{aligned} f(x, y, z) &= (x + z)\,(x' + y)\,(x + y + z') \\ &= (x + yy' + z)\,(x' + y + zz')\,(x + y + z') \\ &= (x + y + z)\,(x + y' + z)\,(x' + y + z)\,(x' + y + z')\,(x + y + z') \end{aligned}$$

The resulting expression is in principal conjunctive normal form.

11.6 Boolean Function Minimization

A Boolean function consists of minterms and/or maxterms. It is obtained from the truth table either in the form of sum of minterms or product of maxterms. While designing the logic circuits, one has to obtain the minimized Boolean expression to optimize the fabrication cost. We will study logic circuits in Chapter 12. Boolean function can be minimized by three methods: using deduction system, Karnaugh map and table method. Deduction system consists of inference rules, while Karnaugh map is a table consisting of cells of minterms or maxterms. Table method is procedural. We will discuss only the first two methods, Deduction and Karnaugh map. For table method, readers may refer any good text on Digital Electronics/Logic Circuit Design.

11.6.1 Deduction Method

The following Boolean laws are used to minimize the Boolean expressions. The complete deduction system can be referred from Chapter 10 (Tables 10.9 and 10.10). These laws are used repeatedly to obtain minimized expression.

1. **Identity Law:** $x + xy = x$
2. **Associative Law:**
   (i) $x + (y + z) = (x + y) + z$
   (ii) $x(yz) = (xy)z$
3. **De Morgan's Law:**
   (i) $(x + y)' = x'\,y'$
   (ii) $(xy)' = x' + y'$

4. **Distributive Law:**
   (i) $x + (yz) = (x + y)(x + z)$
   (ii) $x(y + z) = xy + xz$

EXAMPLE 11.3 Consider the Boolean function,

$$x + xyz + xy'z + x'yz.$$

Steps of minimization are given below. On right hand side of a step, we write the details of the theorem, or a Boolean law, which is applied for deduction.

$$\begin{aligned} & x + xyz + xy'z + x'yz \\ &= x + xz(y + y') + x'yz && \{\text{distributive law}\} \\ &= x + xz + x'yz \\ &= x(1 + z) + x'yz && \{(1 + z) = 1\} \\ &= x + x'yz \end{aligned}$$

This is the minimal expression.

EXAMPLE 11.4 Consider the Boolean function,

$$x + xyz + xy'z + x'yz + yz$$

Steps of minimization are given as follows:

$$\begin{aligned} & x + xyz + xy'z + x'yz + yz \\ &= x + xz(y + y') + x'yz + yz && \{\text{distributive law}\} \\ &= x + xz + x'yz + yz && \{\text{since } y + y' = 1\} \\ &= x + xz + (x' + 1)yz && \{\text{distributive law}\} \\ &= x + xz + yz && \{\text{since } x' + 1 = 1\} \\ &= x(1 + z) + yz && \{1 + z = 1\} \\ &= x + yz \end{aligned}$$

This is the minimal expression.

11.6.2 Karnaugh Map Method

Karnaugh map method was first proposed by E.W. Veitch in 1952. It was further modified by M. Karnaugh in 1953. It is also known as Vietch-Karnaugh method. A Karnaugh map is a table with 2^n (n is the number of variables) cells, where each cell represents a minterm. This method is very simple and straightforward, if the number of variables in the Boolean expression is small, say up to five. Beyond five, it becomes large and complicated. In this section, two, three, four and five variables Karnaugh maps are drawn. Entries in the cells of the Karnaugh map are the minterms of Boolean expressions. If an expression, to be minimized, contains maxterms, or in general, product of sum terms, then it has to be converted into sum of minterms form before applying Karnaugh map method. In a Karnaugh map, arrangement or the relative positions of cells are fixed and cannot be altered; i.e. each cell is fixed for a minterm.

Two-variable Karnaugh map

Karnaugh map for two variables is given in Figure 11.1.

Three-variable Karnaugh map

Karnaugh map for three variables is given in Figure 11.2.

	y'	y
x'	$x'y'$ 0	$x'y$ 1
x	xy' 2	xy 3

Figure 11.1 Two variables map.

	$y'z'$	$y'z$	yz	yz'
x'	$x'y'z'$ 0	$x'y'z$ 1	$x'yz$ 3	$x'yz'$ 2
x	$xy'z'$ 4	$xy'z$ 5	xyz 7	xyz' 6

Figure 11.2 Three variables map.

Four-variable Karnaugh map

Karnaugh map for four variables is given in Figure 11.3.

	$z'w'$	$z'w$	zw	zw'
$x'y'$	$x'y'z'w'$ 0	$x'y'z'w$ 1	$x'y'zw$ 3	$x'y'zw'$ 2
$x'y$	$x'yz'w'$ 4	$x'yz'w$ 5	$x'yzw$ 7	$x'yzw'$ 6
xy	$xyz'w'$ 12	$xyz'w$ 13	$xyzw$ 15	$xyzw'$ 14
xy'	$xy'z'w'$ 8	$xy'z'w$ 9	$xy'zw$ 11	$xy'zw'$ 10

Figure 11.3 Four variables map.

Five-variable Karnaugh map

Karnaugh map for five variables is given in Figure 11.4.

	$z'w't'$	$z'wt'$	zwt'	$zw't'$	$zw't$	zwt	$z'wt$	$z'w't$
$x'y'$	$x'y'z'w't'$ 0	$x'y'z'wt'$ 2	$x'y'zwt'$ 6	$x'y'zw't'$ 4	$x'y'zw't$ 5	$x'y'zwt$ 7	$x'y'z'wt$ 3	$x'y'z'w't$ 1
$x'y$	$x'yz'w't'$ 8	$x'yz'wt'$ 10	$x'yzwt'$ 14	$x'yzw't'$ 12	$x'yzw't$ 13	$x'yzwt$ 15	$x'yz'wt$ 11	$x'yz'w't$ 9
xy	$xyz'w't'$ 24	$xyz'wt'$ 26	$xyzwt'$ 30	$xyzw't'$ 28	$xyzw't$ 29	$xyzwt$ 31	$xyz'wt$ 27	$xyz'w't$ 25
xy'	$xy'z'w't'$ 16	$xy'z'wt'$ 18	$xy'zwt'$ 22	$xy'zw't'$ 20	$xy'zw't$ 21	$xy'zwt$ 23	$xy'z'wt$ 19	$xy'z'w't$ 17

Figure 11.4 Five variables map.

Minimization procedure

1. Enter 1 for each minterm in the Boolean expression in the corresponding cell.
2. Find adjacent cells having 1′s and form pairs, quads, octets etc.
3. Retain only those variables in the term that do not change its nature (i.e. from prime to non prime or from non prime to prime).

EXAMPLE 11.5 Consider the following Boolean expression,

$$F(x, y, z, w) = \Sigma(0, 1, 4, 5, 14, 15)$$

It is in the sum of minterms form and can be written as:

$$F(x, y, z, w) = x'y'z'w' + x'y'z'w + x'yz'w' + x'yz'w + xyzw' + xyzw$$

Since there are four variables in the given expression, therefore, draw a four-variable Karnaugh map and enter 1 in all the cells corresponding to these minterms. The Karnaugh map for this is given in Figure 11.5.

	$z'w'$	$z'w$	zw	zw'
$x'y'$	1	1		
$x'y$	1	1		
xy			1	1
xy'				

Figure 11.5 Four variables Karnaugh map.

Observe that four adjacent 1's form a quad and remaining two 1's form a pair. In the quad, variable w is changing from w' to w and y is changing from y' to y, therefore, these variables will not appear in the term representing the four minterms of the quad. These four 1's will give a minimized term $x'z'$. Similarly, in the pair, variable w is changing from w to w', therefore, this variable will not appear in the term representing the two minterms of the pair. In this case, the minimized term will be xyz. The complete minimized Boolean expression will be given as:

$$F(x, y, z, w) = x'z' + xyz$$

EXAMPLE 11.6 Consider the following Boolean expression,

$$F(x, y, z, w, t) = \Sigma(0, 2, 4, 6, 8, 10, 12, 13, 14, 15, 16, 17, 19, 23, 27, 29, 31).$$

There are five variables in this Boolean expression. It is in the sum of minterms form and can be written as:

$$\begin{aligned} F(x, y, z, w, t) = {} & x'y'z'w't' + x'y'z'wt' + x'y'zw't' + x'y'zwt' + x'yz'w't' \\ & + x'yz'wt' + x'yzw't' + x'yzw't + x'yzwt' + x'yzwt + xy'z'w't' \\ & + xy'z'w't + xy'z'wt + xy'zwt + xyz'wt + xyzw't + xyzwt \end{aligned}$$

A five-variable Karnaugh map with 1 in each cell corresponding to each minterm in the expression is shown in Figure 11.6.

	$z'w't'$	$z'wt'$	zwt'	$zw't'$	$zw't$	zwt	$z'wt$	$z'w't$
$x'y'$	1	1	1	1				
$x'y$	1	1	1	1	1	1		
xy					1	1	1	
xy'	1					1	1	1

Figure 11.6 Five-variable Karnaugh map.

Here the terms 0, 2, 4, 6, 8, 10, 12, and 14 will form an octet giving minimized term $x't'$. A quad is formed by the terms 13, 15, 29 and 31 giving yzt. Another quad is formed by the terms 19, 23, 27 and 31, by overlapping with the quad already formed, giving xwt. Also, to form an octet, quad or pair, wrapping around cells on the extreme edges is allowed. In view of this, cells on the opposite edges are treated as adjacent. Therefore, terms 16 and 17 will form a pair giving $xy'z'w'$. The complete minimized Boolean expression is given as:

$$F(x, y, z, w, t) = x't' + yzt + xwt + xy'z'w'$$

Don't care conditions

There may be instances that a particular minterm when combined with a Boolean expression may not affect the Boolean value (0 or 1) of the expression. Such minterm is said to be the don't care condition for that Boolean expression. The Karnaugh map cells are marked with X corresponding to the given don't care conditions. These don't care conditions can be used to simplify the Boolean expression in a better way. These Xs may be assumed to be 0 or 1 as per the need to form a pair, quad or octet etc.

EXAMPLE 11.7 Consider the Boolean expression:

$$F(x, y, z, w, t) = \Sigma(0, 2, 4, 6, 8, 10, 12, 13, 14, 15, 17, 19, 23, 27, 29, 31)$$

with don't care conditions

$$d(x, y, z, w, t) = \Sigma(21, 25)$$

There are five variables in this expression, and it is in the sum of minterms form. Therefore, a five-variable Karnaugh map is needed for it. Five variables Karnaugh map with corresponding entries of 1 for the minterms of the expression above is given in Figure 11.7. The entries for don't care conditions are indicated by X in the map.

	$z'w't'$	$z'wt'$	zwt'	$zw't'$	$zw't$	zwt	$z'wt$	$z'w't$
$x'y'$	1	1	1	1				
$x'y$	1	1		1	1	1		
xy					1	1	1	X
xy'					X	1	1	1

Figure 11.7 Five variables Karnaugh map.

The Xs may be assigned 1 and combined with the adjacent cells to form octet. The minimized expression is

$$F(x, y, z, w, t) = x't' + yzt + xt.$$

11.7 Solved Problems

PROBLEM 11.1 Minimize the following Boolean expression,

$$F(x, y, z, w) = \Sigma(0, 1, 4, 5, 10, 11, 14, 15).$$

Solution

The given Boolean expression in the sum of minterms form can be written as:

$$F(x, y, z, w) = x'y'z'w' + x'y'z'w + x'yz'w' + x'yz'w + xy'zw' + xy'zw + xyzw' + xyzw$$

Since there are four variables in the given expression, therefore, a four-variable Karnaugh map is needed. It is drawn in Figure 11.8.

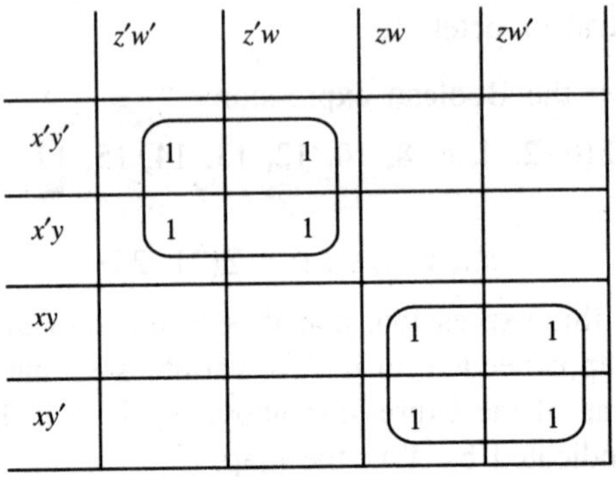

	$z'w'$	$z'w$	zw	zw'
$x'y'$	1	1		
$x'y$	1	1		
xy			1	1
xy'			1	1

Figure 11.8 Four variables Karnaugh map for Problem 11.1.

Observe that there are two quads. In the upper quad, variable w is changing from w' to w and y is changing from y' to y, therefore, these variables will not appear in the term representing the four minterms of the quad. These four 1s will give a minimized term $x'z'$. Similarly, in the lower quad, variable w is changing from w to w' and y is changing from y' to y, therefore, these variables will not appear in the term representing the four minterms of the quad. In this case, the minimized term will be xz. The complete minimized Boolean expression will be given as:

$$F(x, y, z, w) = x'z' + xz.$$

PROBLEM 11.2 Minimize the following Boolean expression,

$$F(x, y, z, w, t) = \Sigma(0, 2, 4, 6, 8, 10, 12, 13, 14, 15, 16, 17, 29, 31).$$

Solution

The given Boolean expression in the sum of minterms form can be written as:

$$\begin{aligned} F(x, y, z, w, t) = {} & x'y'z'w't' + x'y'z'wt' + x'y'zw't' + x'y'zwt' + x'yz'w't' \\ & + x'yz'wt' + x'yzwt' + x'yzw't' + x'yzw't + x'yzwt \\ & + xyzw't + xyzwt + xy'z'w't' + xy'z'w't \end{aligned}$$

Now, draw a five-variable Karnaugh map and enter 1 in the cells corresponding to these minterms. This map is drawn in Figure 11.9.

Here the terms 0, 2, 4, 6, 8, 10, 12, and 14 will form an octet giving minimized term $x't'$. A quad is formed by the terms 13, 15, 29 and 31 giving yzt. Also, to form an octet, quad or pair overlapping of the adjacent cells or wrapping around cells on the extreme edges can be done. Cells on the edges are treated as adjacent. Therefore, terms 16 and 17 will form a pair giving $xy'z'w'$. The complete minimized Boolean expression is given as:

$$F(x, y, z, w, t) = x't' + yzt + xy'z'w'$$

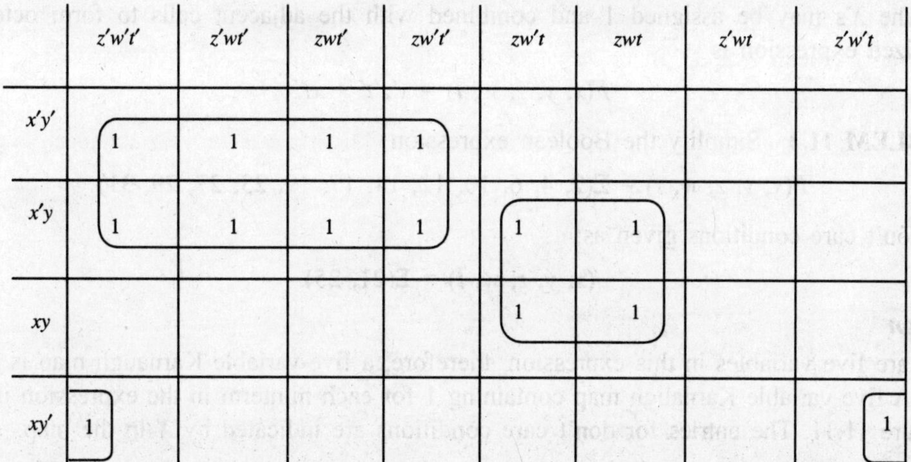

	$z'w't'$	$z'wt'$	zwt'	$zw't'$	$zw't$	zwt	$z'wt$	$z'w't$
$x'y'$	1	1	1	1				
$x'y$	1	1	1	1	1	1		
xy					1	1		
xy'	1							1

Figure 11.9 Five variables Karnaugh map for Problem 11.2.

PROBLEM 11.3 Simplify the Boolean function:

$$F(x, y, z, w, t) = \Sigma(0, 2, 8, 10, 17, 19, 23, 27, 29, 31)$$

with don't care conditions $d(x, y, z, w, t) = \Sigma(21,25)$.

Solution

The given Boolean function is:

$$F(x, y, z, w, t) = \Sigma(0, 2, 8, 10, 17, 19, 23, 27, 29, 31).$$

There are five variables in this expression. Therefore, a five-variable Karnaugh map is needed here. This map is drawn in Figure 11.10 with corresponding entries of 1 for each of the minterm appearing in the expression. Further, the entries for don't care conditions are indicated by X in the map.

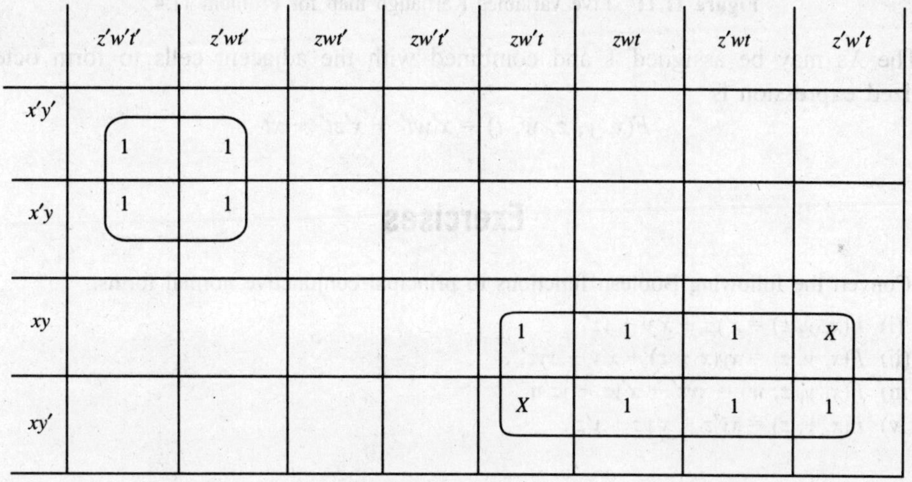

	$z'w't'$	$z'wt'$	zwt'	$zw't'$	$zw't$	zwt	$z'wt$	$z'w't$
$x'y'$	1	1						
$x'y$	1	1						
xy					1	1	1	X
xy'					X	1	1	1

Figure 11.10 Two variables Karnaugh map for Problem 11.3.

The Xs may be assigned 1 and combined with the adjacent cells to form octet. The minimized expression is

$$F(x, y, z, w, t) = x'z't' + xt.$$

PROBLEM 11.4 Simplify the Boolean expression

$$F(x, y, z, w, t) = \Sigma(2, 4, 6, 10, 12, 14, 17, 19, 23, 27, 29, 31)$$

with don't care conditions given as:

$$d(x, y, z, w, t) = \Sigma(21, 25).$$

Solution

There are five variables in this expression, therefore, a five-variable Karnaugh map is needed for it. A five-variable Karnaugh map containing 1 for each minterm in the expression is given in Figure 11.11. The entries for don't care conditions are indicated by X in the map.

	$z'w't'$	$z'wt'$	zwt'	$zw't'$	$zw't$	zwt	$z'wt$	$z'w't$
$x'y'$		1	1	1				
$x'y$		1	1	1				
xy					1	1	1	X
xy'					X	1	1	1

Figure 11.11 Five variables Karnaugh map for Problem 11.4.

The Xs may be assigned 1 and combined with the adjacent cells to form octet. The minimized expression is

$$F(x, y, z, w, t) = x'wt' + x'zt' + xt.$$

Exercises

11.1 Convert the following Boolean functions to principal conjunctive normal forms.

(i) $F(x, y, z) = x'yz + x'y + yz'$

(ii) $F(x, y, z) = xy(x + z) + x'y + xyz'$

(iii) $F(x, y, z, w) = xyz' + x'yz + yz'w$

(iv) $F(x, y, z) = xy'z + x'yz + y'z'$.

11.2 Convert the following Boolean functions to principal disjunctive normal forms.

(i) $F(x, y, z, t) = xy'z + x'y + yz' + xt$
(ii) $F(x, y, z, t, w) = xyt + x'yw + xyz'$
(iii) $F(x, y, z, t) = xyz' + x'yz + yz' + xt$
(iv) $F(x, y, z, w) = xy'z + x'yz + y'z'w.$

11.3 Minimize the following Boolean expressions using deduction system.

(i) $F(x, y, z, t, w) = x'yzw + xy'zt + x'yzwt + yz' + xy'zt' + yzt$
(ii) $F(x, y, z, t, w) = xyt + x'ywt + xyz' + x'yzwt + yz' + xyt'$
(iii) $F(x, y, z, t) = x'yzt + yz' + xt + xyz' + x'yz + y'z' + xt$
(iv) $F(x, y, z, w, t) = x'y'z + x'yz + y'z'w + x'yzwt + yz' + x'yzwt'$
(v) $F(x, y, z, t) = xy'z + x'y'z + x'yzt + y'z' + xyt + yz' + xt$
(vi) $F(x, y, z, w, t) = xyt + x'yw + xyz' + xy'z + x'y + x'yzt + yz' + xyt'$
(vii) $F(x, y, z, t) = xyz' + x'yz' + yz' + xt + xy'z + x'y + x'yzt + yz' + xyt$
(viii) $F(x, y, z, w, t) = xy'z + x'yz + y'z'w + x'yzwt + xyt' + xyt + xyzwt.$

11.4 Minimize the following Boolean expressions by Karnaugh map.

(i) $F(x, y, z, w) = \Sigma(0, 2, 4, 10, 12, 13, 14, 15)$
(ii) $F(x, y, z, w, t) = \Sigma(0, 2, 4, 10, 12, 13, 14, 15, 16, 23, 27, 29, 31)$
(iii) $F(x, y, z, w, t) = \Sigma(1,10, 12, 13, 14, 15, 16, 17, 23, 29, 31)$
(iv) $F(x, y, z, w, t) = \Sigma(0, 3, 4, 10, 12, 13, 15, 16, 17, 19, 23, 27, 29, 31).$

11.5 Minimize the following Boolean expression by Karnaugh map. Use don't care conditions.

(i) $F(x, y, z, w) = \Sigma(0, 2, 3, 4, 10, 12, 13, 14, 15),$
$d(x, y, z, w, t) = \Sigma(1, 5, 11)$
(ii) $F(x, y, z, w, t) = \Sigma(0, 2, 4, 10, 12, 13, 14, 15, 17, 23, 29, 31),$
$d(x, y, z, w, t) = \Sigma(21, 25)$
(iii) $F(x, y, z, w, t) = \Sigma(1, 10, 14, 15, 16, 18, 23, 26, 27, 31),$
$d(x, y, z, w, t) = \Sigma(0, 11, 22, 24)$
(iv) $F(x, y, z, w, t) = \Sigma(0, 3, 13, 15, 16, 17, 19, 23, 28, 29, 31),$
$d(x, y, z, w, t) = \Sigma(1, 27, 30).$

12

Logic Gates and Circuit Design

12.1 Introduction

In Chapter 11, we discussed the minimization of Boolean functions, which is an important part of circuit design. In this chapter, we will introduce logic gates, and develop a procedure to implement the logic circuits using gates, which is our ultimate goal. Logic gates are the basic building blocks of electronic circuits. We will also study the implementation of logic gates using universal gates.

12.2 Logic Gate

A logic gate is a fundamental atomic component of electronic circuit. It is called atomic, because an electronic circuit is fabricated with gates. It is a device that accepts some input signals and produces one output. The principles of two-valued Boolean algebra, which have been discussed in Chapter 10, are used in signal processing. Further, there is a lot of similarity between mathematical logic and Boolean algebra. A signal (input or output) may be at one of the two levels 0 and 1. Here 0 represents low and 1 represents high. These are also called binary digits or bits. There are three basic logic gates, termed NOT, AND and OR. Other gates described in the following sections, can be realized in terms of these gates.

Lines entering in the gate symbol from the left are considered as input lines and a single line on the right is the output line. A gate is a logic component consisting of diodes, transistors, capacitors and resistors. In reality, it is fabricated on a semiconductor chip by doping it with a desirable impurity to modify its properties in a controlled manner. An IC (Integrated circuit) chip contains several gates. A chip can be purchased from the electronic market at a nominal price. Almost every circuit can be implemented using gates.

12.3 Logic Gate Symbols

Each gate is assigned a special symbol so that it can be identified in a logic circuit. Here, in this section, we will discuss all these symbols along with their truth tables.

12.3.1 NOT Gate

Figure 12.1 shows a NOT gate symbol. It is also called an inverter. NOT gate has only one input and one output.

Truth table (Table 12.1) gives the complete definition of NOT gate. Here A is the input and A' is the output. If input is low then output will be high, and if input is high then output will be low.

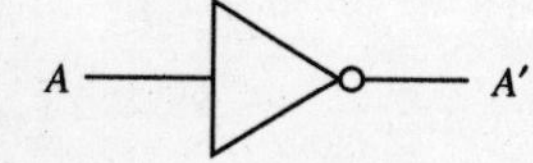

Figure 12.1 Symbol for NOT.

Table 12.1 Truth Table for NOT

A	A'
0	1
1	0

12.3.2 AND Gate

Figure 12.2 shows a symbol of two input AND gate. It has one output. Truth table (Table 12.2) gives the complete definition of AND gate. Here A and B are the inputs and $A \cdot B$ (or simply AB) is the output. If any of the inputs is low then the output will be low. The output will be high only if all the inputs are high. An AND gate may also have three or more inputs. Theoretically, it may have any number of inputs.

Figure 12.2 Two input AND gate.

Table 12.2 Truth Table for AND

A	B	$A \cdot B$
0	0	0
0	1	0
1	0	0
1	1	1

12.3.3 OR Gate

Figure 12.3 shows a two input OR gate symbol.

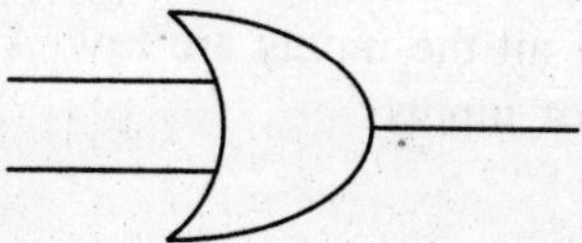

Figure 12.3 Two input OR gate.

Truth table (Table 12.3) gives the complete definition of OR gate. Here A and B are the inputs and $A + B$ is the output. If any of the inputs is high then the output will be high, and output will be low only if all the inputs are low. Like AND gate, an OR gate may have any number of inputs.

Table 12.3 Truth Table for OR

A	B	$A + B$
0	0	0
0	1	1
1	0	1
1	1	1

12.3.4 NAND Gate

Figure 12.4 shows a two-input NAND gate symbol. Truth table (Table 12.4) gives the complete definition of NAND gate. Here A and B are the inputs and $(A \cdot B)'$ (or simply $(AB)'$) is the output. If any of the inputs is low then the output will be high. The output will be low if all the inputs are high.

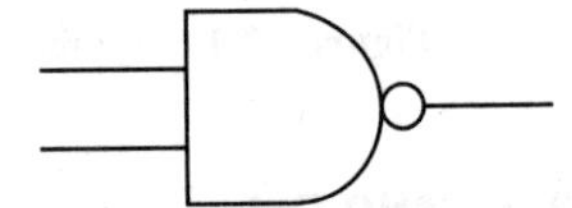

Figure 12.4 Two input NAND gate.

Table 12.4 Truth Table for NAND

A	B	$(A \cdot B)'$
0	0	1
0	1	1
1	0	1
1	1	0

A NAND gate is the complement of AND gate, i.e., AND gate followed by NOT gate. Like AND gate, it may have any number of inputs.

12.3.5 NOR Gate

Figure 12.5 shows a two-input NOR gate symbol.

When output signal of an OR gate is followed by a NOT gate, a NOR gate is constructed. Truth table (Table 12.5) gives the complete definition of NOR gate. Here A and B are the inputs and $(A + B)'$ is the output. If any of the inputs is high then the output will be low, and the output will be high only if all the inputs are low. A NOR gate may have any number of inputs.

Figure 12.5 Two input NOR gate.

Table 12.5 Truth Table for NOR

A	B	$(A + B)'$
0	0	1
0	1	0
1	0	0
1	1	0

12.3.6 EX-OR Gate

It is also abbreviated as **X-OR** gate. Figure 12.6 shows a two-input EX-OR gate symbol. Truth table (Table 12.6) gives the complete definition of EX-OR gate. Here A and B are the inputs and $A \oplus B$ is the output. The equivalent Boolean expression for $A \oplus B$ is $A'B + AB'$.

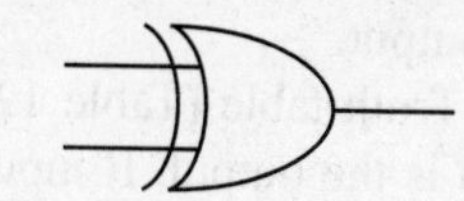

Figure 12.6 Two input EX-OR gate.

That is, $$A \oplus B = A'B + AB'.$$

Table 12.6 Truth Table for EX-OR

A	B	$A \oplus B$
0	0	0
0	1	1
1	0	1
1	1	0

When the number of inputs in an EX-OR gate is more than two then its definition is slightly modified. It is then considered as an odd function, i.e., its output is 1 if an odd number of inputs are 1. Thus, in a three-input EX-OR gate, the output is high if either one or all inputs are high, otherwise it will be low.

12.3.7 EX-NOR Gate

It is also known as **X-NOR** gate. Figure 12.7 shows the symbol of EX-NOR gate. Truth table (Table 12.7) gives the complete definition of EX-NOR gate. Here A and B are the inputs and $(A \oplus B)'$ is the output. The equivalent Boolean expression for $(A \oplus B)'$ is $A'B' + AB$.

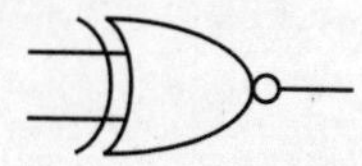

Figure 12.7 Two input EX-NOR gate.

That is, $$(A \oplus B)' = A'B' + AB.$$

Table 12.7 Truth Table for EX-NOR

A	B	$(A \oplus B)'$
0	0	1
0	1	0
1	0	0
1	1	1

Like EX-OR gate, the definition of EX-NOR gate is modified when the number of inputs exceeds two. In that case, it behaves as an even function, i.e., its output is 1 if an even number of inputs are 0. Thus, in a three-input EX-NOR gate, output is high if none of the inputs are 0 (all inputs are 1), or if two of its inputs are zero. EX-NOR gate is also called **equivalence gate**. It may be observed that EX-OR and EX-NOR gates have complemented outputs when number of inputs are even, but have same outputs when number of inputs are odd.

12.3.8 Buffer Gate

Figure 12.8 shows a buffer gate symbol. Like NOT gate, a buffer gate has only one input and one output.

Truth table (Table 12.8) gives the complete definition of buffer gate. Here A is the input and B is the output. If input is low then output will be low, and if input is high then output will be high, i.e., there is no change in the signal. Its main purpose is to create delay in the processing.

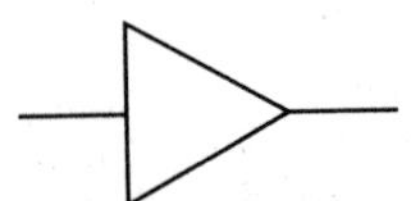

Figure 12.8 Buffer gate.

Table 12.8 Truth Table for Buffer

A	B
0	0
1	1

12.4 Logic Circuits

A logic circuit is a well defined and well formed connected structure whose basic components are logic gates. It is also called combinational circuit. A combinational circuit transforms binary input data to a desirable form. Figure 12.9 shows a general block diagram of a logic circuit whose actual implementation with gates may depend on a specific problem.

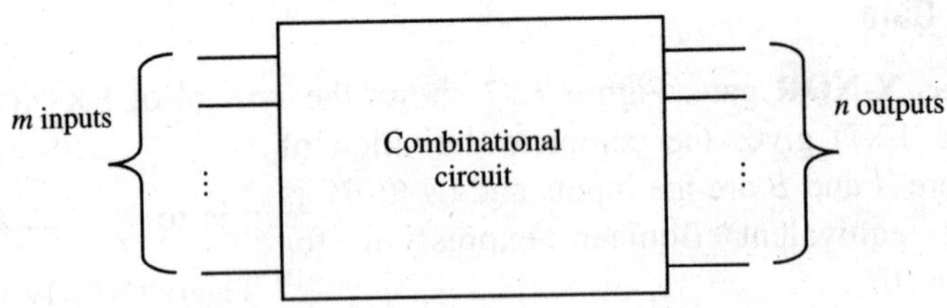

Figure 12.9 Combinational circuit.

There are m inputs coming from some source and n outputs going to some external destination. Each input and output may have value 0 or 1. Different input combinations will produce different desirable output combinations. The complete behaviour of a combinational circuit can be described by a truth table showing the binary relationship between its m inputs and n outputs. There are 2^m possible combinations of m binary inputs for which the outputs are specified.

Figure 12.10 is an example of a combinational circuit whose input and output relationship is given in Table 12.9.

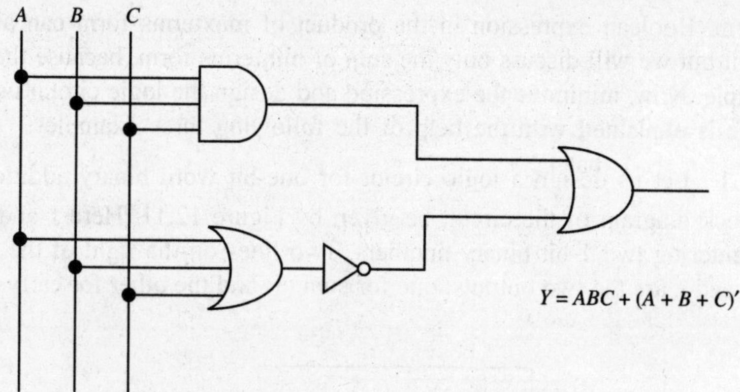

Figure 12.10 Combinational circuit.

Table 12.9 Truth Table

Input(*A*, *B*, *C*)			Output(*Y*)
A	*B*	*C*	*Y*
0	0	0	1
0	0	1	0
0	1	0	0
0	1	1	0
1	0	0	0
1	0	1	0
1	1	0	0
1	1	1	1

Here *A*, *B* and *C* are the inputs, and *Y* is the output. The construction of a truth table and its corresponding logic circuit is explained with the help a procedure given below.

12.5 Designing Logic Circuits

A logic circuit can be designed by the following procedure:

1. State and analyze the problem for which the circuit is to be constructed.
2. Construct the truth table.
3. Obtain the Boolean expression from the truth table.
4. Minimize the Boolean expression.
5. Design the circuit using gates.

The problem for which the circuit is to be designed must be analyzed, i.e., a relationship between input and output must be expressed by a mathematical function. Now, since truth table is the primary requirement of a logic circuit, therefore, transform this function into a truth table. In view of this, if it is not possible to interpret the problem in terms of a truth table then logic circuit cannot be constructed. From the truth table, obtain the Boolean expression in the sum of minterms form. Boolean expression in the product of maxterms form can also be used to design the circuit but we will discuss only the sum of minterms form, because the other is quite similar in principle. Now, minimize the expression and design the logic circuit using gates. The whole procedure is explained with the help of the following three examples.

EXAMPLE 12.1 Let us design a logic circuit for one-bit word binary addition.

Let the block diagram of the circuit be given by Figure 12.11. Here x and y are the two input lines for entering two 1-bit binary numbers. Two lines on the right of the block diagram designated by s and c are the two outputs: one for sum (s) and the other for carry (c). It is called half adder.

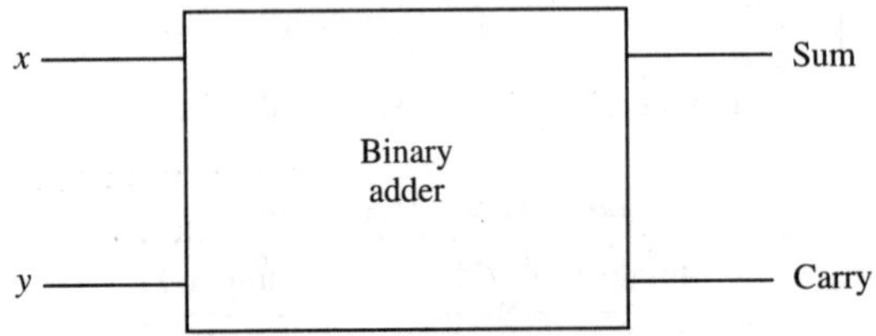

Figure 12.11 Half adder.

1. **Analysis of the problem:** Consider the addition of two one-bit binary numbers x and y. Since there are two variables x and y, therefore, the number of possible binary combinations will be four (for $n = 2$, $2^n = 4$).

   Now, compute the sum s and carry c for $x + y$. It is ordinary arithmetic addition. In this chapter, we are taking symbol + for this operation, while + is usual OR operator.

$x + y =$	sum(s),	carry(c)
$0 + 0 =$	0	0
$0 + 1 =$	1	0
$1 + 0 =$	1	0
$1 + 1 =$	1	1

2. **Construction of the truth table:** The above addition can be summarized in a truth table given by Table 12.10.

Table 12.10 Truth Table for Half Adder

x	y	s	c
0	0	0	0
0	1	1	0
1	0	1	0
1	1	0	1

3. **Boolean expression:** Since there are two output functions s and c, therefore, there will be two Boolean expressions one for s and another for c.

   Sum of minterms expression for s will be

$$s = x'y + xy'.$$

It is obtained by adding the minterms corresponding to 1s in column containing s. Two minterms $x'y$ and xy' which have value 1 in column represented by s. Thus, the function s is

$$x'y + xy'.$$

Further, there is only one minterm xy, which has 1 in column headed by c, therefore, function c is xy, i.e., $c = xy$. Finally, the output Boolean functions for s and c will be:

$$s = x'y + xy'$$

$$c = xy$$

4. **Minimization of the Boolean expression:** Two Boolean expressions for s and c, given above, are already minimized, therefore, go to the final step, i.e., gate implementation.

5. **Designing the circuit using gates:** There are two input variables x and y, therefore, draw four input lines one each for x, x', y and y'. In fact, lines for x' and y' are obtained by using NOT gates with x and y. Now, use three AND gates, one each for three minterms $x'y$, xy' and xy. The output lines from two AND gates implementing minterms $x'y$ and xy' are given as the input lines to an OR gate producing the output sum (s). The output carry (c) is obtained from the AND gate producing xy. The complete circuit is given in Figure 12.12. This circuit is also known as half adder circuit.

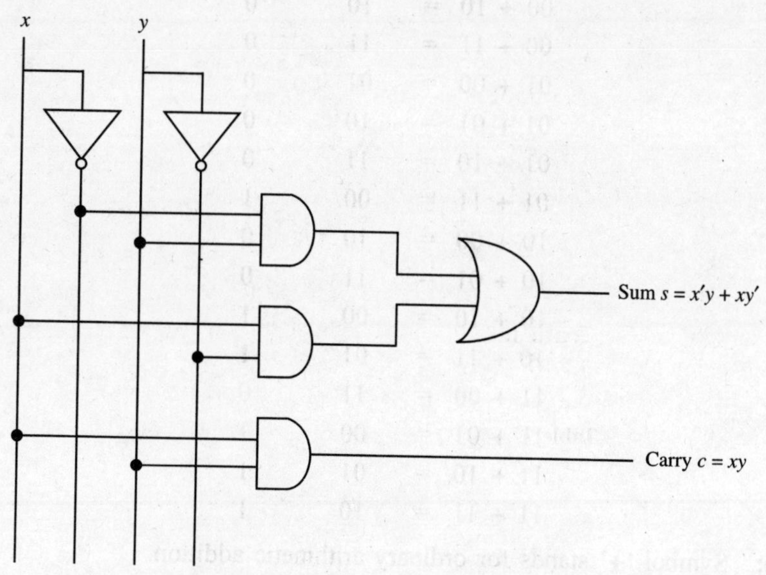

Figure 12.12 Logic circuit for half adder.

EXAMPLE 12.2 Let us consider to design a logic circuit for two-bit word binary addition.

Let the block diagram of the circuit be given in the Figure 12.13. Since each of the two words needs two input lines to enter the data separately, therefore, four input lines (a, b, c and d)

are considered. There will be three output lines on the right of the block diagram, two for representing sum (s) and one for final carry (c).

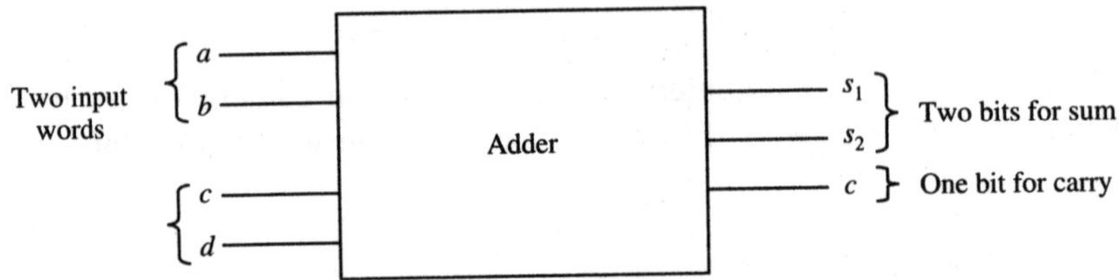

Figure 12.13 Two-bit word adder.

1. **Analysis of the problem:** Consider the addition of two 2-bit binary numbers x and y, each with two bits. Since there are two words x and y, each consisting of the two variables (making four variables in total), therefore the number of possible binary combinations will be sixteen (for $n = 4$, $2^n = 16$). These sixteen combinations are as follows:

$x + y$	=	Sum(s)	Carry(c)
00 + 00	=	00	0
00 + 01	=	01	0
00 + 10	=	10	0
00 + 11	=	11	0
01 + 00	=	01	0
01 + 01	=	10	0
01 + 10	=	11	0
01 + 11	=	00	1
10 + 00	=	10	0
10 + 01	=	11	0
10 + 10	=	00	1
10 + 11	=	01	1
11 + 00	=	11	0
11 + 01	=	00	1
11 + 10	=	01	1
11 + 11	=	10	1

Note: Symbol '+' stands for ordinary arithmetic addition.

2. **Construction of the truth table:** The above addition can be summarized in a truth table given in Table 12.11. Here word x is represented by two bits x_1 and x_2 (i.e. x_1x_2), and y is represented by y_1 and y_2 (i.e. y_1y_2). The output is given by one-bit word carry c and two-bit word sum s_1 and s_2.

Table 12.11 Truth Table for Two-bit Word Adder

x_1	x_2	y_1	y_2	*Sum* s_1s_2	*Carry* c
0	0	0	0	00	0
0	0	0	1	01	0
0	0	1	0	10	0
0	0	1	1	11	0
0	1	0	0	01	0
0	1	0	1	10	0
0	1	1	0	11	0
0	1	1	1	00	1
1	0	0	0	10	0
1	0	0	1	11	0
1	0	1	0	00	1
1	0	1	1	01	1
1	1	0	0	11	0
1	1	0	1	00	1
1	1	1	0	01	1
1	1	1	1	10	1

3. **Boolean expression:** Since there are three output functions s_1, s_2 and c, therefore, there will be three Boolean expressions, one each for s_1, s_2 and c.

   The sum of minterms expressions for each will be written as:

$$s_1 = x_1'x_2'y_1y_2' + x_1'x_2'y_1y_2 + x_1'x_2y_1'y_2 + x_1'x_2y_1y_2' + x_1x_2'y_1'y_2' + x_1x_2'y_1'y_2 + x_1x_2y_1'y_2' + x_1x_2y_1y_2,$$

$$s_2 = x_1'x_2'y_1'y_2 + x_1'x_2'y_1y_2 + x_1'x_2y_1'y_2' + x_1'x_2y_1y_2' + x_1x_2'y_1'y_2 + x_1x_2'y_1y_2 + x_1x_2y_1'y_2' + x_1x_2y_1y_2',$$

$$c = x_1'x_2y_1y_2 + x_1x_2'y_1y_2' + x_1x_2'y_1y_2 + x_1x_2y_1'y_2 + x_1x_2y_1y_2' + x_1x_2y_1y_2.$$

   These Boolean expressions are obtained by adding the minterms corresponding to 1*s* in columns for s_1, s_2 and c.

4. **Minimization of the Boolean expressions:** Three Boolean expressions obtained above may be minimized by Karnaugh map method, which has already been discussed in Chapter 11. Construct four variables Karnaugh map for each expression.

   Karnaugh map for s_1 is given in Figure 12.14.
   Minimized expression for s_1 will be written as:

$$s_1 = x_1'x_2'y_1 + x_1'x_2y_1'y_2 + x_1'y_1y_2' + x_1y_1'y_2' + x_1x_2'y_1' + x_1x_2y_1y_2$$

   Karnaugh map for s_2 is given in Figure 12.15.

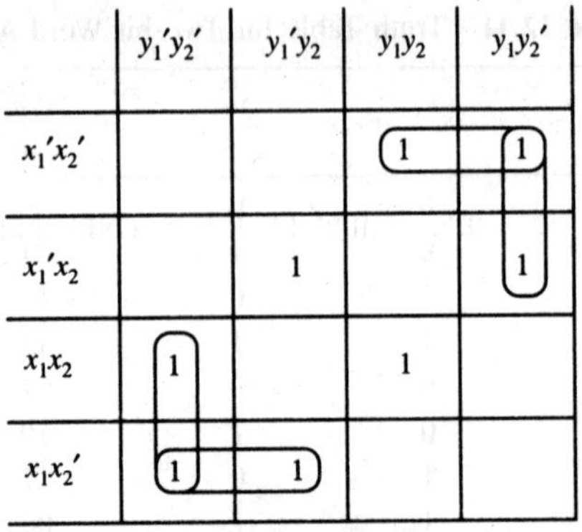

	$y_1'y_2'$	$y_1'y_2$	y_1y_2	y_1y_2'
$x_1'x_2'$			1	1
$x_1'x_2$		1		1
x_1x_2	1		1	
x_1x_2'	1	1		

Figure 12.14 Karnaugh map for s_1.

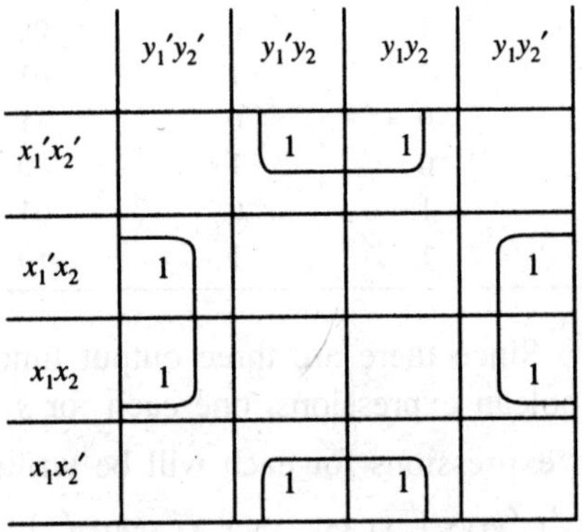

	$y_1'y_2'$	$y_1'y_2$	y_1y_2	y_1y_2'
$x_1'x_2'$		1	1	
$x_1'x_2$	1			1
x_1x_2	1			1
x_1x_2'		1	1	

Figure 12.15 Karnaugh map for s_2.

Minimized expression for s_2 will be:

$$s_2 = x_2'y_2 + x_2y_2'$$

Karnaugh map for c is given in Figure 12.16.

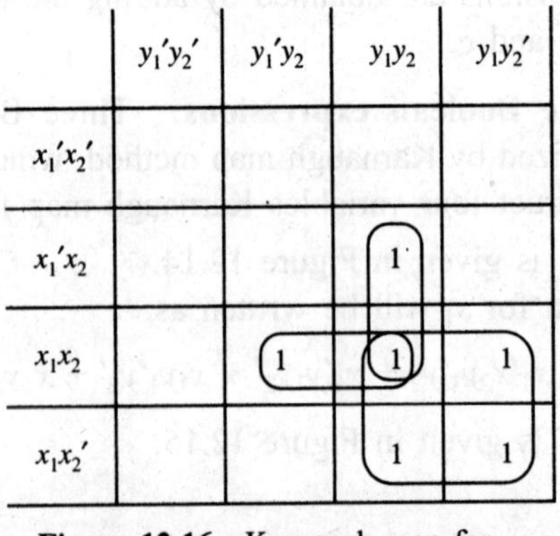

	$y_1'y_2'$	$y_1'y_2$	y_1y_2	y_1y_2'
$x_1'x_2'$				
$x_1'x_2$			1	
x_1x_2		1	1	1
x_1x_2'			1	1

Figure 12.16 Karnaugh map for c.

Minimized expression for c will be written as:

$$c = x_2y_1y_2 + x_1x_2y_2 + x_1y_1$$

5. **Designing the circuit using gates:** There are four input variables, therefore, draw four input lines, one each for x_1, x_2, y_1 and y_2. Obtain lines for x_1', x_2', y_1' and y_2' by using NOT gates with x_1, x_2, y_1, and y_2, respectively. Realize expression for s_1 by using six AND gates and one OR gate. For s_2, use two AND gates and one OR gate. Expression for c can be implemented by three AND and one OR gate. The complete circuit is given in Figure 12.17.

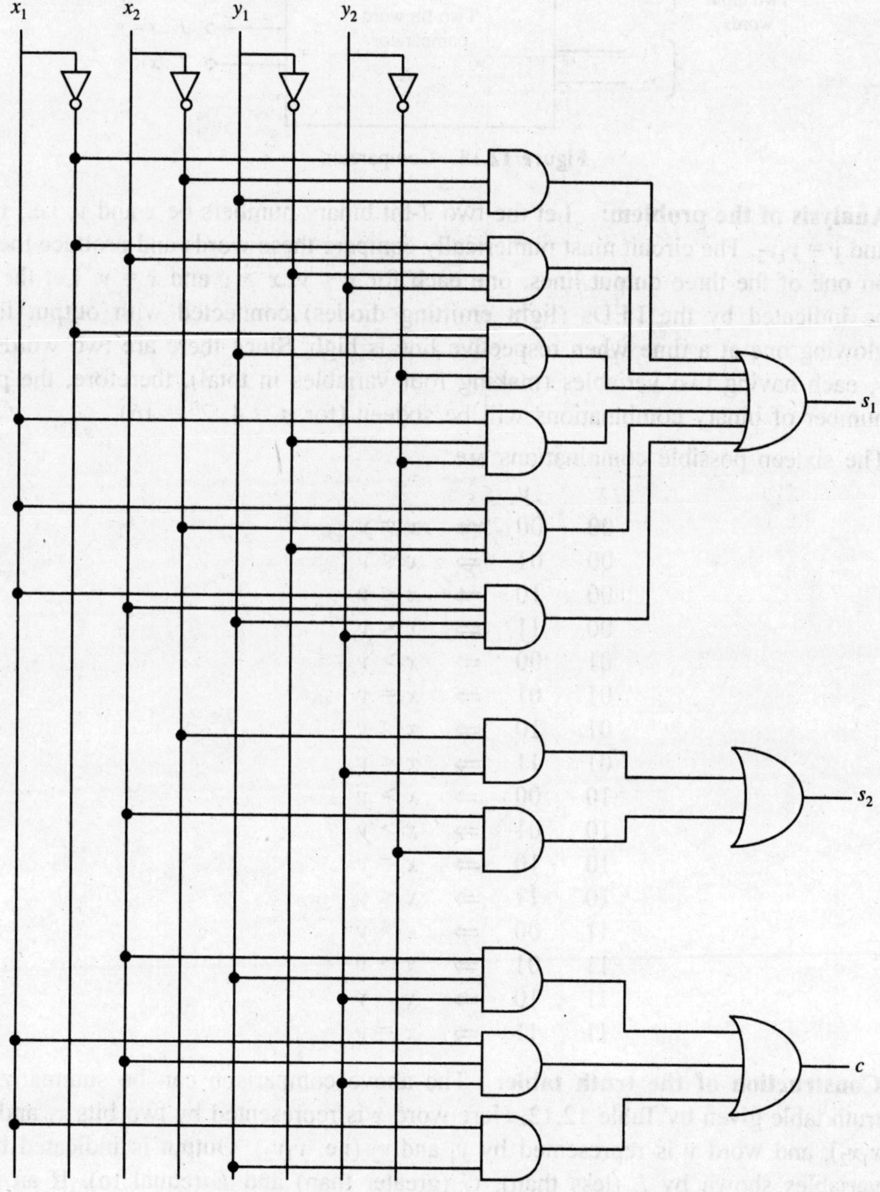

Figure 12.17 Logic circuit for two-bit word adder

EXAMPLE 12.3 Let us design a logic circuit for two-bit word binary comparison.

Suppose, the block diagram of the circuit is given in Figure 12.18. Here x_1, x_2, y_1 and y_2 are the four input lines for entering two 2- bit binary numbers (assuming separate lines for each word). There will be three output lines (one each for <, > and =).

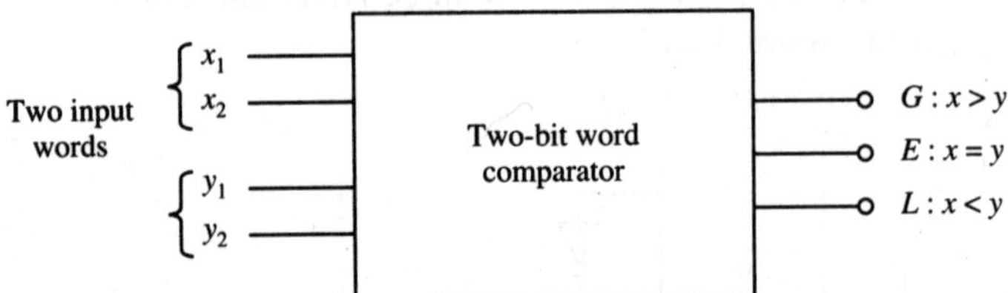

Figure 12.18 Comparator.

1. **Analysis of the problem:** Let the two 2-bit binary numbers be x and y, i.e., $x = x_1x_2$ and $y = y_1y_2$. The circuit must numerically compare these words and produce the output on one of the three output lines, one each for $x < y$, $x > y$ and $x = y$. Let the output be indicated by the LEDs (light emitting diodes) connected with output lines by glowing one at a time when respective line is high. Since there are two words x and y, each having two variables (making four variables in total), therefore, the possible number of binary combinations will be sixteen (for $n = 4$, $2^n = 16$).

   The sixteen possible combinations are:

x	y		
00	00	$\Rightarrow$	$x = y$
00	01	$\Rightarrow$	$x < y$
00	10	$\Rightarrow$	$x < y$
00	11	$\Rightarrow$	$x < y$
01	00	$\Rightarrow$	$x > y$
01	01	$\Rightarrow$	$x = y$
01	10	$\Rightarrow$	$x < y$
01	11	$\Rightarrow$	$x < y$
10	00	$\Rightarrow$	$x > y$
10	01	$\Rightarrow$	$x > y$
10	10	$\Rightarrow$	$x = y$
10	11	$\Rightarrow$	$x < y$
11	00	$\Rightarrow$	$x > y$
11	01	$\Rightarrow$	$x > y$
11	10	$\Rightarrow$	$x > y$
11	11	$\Rightarrow$	$x = y$

2. **Construction of the truth table:** The above comparison can be summarized in a truth table given by Table 12.12. Here word x is represented by two bits x_1 and x_2 (i.e. x_1x_2), and word y is represented by y_1 and y_2 (i.e. y_1y_2). Output is indicated by three variables shown by L (less than), G (greater than) and E (equal to). If an LED is glowing, then it means respective output is 1. Only one out of the three LEDs will glow at a time.

Table 12.12 Truth Table for Comparator

x_1	x_2	y_1	y_2	L	G	E
0	0	0	0	0	0	1
0	0	0	1	1	0	0
0	0	1	0	1	0	0
0	0	1	1	1	0	0
0	1	0	0	0	1	0
0	1	0	1	0	0	1
0	1	1	0	1	0	0
0	1	1	1	1	0	0
1	0	0	0	0	1	0
1	0	0	1	0	1	0
1	0	1	0	0	0	1
1	0	1	1	1	0	0
1	1	0	0	0	1	0
1	1	0	1	0	1	0
1	1	1	0	0	1	0
1	1	1	1	0	0	1

3. **Boolean expression:** Since there are three output functions L, G and E, therefore, there will be three Boolean expressions, one each for L, G and E.
The sum of minterms expressions for each will be:

$$L = x_1'x_2'y_1'y_2 + x_1'x_2'y_1y_2' + x_1'x_2'y_1y_2 + x_1'x_2y_1y_2' + x_1'x_2y_1y_2 + x_1x_2'y_1y_2$$

$$G = x_1'x_2y_1'y_2' + x_1x_2'y_1'y_2' + x_1x_2'y_1'y_2 + x_1x_2y_1'y_2' + x_1x_2y_1'y_2 + x_1x_2y_1y_2'$$

$$E = x_1'x_2'y_1'y_2' + x_1'x_2y_1'y_2 + x_1x_2'y_1y_2' + x_1x_2y_1y_2$$

These Boolean expressions are obtained by adding the minterms corresponding to 1s in the columns of L, G and E.

4. **Minimization of the Boolean expressions:** Three Boolean expressions obtained above may be minimized by Karnaugh map method. Construct four variables Karnaugh map for each expression.
Karnaugh map for L is given in Figure 12.19.

	$y_1'y_2'$	$y_1'y_2$	y_1y_2	y_1y_2'
$x_1'x_2'$		1	1	1
$x_1'x_2$			1	1
x_1x_2				
x_1x_2'			1	

Figure 12.19 Karnaugh map for L.

Minimized expression for L will be written as:

$$L = x_1'y_1 + x_1'x_2'y_2 + x_2'y_1y_2$$

Karnaugh map for G is given in Figure 12.20.

	$y_1'y_2'$	$y_1'y_2$	y_1y_2	y_1y_2'
$x_1'x_2'$				
$x_1'x_2$	1			
x_1x_2	1	1		1
x_1x_2'	1	1		

Figure 12.20 Karnaugh map for G.

Minimized expression for G will be:

$$G = x_1y_1' + x_2y_1'y_2' + x_1x_2y_2'$$

Karnaugh map for E is given in Figure 12.21.

	$y_1'y_2'$	$y_1'y_2$	y_1y_2	y_1y_2'
$x_1'x_2'$	1			
$x_1'x_2$		1		
x_1x_2			1	
x_1x_2'				1

Figure 12.21 Karnaugh map for E.

It is clear from the Karnaugh map that no further minimization of the expression E is possible. Therefore, it remains unchanged.

Hence, $$E = x_1'x_2'y_1'y_2' + x_1'x_2y_1'y_2 + x_1x_2'y_1y_2' + x_1x_2y_1y_2.$$

5. **Design the circuit using gates:** There are four input variables, therefore, draw four input lines one each for x_1, x_2, y_1 and y_2. Obtain lines for x_1', x_2', y_1' and y_2' by using NOT gates with x_1, x_2, y_1, and y_2, respectively. Realize expression for L by using three AND gates and one OR gate. Also, for G, use three AND gates and one OR gate. Expression for E can be implemented by four AND and one OR gate. The complete circuit is given in Figure 12.22.

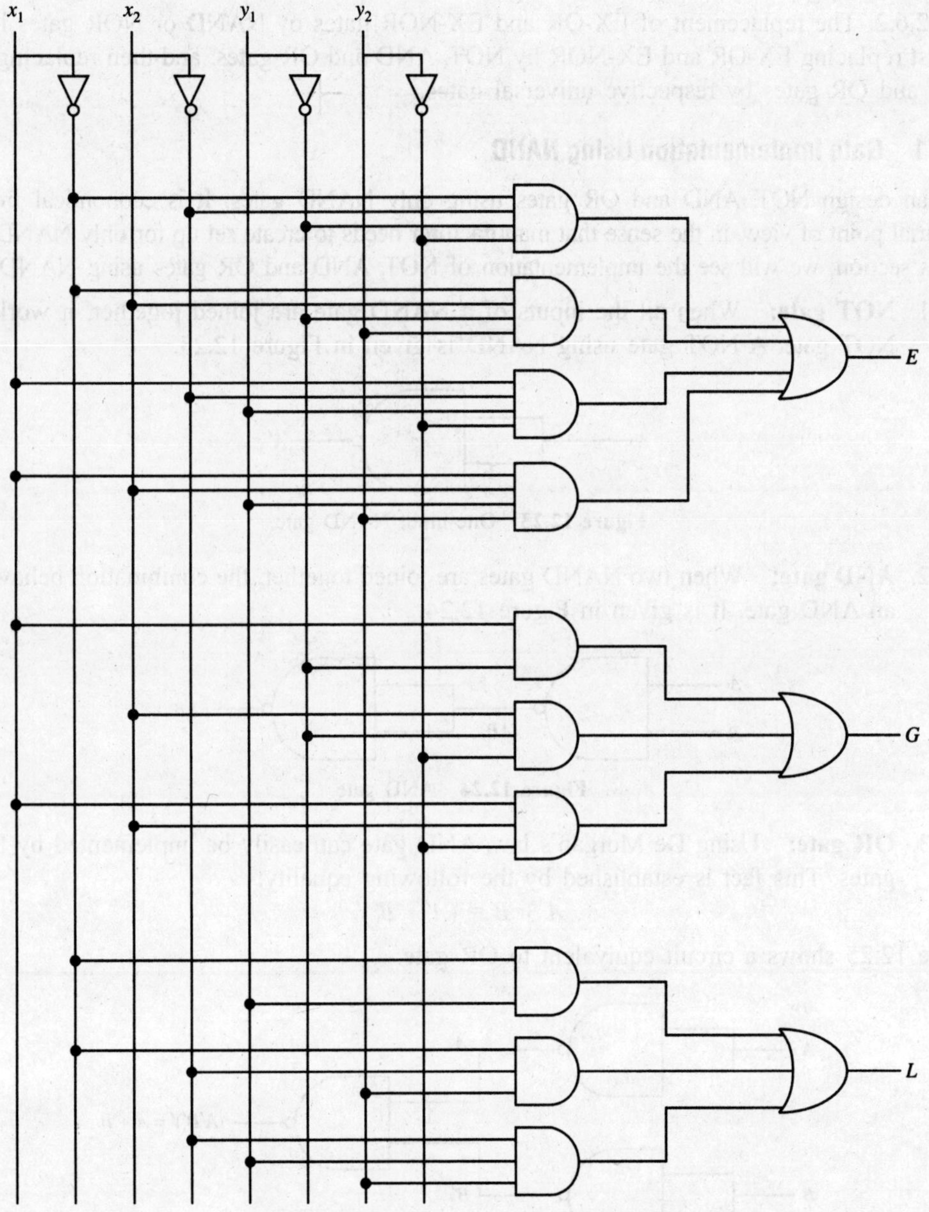

Figure 12.22 Logic circuit for comparator.

12.6 Universal Gates

A logic circuit can be designed by using only NAND gates, i.e., every gate in the circuit can be replaced by a combination of NAND gates. Due to this property, NAND gate is called universal gate. NOR gate is also called universal gate because of its similar property. The procedure to design a circuit using only NAND gates or only NOR gates is to replace the AND, OR and NOT gates by NAND or NOR gates. This replacement is shown in Sections 12.6.1 and 12.6.2. The replacement of EX-OR and EX-NOR gates by NAND or NOR gates is done by first replacing EX-OR and EX-NOR by NOT, AND and OR gates, and then replacing NOT, AND and OR gates by respective universal gates.

12.6.1 Gate Implementation Using NAND

We can design NOT, AND and OR gates using only NAND gates. It is economical from the industrial point of view, in the sense that manufacturer needs to create set up for only NAND gates. In this section, we will see the implementation of NOT, AND and OR gates using NAND gates.

1. **NOT gate:** When all the inputs of a NAND gate are joined together, it works as a NOT gate. A NOT gate using NAND is given in Figure 12.23.

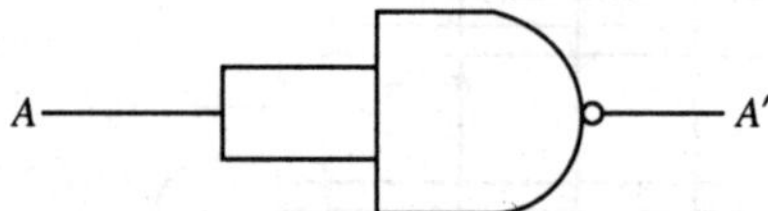

Figure 12.23 One-input NAND gate.

2. **AND gate:** When two NAND gates are joined together, the combination behaves like an AND gate. It is given in Figure 12.24.

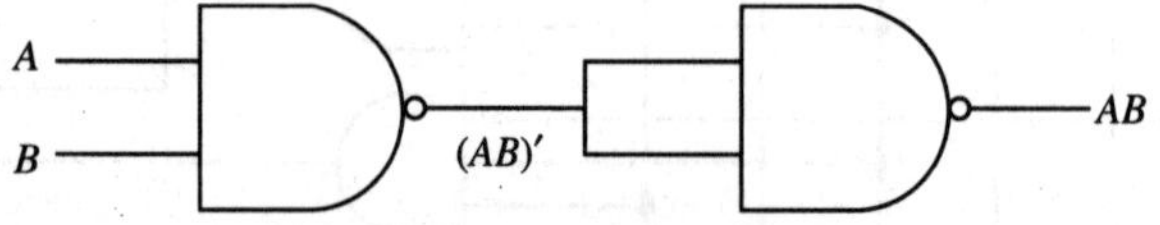

Figure 12.24 AND gate.

3. **OR gate:** Using De Morgan's law, AND gate can easily be implemented by NAND gates. This fact is established by the following equality:

$$A + B = (A' \cdot B')'$$

Figure 12.25 shows a circuit equivalent to OR gate.

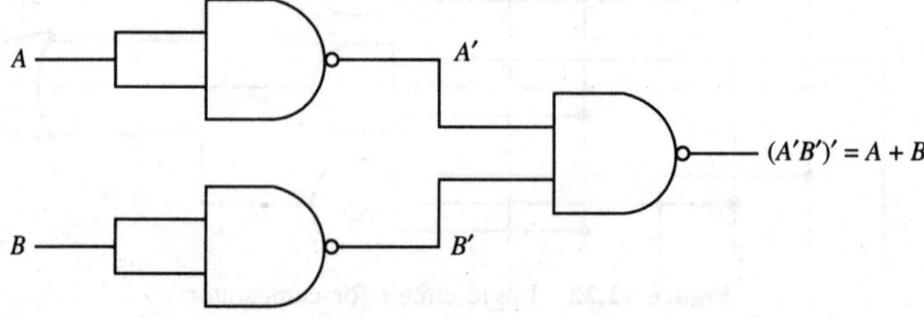

Figure 12.25 OR gate.

12.6.2 Gate Implementation Using NOR Gate

Like NAND gate implementation, we can also design NOT, AND and OR gates using NOR gates only. In this section, we will see the implementation of NOT, AND and OR gates using NOR gates.

1. **NOT gate:** When all the inputs of a NOR gate are joined together, it works as a NOT gate. A NOT gate using NOR gate is given in Figure 12.26.

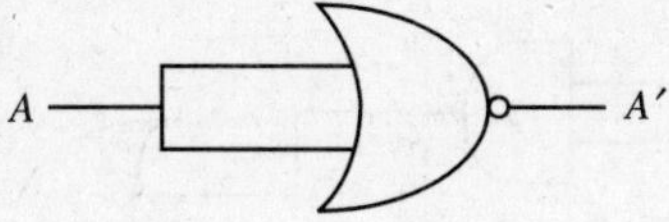

Figure 12.26 NOT gate.

2. **OR gate:** When two NOR gates are joined together as given in Figure 12.27, the combination behaves like an OR gate.

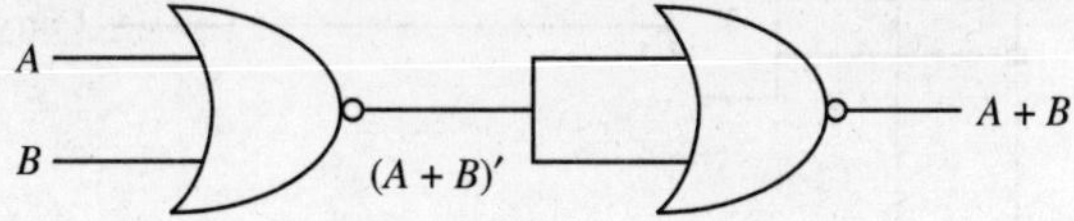

Figure 12.27 OR gate.

3. **AND gate:** Using De Morgan's law, the following equality establishes the implementation of AND gate by NOR gates.

$$A \cdot B = (A' + B')'$$

Figure 12.28 shows a circuit equivalent to AND gate.

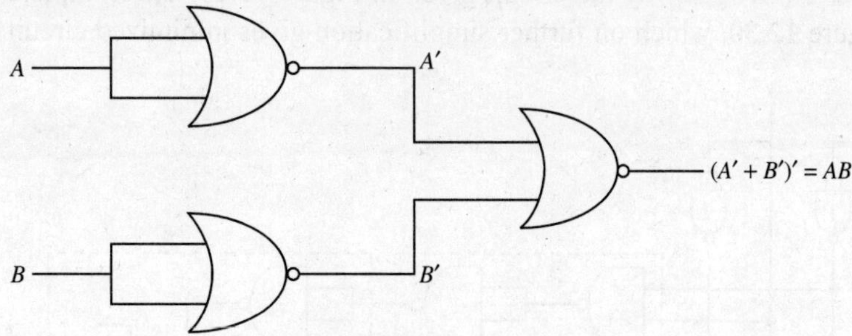

Figure 12.28 AND gate.

12.6.3 Circuit Designing Using Universal Gates

First of all, design the circuit by using AND, OR and NOT gates and then replace each gate by its NAND or NOR gates equivalent form. The procedure can be understood by the following example.

EXAMPLE 12.4 Consider the circuit of half adder given in Figure 12.12. The figure is redrawn here (Figure 12.29), and the gates are numbered to facilitate the discussion further.

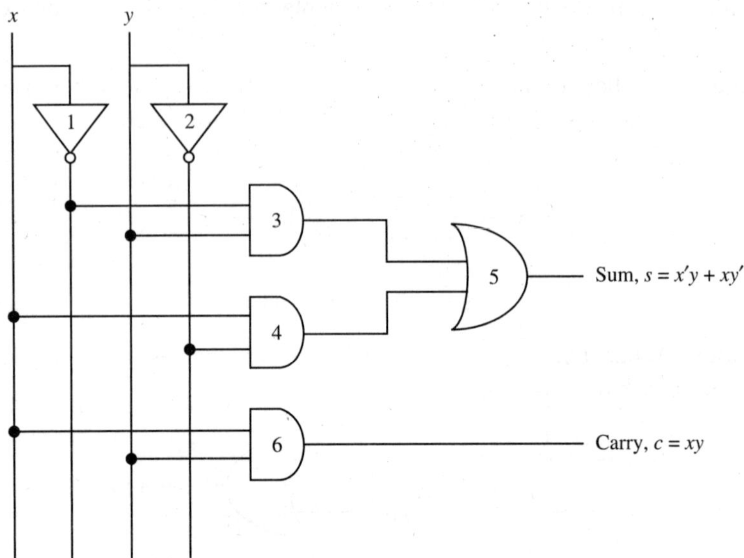

Figure 12.29 Half adder.

1. **Implementation by NAND gates:** Replace gate number 1 and 2 in the above circuit (Figure 12.29) by the NAND gates with two inputs of each gate joined together. Further, replace gate number 3, 4 and 6 by their NAND gates equivalent circuit. This can be done by replacing each gate by two NAND gates joined together. Replace gate number 5 (OR gate) by the circuit given in Figure 12.25. The complete circuit is given in Figure 12.30, which on further simplification gives minimized circuit (Figure 12.31).

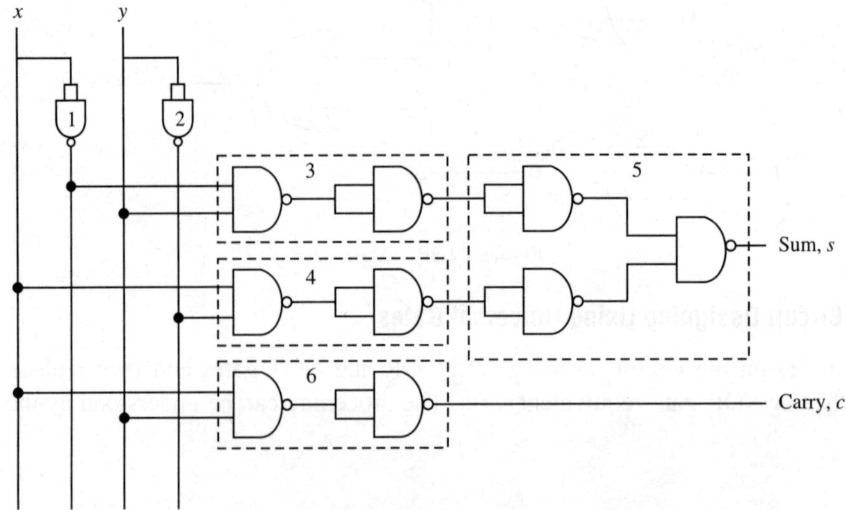

Figure 12.30 Circuit using NAND gates.

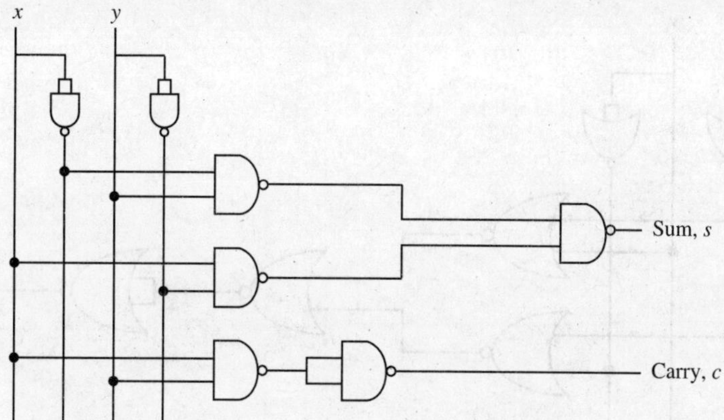

Figure 12.31 Circuit using NAND gates : minimized.

2. **Implementation by NOR gates:** Replace gate number 1 and 2 in the circuit given in Figure 12.29 by the NOR gates with two inputs of each gate joined. Further, replace gate number 5 by its NOR gates equivalent circuit. This can be done by replacing OR gate by two NOR gates joined together. Replace gate number 3, 4 and 6 (AND gates) by using the circuit given in Figure 12.28. The complete circuit is given in the Figure 12.32. On further simplification and little modification, it is transformed into the circuit shown in Figure 12.33. Note that, in the minimized figure, inputs for carry has been taken directly from the lines x' and y'.

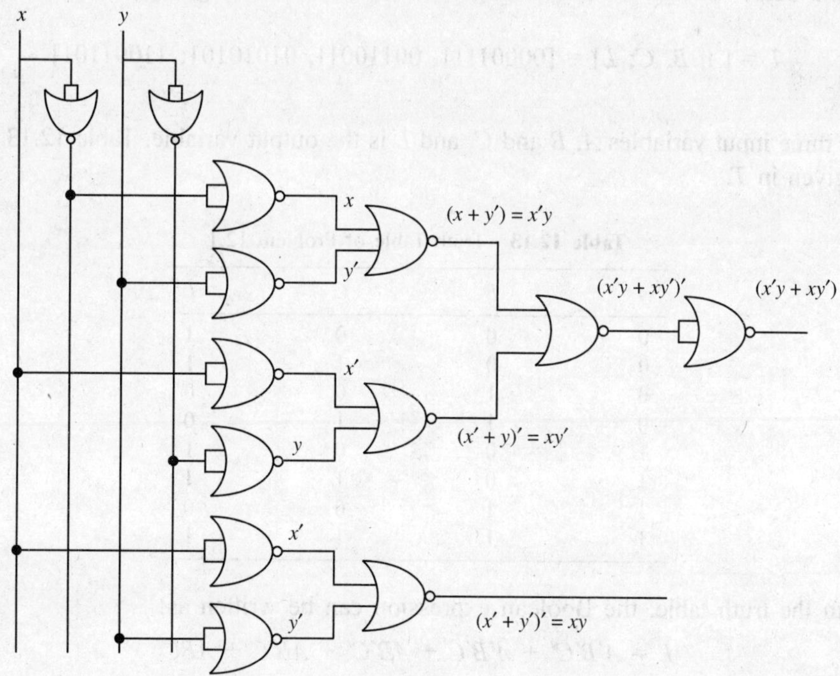

Figure 12.32 Circuit using NOR gates.

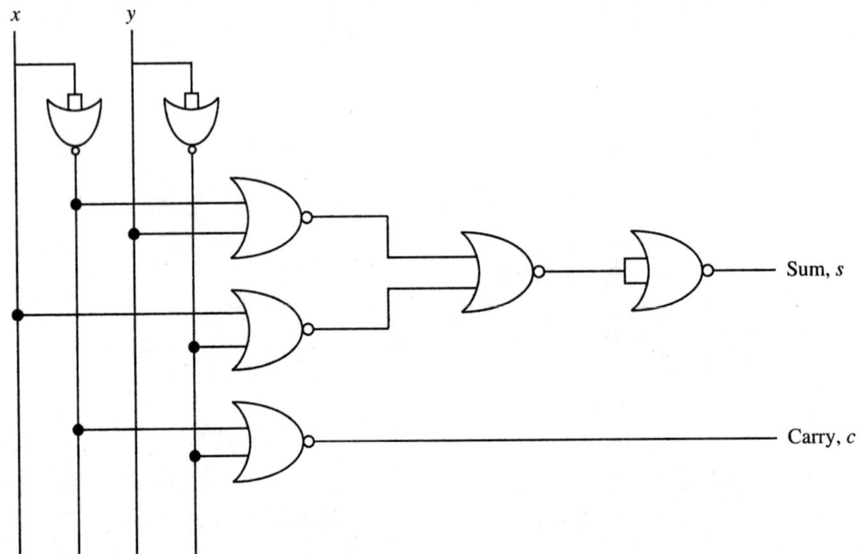

Figure 12.33 Circuit using NOR gates : minimized.

12.7 Solved Problems

PROBLEM 12.1 Design a minimal logic circuit for the data given in the following truth table.

$$T = [A, B, C; L] = [00001111, 00110011, 01010101; 11001101]$$

Solution

There are three input variables A, B and C, and L is the output variable. Table 12.13 represents the data given in T.

Table 12.13 Truth Table of Problem 12.1

A	B	C	L
0	0	0	1
0	0	1	1
0	1	0	0
0	1	1	0
1	0	0	1
1	0	1	1
1	1	0	0
1	1	1	1

From the truth table, the Boolean expression can be written as:

$$L = A'B'C' + A'B'C + AB'C' + AB'C + ABC.$$

This can be minimized by Karnaugh map shown in Figure 12.34. The minimized Boolean expression is

$$L = B' + AC$$

and the minimized logic circuit is given in Figure 12.35.

	$B'C'$	$B'C$	BC	BC'
A'	1	1		
A	1	1	1	

Figure 12.34 Karnaugh map.

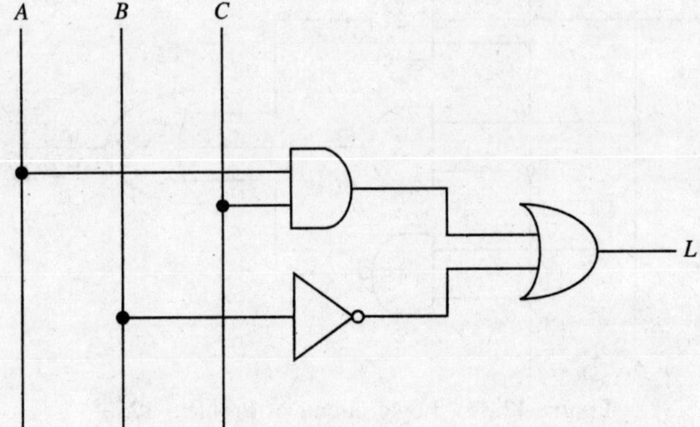

Figure 12.35 Logic circuit for Problem 12.1.

PROBLEM 12.2 Obtain the output Boolean expression for the logic circuit shown in Figure 12.36.

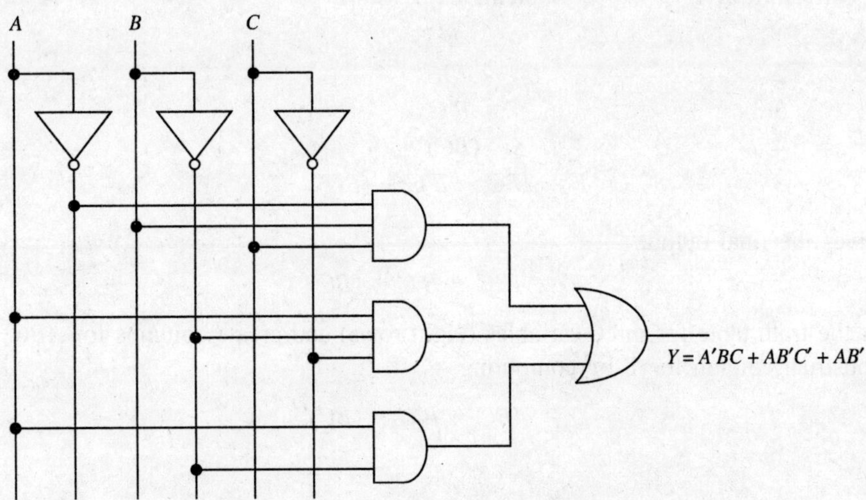

Figure 12.36 Logic circuit of Problem 12.2.

Solution

The outputs of the first, second and third AND gates are $A'BC$, $AB'C'$, and AB' respectively. Therefore, output $y = A'BC + AB'C' + AB'$, which on minimization becomes

$$y = A'BC + AB'.$$

PROBLEM 12.3 Construct the truth table for the logic circuit in Figure 12.37.

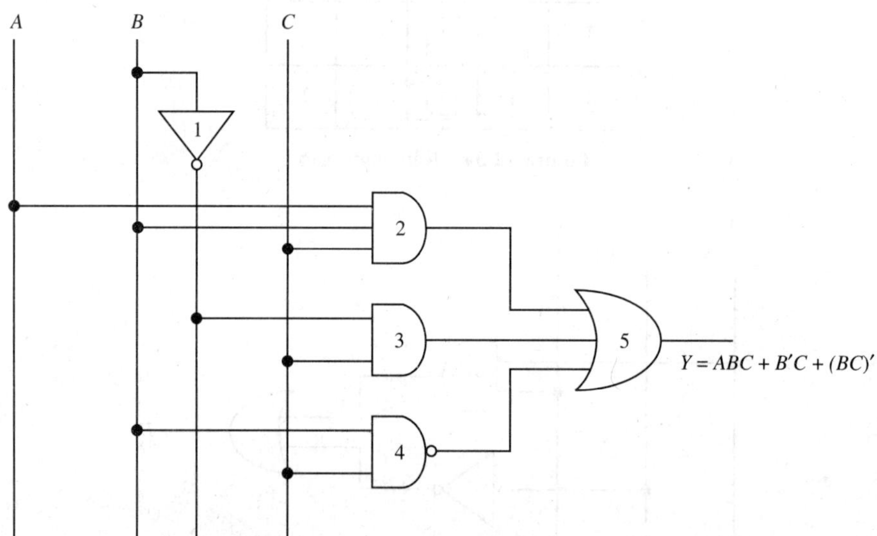

Figure 12.37 Logic circuit of Problem 12.3.

Solution

The output expressions of all the gates are listed below.

Gate number	Boolean expression
1	B'
2	ABC
3	$B'C$
4	$(BC)'$
5	$Y = ABC + B'C + (BC)'$

Gate 5 gives the final output

$$Y = ABC + B'C + (BC)'.$$

Construct the truth table for three variables (eight rows) and create columns for ABC, $B'C$ and $(BC)'$. Construct column for Y by computing

$$Y = ABC + B'C + (BC)'.$$

The complete truth table is given in Table 12.14.

Table 12.14 Truth Table for Problem 12.3

A	B	C	ABC	B'	$B'C$	BC	$(BC)'$	Y
0	0	0	0	1	0	0	1	1
0	0	1	0	1	1	0	1	1
0	1	0	0	0	0	0	1	1
0	1	1	0	0	0	1	0	0
1	0	0	0	1	0	0	1	1
1	0	1	0	1	1	0	1	1
1	1	0	0	0	0	0	1	1
1	1	1	1	0	0	1	0	1

PROBLEM 12.4 Draw the logic circuits for the following Boolean expressions.

(i) $Y = ABC + A'B'C + B'C'$

(ii) $Y = A'B + B'C + ABC'$

Solution

Both expressions are in the sum of products form. The circuits for both the expressions are given in Figure 12.38(a and b).

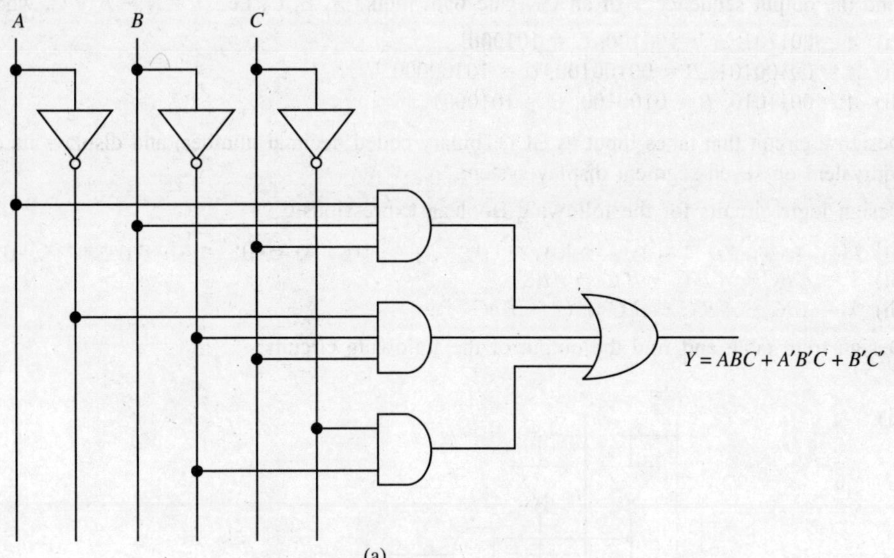

(a)

Figure 12.38 (*Contd.*)

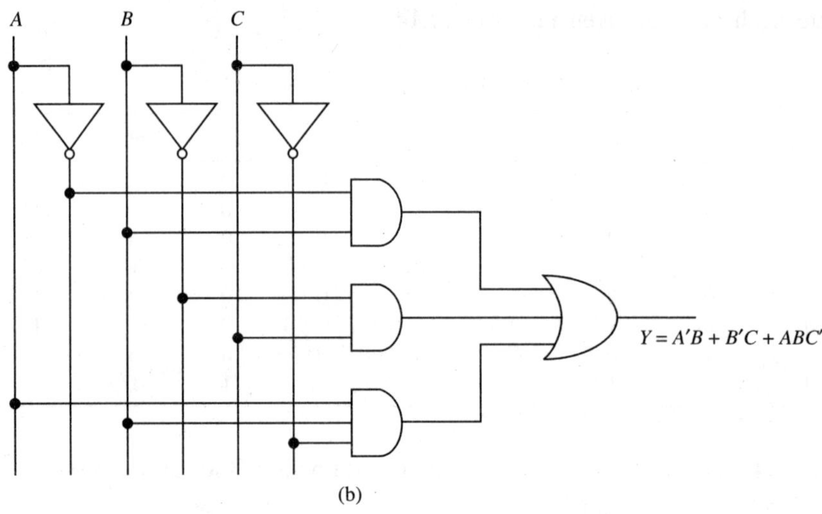

(b)

Figure 12.38 Logic circuits for Problem 12.4.

Exercises

12.1 Implement EX-OR and EX-NOR gates with AND, OR and NOT gates.

12.2 Find the output sequence Y of an OR gate with inputs A, B, C, i.e., $Y = A + B + C$, where

(i) $A = 001101$; $B = 100100$, $C = 101000$.
(ii) $A = 00100101$, $B = 00100100$, $C = 10100000$.
(iii) $A = 0011010$, $B = 0100100$, $C = 1010001$.

12.3 Design a circuit that takes input as BCD (binary coded decimal number) and displays its decimal equivalent on seven segment display system.

12.4 Design logic circuits for the following Boolean expressions.

(i) $f(x_1, x_2, y_1, y_2) = x_1'x_2y_1'y_2 + x_1'x_2'y_1y_2 + x_1'x_2'y_1y_2 + x_1'x_2y_1y_2' + x_1'x_2y_1y_2 + x_1x_2'y_1y_2$
(ii) $Y = A'BC + A'B'C + B'C' + ABC'$
(iii) $Y = ABC + A'B'C + B'C' + (A + B')C$

12.5 Design truth table and find the output of the following circuits.

(i)

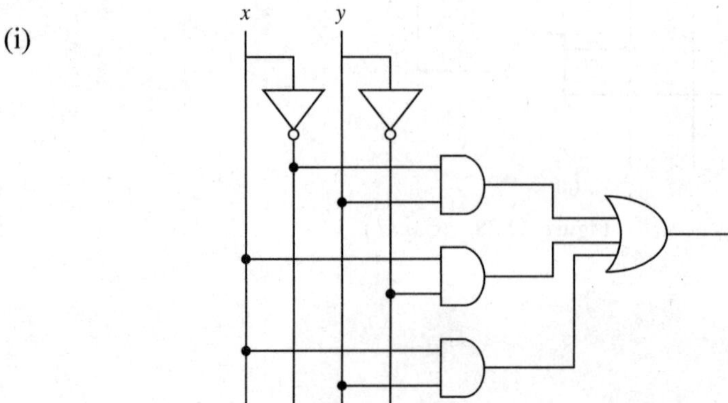

Figure 12.39 Logic circuit.

(ii)

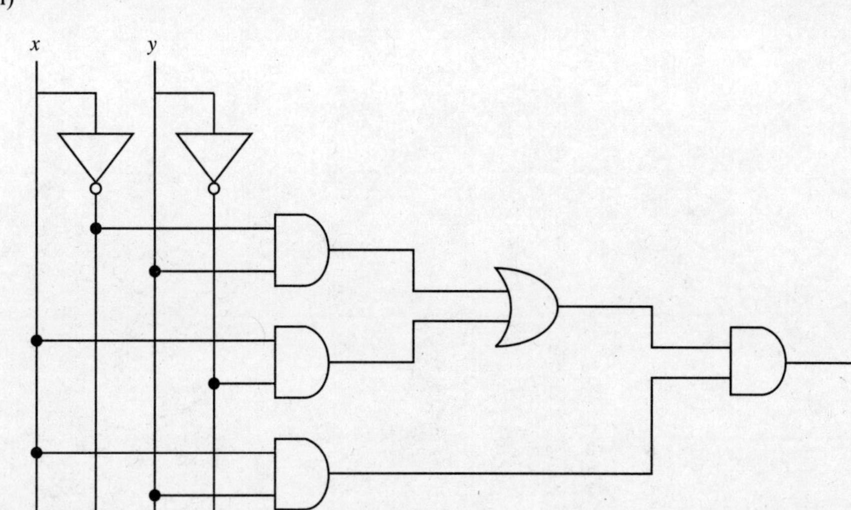

Figure 12.40 Logic circuit.

12.6 Design a minimal logic circuit that takes seven bit-word input and determines the parity: even parity (even number of 1s) and odd parity (odd number of 1s).

12.7 Design a circuit using NAND gates to implement the following circuits:

(i) $Y = ABCD + A'B'CD + (A + D)B'C'$
(ii) $Y = A'BCD' + AB'CD + B'C' + D$

12.8 Design logic circuits using only NOR gates for the following Boolean expressions.

(i) $Y = ABCD' + A'B'CD + (A + B)C'D$
(ii) $Y = A'BC'D' + AB'CD + B'C' + D'$

12.9 Find the output Boolean expression Y for the following truth table. Also, design the logic circuit for it.

Table 12.15 Truth Table

A	B	C	Y
0	0	0	0
0	0	1	1
0	1	0	1
0	1	1	0
1	0	0	1
1	0	1	1
1	1	0	0
1	1	1	1

Part D
GRAPH THEORY

13

Directed and Undirected Graphs

13.1 Introduction

Graph is a great mathematical tool with its applications in several areas such as computer science, communication, physical science, networking and so on. Euler (1707–1783) was the first to introduce the concept of graph theory in 1736. He wrote the first paper on graph theory while solving a very famous problem 'Königsberg bridge problem'. The Königsberg bridge problem is a notable historical problem in mathematics. In the eighteenth century, Königsberg town (formerly in East Prussia but now in Russia and known as Kaliningrad) was situated on two banks and two islands of the river Pregel, connected to one another by seven bridges as shown in Figure 13.1. The problem was to find a path starting at any place and ending at the same, moving through the city and crossing each bridge once and only once. Did such a path exist? People out of fun tried to find the solution but in vain.

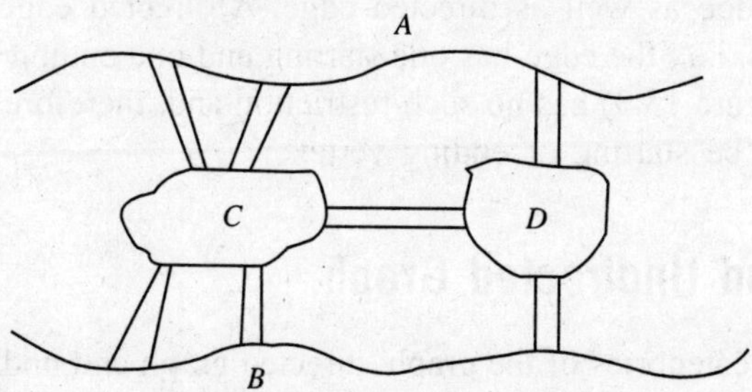

Figure 13.1 Königsberg town.

The problem was sent to the then famous mathematician Euler with the map of the city. Euler analyzed the problem after redrawing it on a paper with a simple diagram. He named the diagram 'graph'. He also answered that no such path/walk was possible. Details related to his deliberations will be discussed in Chapter 15.

13.2 Graph and its Terminology

In this section, we will define three basic terms of graph theory: graph, vertex and edge.

Graph

A graph G is a mathematical structure consisting of two sets: set of vertices and set of edges. Formally,

$$G = (V, E)$$

V: set of vertices

E: set of edges

Vertex

It is a point (drawn as a small circle), representing a physical entity like city, house, or an office. It is also called a **node**. Name of the vertex is generally written inside or outside the small circle. Some symbols which are in common use are:

$$\circ^a, \quad \bullet^a \quad \text{and} \quad \textcircled{a}$$

where a is the name of the vertex.

In this text, just for convenience, we will use first symbol for showing vertices in a graph, while the third one will be used in some trees, specially expression trees in Chapter 14.

Edge

A link joining two vertices is called an edge. It is represented by a pair of two vertices called the end points of the edge, e.g., (e, f). Here 'e' and 'f' are the two end points of the edge. Another alternate notation, which is more suitable to use is e-f. An edge may be directed or undirected. If it has an arrow on it then it is directed, otherwise undirected.

An edge of a graph, which joins a vertex to itself, is called a **loop**. A loop is shown in Figure 13.2 at vertex b. The direction of the loop has no significance, since it may be considered as both an undirected edge as well as directed edge. A directed edge (Figure 13.2) joins an **ordered pair of vertices**, i.e., the edge has one starting and one ending vertex clearly specified. The undirected edge (Figure 13.3) has no such restriction and, therefore, any of the two vertices joined by the edge may be starting or ending vertex.

13.3 Directed and Undirected Graph

There are two important categories of the graph: directed graph and undirected graph. However, the study of undirected graph is more general. Further, readers will find that most of the algorithms of graph theory are common to both. In this section, we will define directed and undirected graph.

13.3.1 Directed Graph

A graph $G = (V, E)$ is directed, if each edge of the graph is directed. An arrow on it shows the direction of the edge. The arrow mark on the edge may be either at the end of the edge or somewhere at the middle. Both are in common use. For example,

o——→o or o——→——o

Directed graph is also called a digraph. The graph given in Figure 13.2 is a directed graph.

EXAMPLE 13.1 Consider a directed, graph $G = (V, E)$ with set of vertices V, and set of edges E, given by:

$$V = \{a, b, c, d\}$$

$$E = \{(a, b), (a, d), (b, b), (b, c), (c, a), (d, c)\}$$

The pictorial representation of this graph is shown in Figure 13.2.

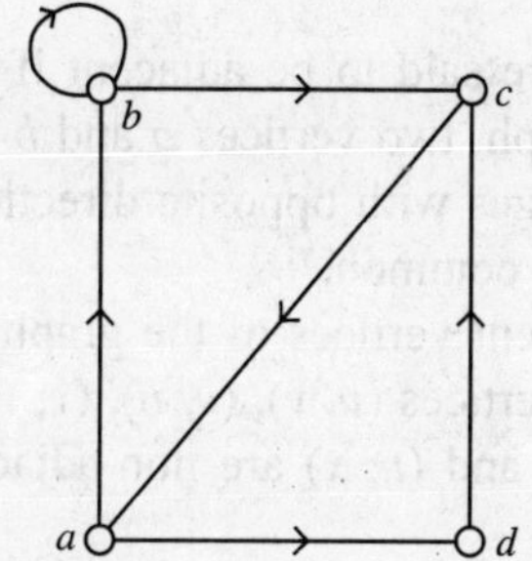

Figure 13.2 Directed graph.

13.3.2 Undirected Graph

If all the edges are undirected then it is called undirected graph or simply a graph. The graph given in Figure 13.3 is an undirected graph. If however, a graph contains both directed and undirected edges, then it is called a **mixed graph.**

EXAMPLE 13.2 Consider an undirected graph $G = (V, E)$ with set of vertices V, and set of edges E, given by:

$$V = \{a, b, c, d\}$$

$$E = \{(a, b), (a, d), (b, c), (c, a), (c, d)\}$$

The pictorial representation of the graph is shown in Figure 13.3.

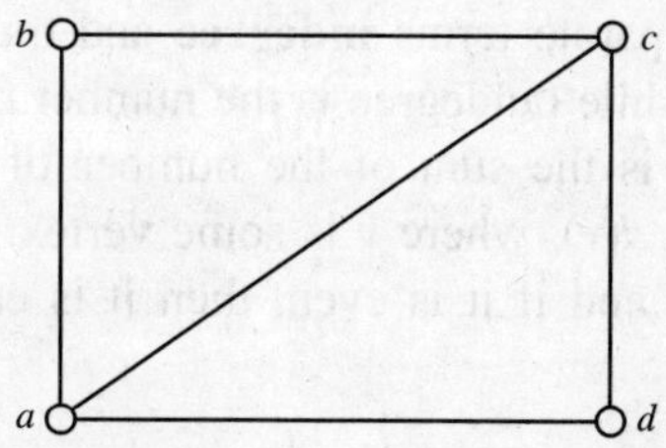

Figure 13.3 Undirected graph.

13.4 Path

An alternating connected sequence of vertices and edges is called a path. A path starts and ends at a vertex. The two vertices, first and last, with which the path begins and ends, respectively, are called terminal vertices. If a path starts and ends at the same vertex then it is termed as **closed path**. A closed path is called a **circuit** or a **cycle**. In a directed graph, a path is a **connected directed sequence** of vertices and edges. For example, consider the undirected graph given in Figure 13.3.

Here edge sequence $< (a, b), (b, c), (c, d) >$ is a path and $< (a, b), (b, c), (c, d), (d, a) >$ is a circuit of the graph. Other appropriate notations for path and circuit are *a-b-c-d* and *a-b-c-d-a*, respectively.

13.5 Adjacent Vertices and Adjacent Edges

Two vertices a and b of a graph are said to be adjacent if there is a direct edge joining these vertices. In the case of directed graph, two vertices a and b of the graph are said to be adjacent if there are at least two directed edges with opposite directions between them. **Adjacent edges** are those that have an endpoint in common.

For example, u and v are the adjacent vertices of the graph given by Figure 13.4(a). However, in Figure 13.4(b), ordered pair of vertices (u, v), (v, u), (v, w), (w, x) and (x, u) are all adjacent, while (w, v), (x, w), (u, w), (w, u) and (u, x) are non-adjacent.

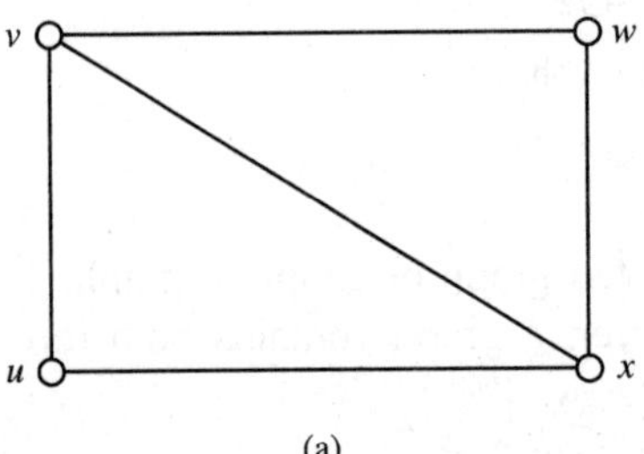

(a)

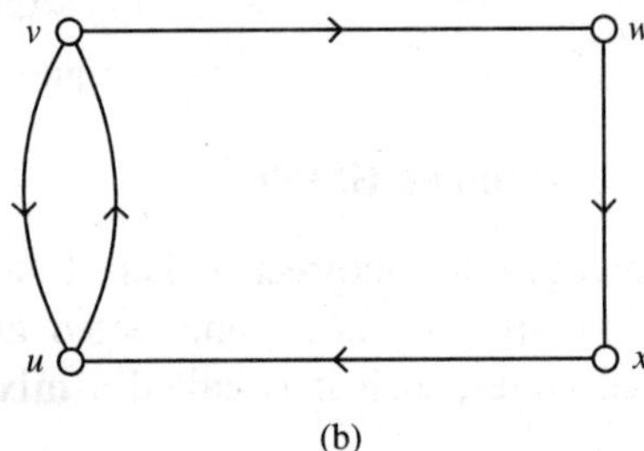

(b)

Figure 13.4 Adjacent and non-adjacent vertices.

13.6 Degree

The number of edges incident on a vertex is called the degree of a vertex. In the case of a directed graph, there are two separate terms **indegree** and **outdegree**. Indegree of a vertex is the number of incoming edges while outdegree is the number of outgoing edges. Hence, degree of a vertex in a directed graph is the sum of the number of incoming edges and number of outgoing edges. It is denoted by $d(v)$, where v is some vertex. If the degree of a vertex is odd, then it is called an **odd vertex,** and if it is even, then it is called an **even vertex**.

13.7 Multi Graph

If in a directed or undirected graph, a pair of vertices is joined by more than one edge, it is called a multi graph and the edges that join the vertices are called parallel edges. However, in the case of directed graph, the two possible directed edges joining a pair of vertices will not be treated as parallel if they have opposite direction, and consequently, such graph will not be a multi graph. The graphs given in Figures 13.5(a and b) are multi graphs, while the graph given in Figure 13.5(c) has no parallel edges, and therefore, it is not a multi graph.

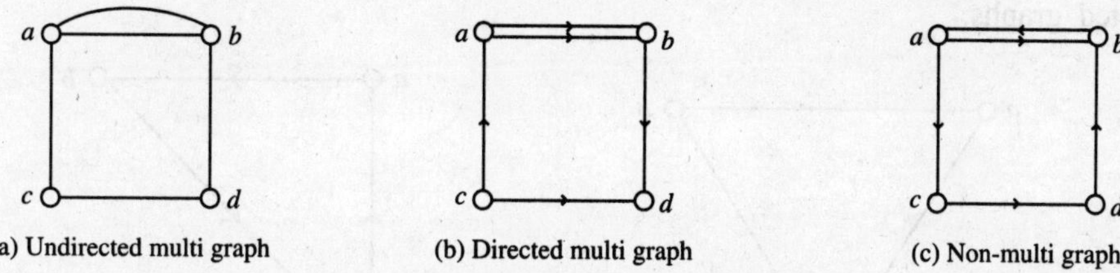

Figure 13.5 Multi and non-multi graphs.

13.8 Simple Graph

A graph is said to be simple if it has no loops and no parallel edges. However, in the case of directed graph, if more than one edges (non parallel) exist between two vertices, it is called **pseudo simple digraph**. All graphs given in Figure 13.6 are simple, while the graph given in Figure 13.7 is pseudo simple digraph.

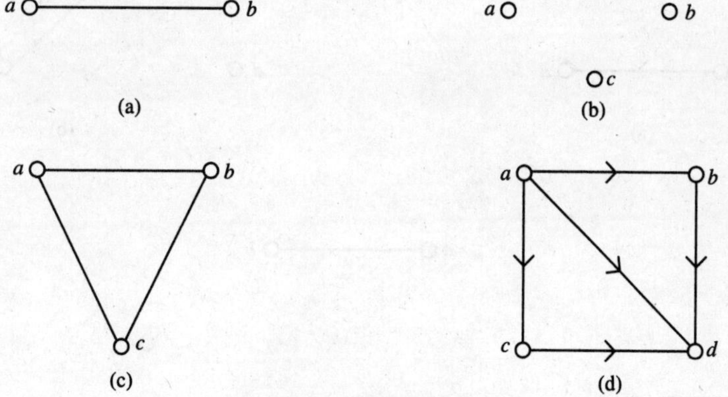

Figure 13.6 Simple graph.

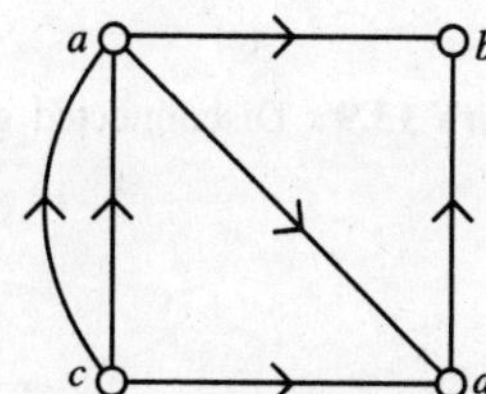

Figure 13.7 Pseudo simple digraph.

13.9 Connected and Disconnected Graph

A graph $G = (V, E)$ is said to be connected if every pair of vertices is connected, i.e., there is a path between each pair of vertices. All the graphs given in Figure 13.8 are connected, while the graphs in Figure 13.9 are disconnected. A directed graph has two types of connectivity: **strong connectivity** and **weak connectivity**. A directed graph is **strongly connected** if there is a directed path between every two of its vertices. Further, a directed graph is **weakly connected** if its corresponding undirected graph is connected. Figure 13.10 shows strongly and weakly connected graphs.

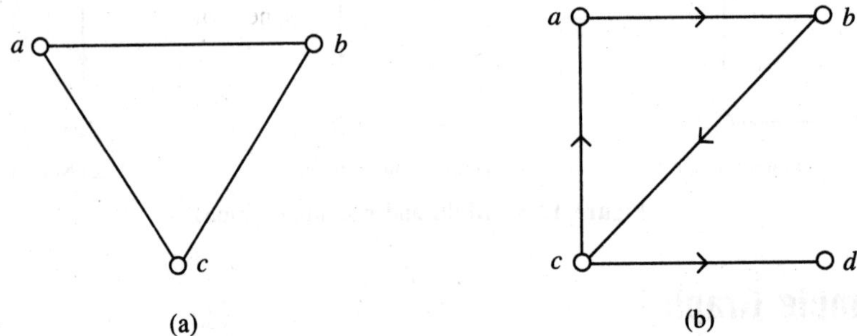

Figure 13.8 Connected graph.

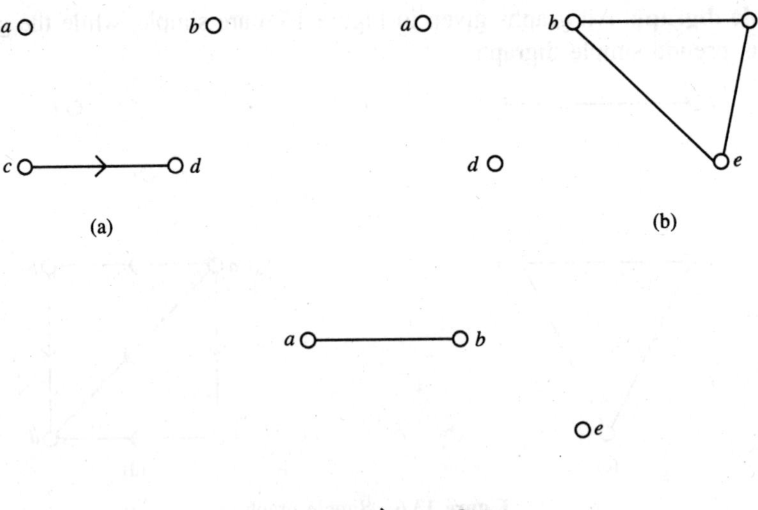

Figure 13.9 Disconnected graph.

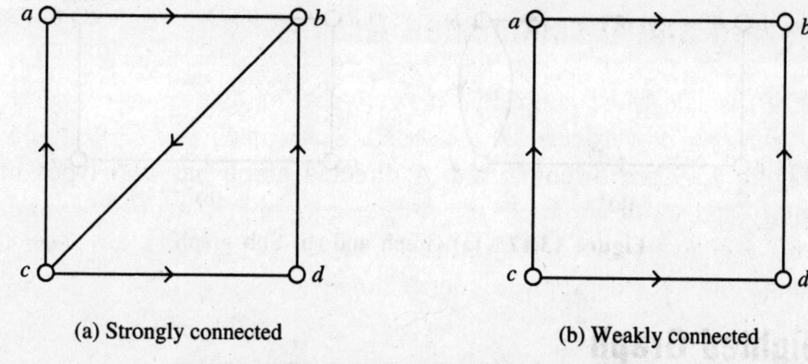

(a) Strongly connected (b) Weakly connected

Figure 13.10 Weak and strong connectivity.

13.10 Complete Graph

A graph, in which every pair of distinct vertices is adjacent, is called a complete graph. The graphs given in Figure 13.11 are all complete graphs.

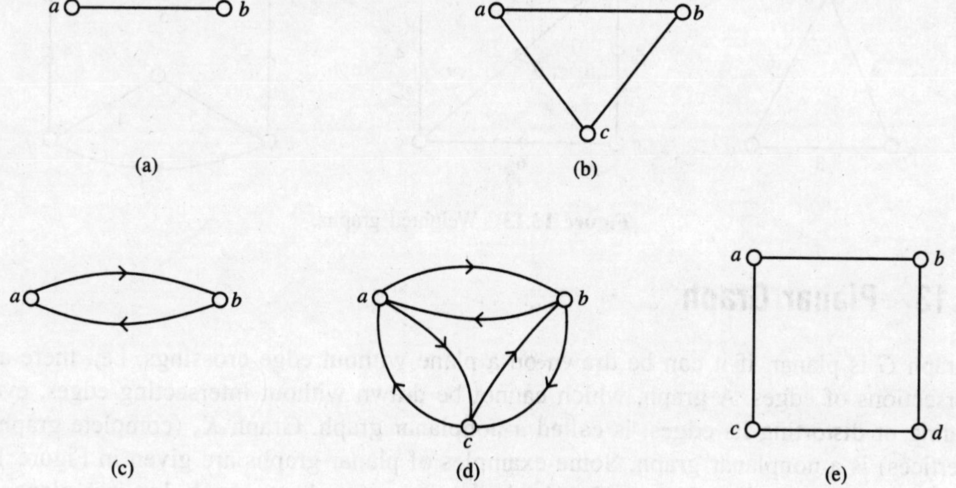

(a) (b) (c) (d) (e)

Figure 13.11 Complete graphs.

13.11 Sub Graph

Let $G = (V, E)$ be a graph, and let $V_1 \subseteq V$ and $E_1 \subseteq E$, then $G_1 = (V_1, E_1)$ is called a sub graph of G. Obviously, a sub graph of a graph can be obtained by deleting some edges, or vertices (or both) in the graph. The graph given in Figure 13.12(b) is a sub graph of the graph given in Figure 13.12(a).

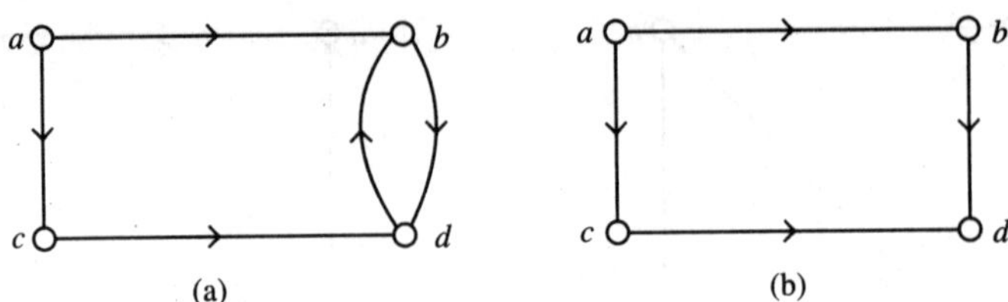

Figure 13.12 (a) Graph and (b) Sub graph.

13.12 Weighted Graph

A graph is called weighted, if a weight is assigned to each edge of the graph. This weight may represent some physical quantity like distance, number of parallel edges etc., between two vertices. In general, it may be anything like a parameter, or even a tag for some entity. Some examples of weighted graphs are given in Figure 13.13.

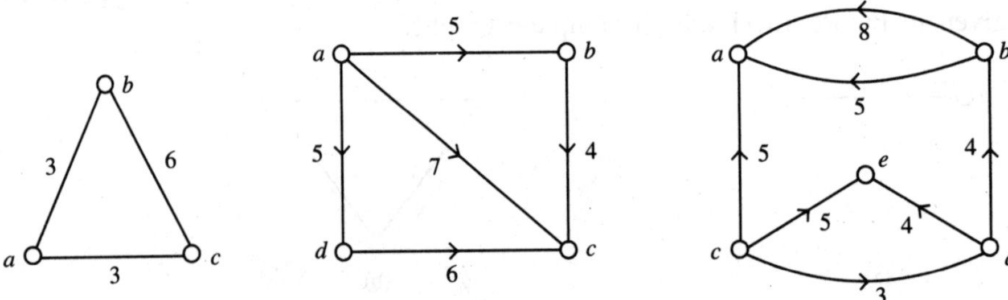

Figure 13.13 Weighted graphs.

13.13 Planar Graph

A graph G is planar, if it can be drawn on a plane without edge crossings, i.e., there are no intersections of edges. A graph, which cannot be drawn without intersecting edges, even by bending or distorting its edges, is called a nonplanar graph. Graph K_5 (complete graph with 5 vertices) is a nonplanar graph. Some examples of planar graphs are given in Figure 13.14. Observe that the graph in Figure 13.14(a) looks like a nonplanar graph, but it is planar since it can be redrawn as given in Figure 13.14(b). A particular planar representation of a planar

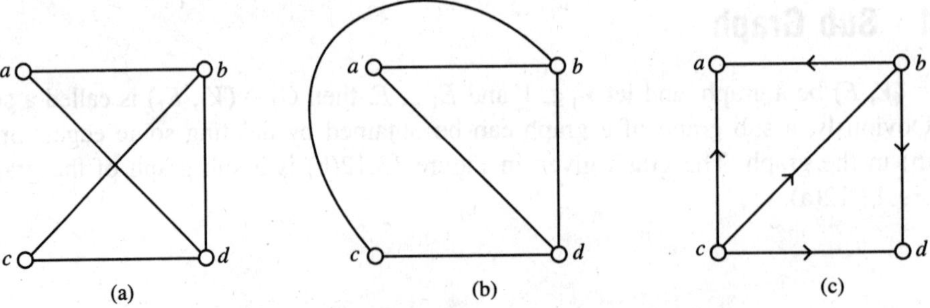

Figure 13.14 Planar graphs.

graph is called a **map**. A map is connected if the graph is connected. A map of a graph is divided into the closed sub graphs called **regions.** The **degree of a region** is the number of edges, which border the region. If a graph contains only one vertex then the number of regions is considered as one, which is in fact the total outer region. Examples of nonplanar graphs are given in Figure 13.15.

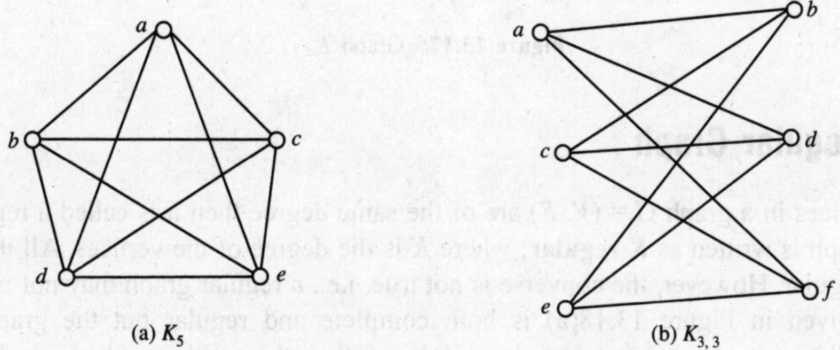

Figure 13.15 Non-planar graphs.

13.14 Bipartite Graph

If the vertex set V of a graph $G = (V, E)$ can be partitioned into two disjoint subsets, say V_1 and V_2, in such a way that each edge in E connects a vertex of V_1 to a vertex of V_2, then the graph G is called a bipartite graph. The graph given in Figure 13.15(b) is a bipartite graph. Some more examples of bipartite graphs are given in Figure 13.16.

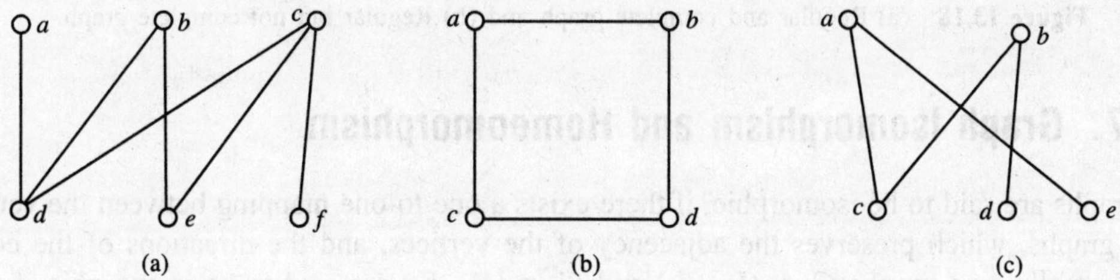

Figure 13.16 Bipartite graphs.

13.15 Complete Bipartite Graph

A bipartite graph is complete if every vertex in one of the bipartition subsets is connected to every vertex in the other bipartition subset. A bipartite graph is denoted by $K_{m,n}$, where m and n are the number of vertices in two respective bipartition subsets V_1 and V_2. Generally, m is greater than n. A complete bipartite graph $K_{4,3}$ is shown in Figure 13.17. The graph given in Figure 13.15(b) is also a complete bipartite graph.

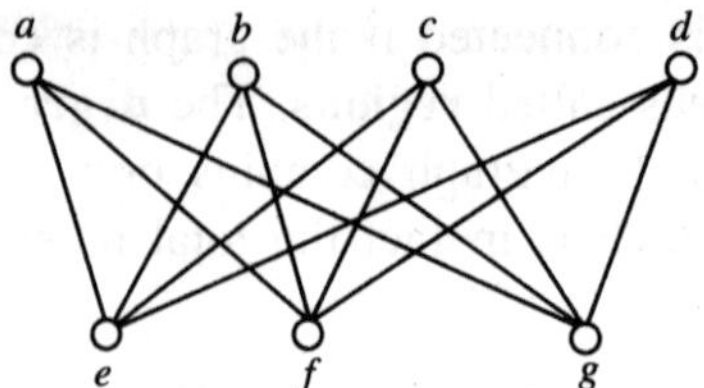

Figure 13.17 Graph $K_{4,3}$

13.16 Regular Graph

If all the vertices in a graph $G = (V, E)$ are of the same degree then it is called a regular graph. A regular graph is written as ***K*-regular**; where K is the degree of the vertices. All the complete graphs are regular. However, the converse is not true, i.e., a regular graph may not be complete. The graph given in Figure 13.18(a) is both complete and regular but the graph given in Figure 13.18(b) is regular but not complete. If $K = 0$ then the graph is called a **null graph**. In a null graph, there are no edges.

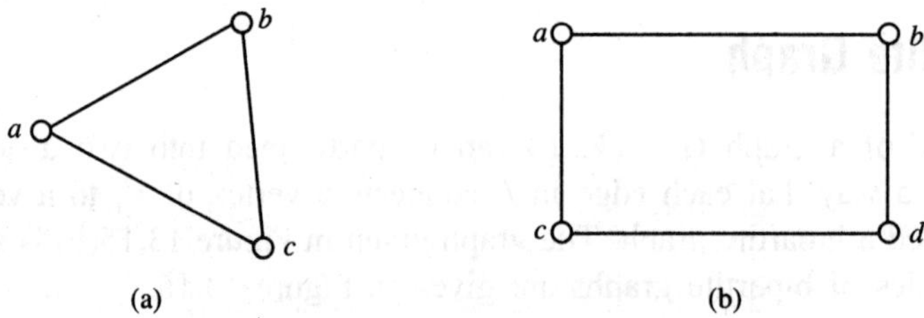

Figure 13.18 (a) Regular and complete graph and (b) Regular but not complete graph.

13.17 Graph Isomorphism and Homeomorphism

Two graphs are said to be isomorphic, if there exists a one-to-one mapping between the vertices of the graphs, which preserves the adjacency of the vertices, and the directions of the edges.

Formally, two graphs $G_1 = (V_1, E_1)$ and $G_2 = (V_2, E_2)$ are said to be isomorphic if there exists a one-to-one mapping $f: V_1 \rightarrow V_2$ between the vertices of the graphs, such that the adjacency of the vertices and the directions of the edges are preserved.

The undirected graphs given in Figure 13.19 are isomorphic. The directed graphs given in Figure 13.20 are also isomorphic. Further, two graphs G_1 and G_2 are said to be **homeomorphic** if there is an isomorphism from a subdivision of G_1 to a subdivision of G_2. In other words, if by subdividing the edges of the graphs G_1 and G_2, the two graphs become isomorphic, then these graphs are called homeomorphic graphs. The subdivision of a graph is obtained by cutting an edge of the graph at some point and subsequently adding a vertex there. Also, two graphs are homeomorphic if they can be obtained from the same graph by dividing edges using additional vertices. Graphs G_1 and G_2 given in Figure 13.21 are homeomorphic, because they can be obtained from the same graph G_3 by dividing edges (a, b) and (c, d), respectively. Also, the subdivision of an edge (c, d) in G_1 and subdivision of an edge (a, b) in G_2, results in an isomorphism of G_1 and G_2.

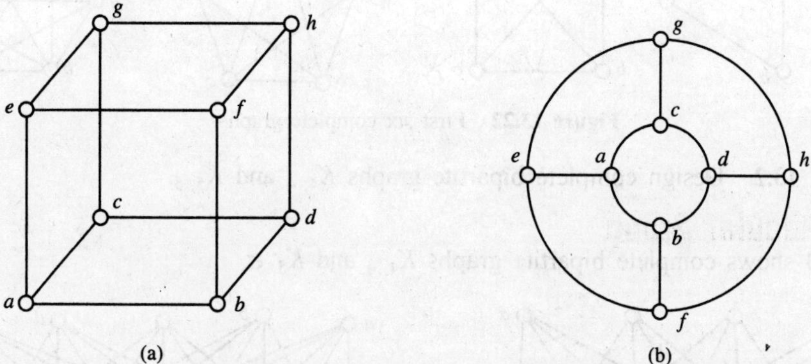

Figure 13.19 Isomorphic undirected graphs.

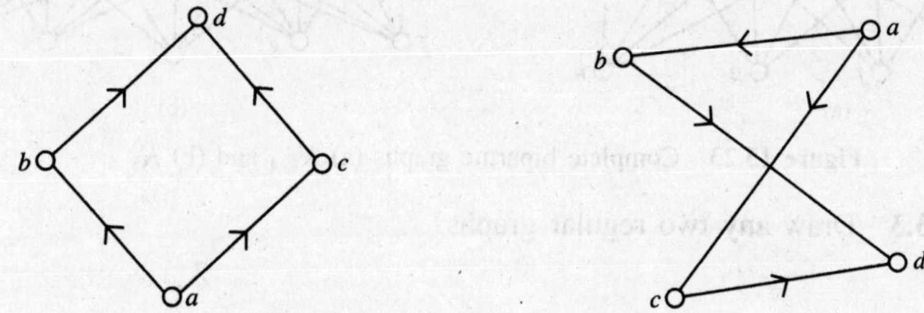

Figure 13.20 Isomorphic directed graphs.

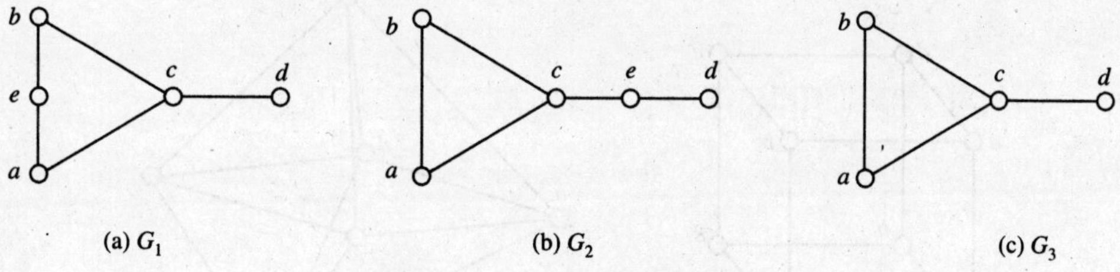

Figure 13.21 Homeomorphism.

13.18 Solved Problems

PROBLEM 13.1 Draw the first six complete graphs with number of vertices ≤ 6.

Solution

Figure 13.22 shows the first six complete graphs.

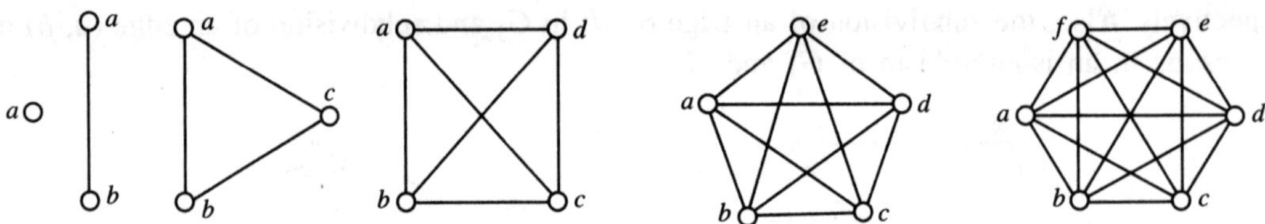

Figure 13.22 First six complete graphs.

PROBLEM 13.2 Design complete bipartite graphs $K_{4,\,4}$ and $K_{5,\,5}$.

Solution

Figure 13.23 shows complete bipartite graphs $K_{4,\,4}$ and $K_{5,\,5}$.

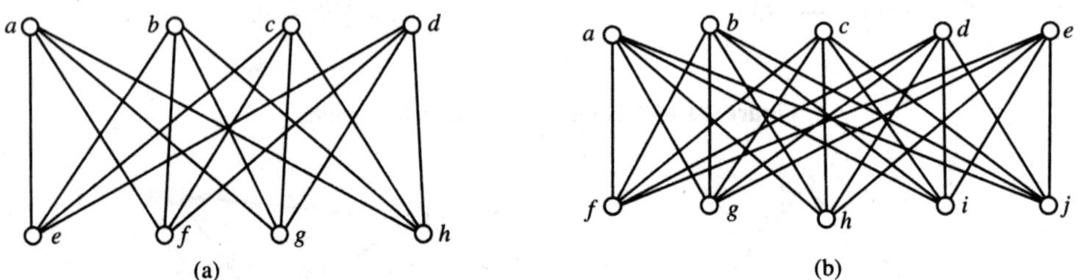

Figure 13.23 Complete bipartite graphs (a) $K_{4,\,4}$ and (b) $K_{5,\,5}$.

PROBLEM 13.3 Draw any two regular graphs.

Solution

Two regular graphs are drawn in Figure 13.24.

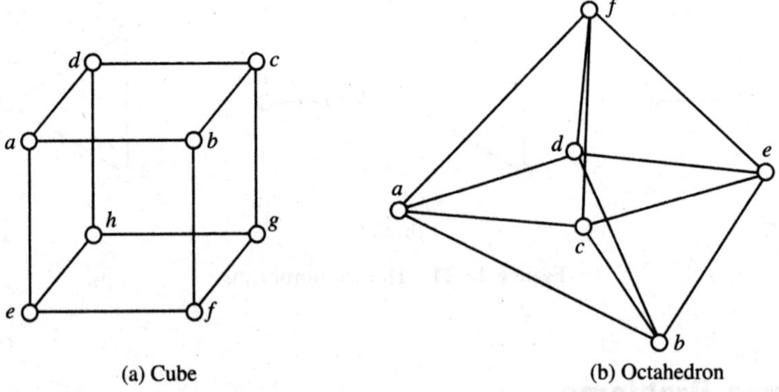

Figure 13.24 Regular graphs.

PROBLEM 13.4 Draw Petersen graph.

Solution

Graph given in Figure 13.25 is the Petersen graph.

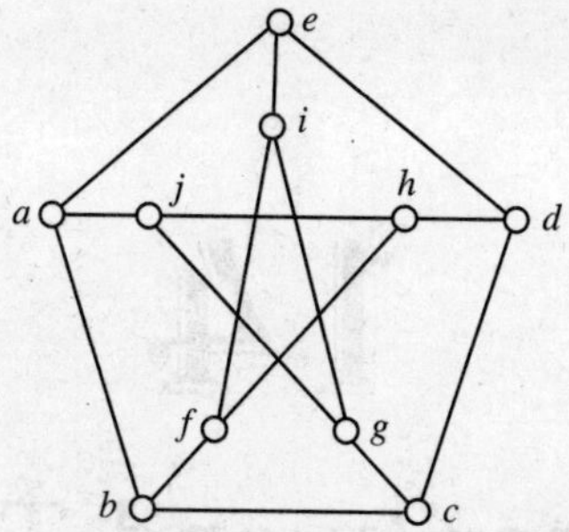

Figure 13.25 Petersen graph.

PROBLEM 13.5 Draw any two isomorphic graphs.

Solution

Graphs given in Figure 13.26 are isomorphic graphs.

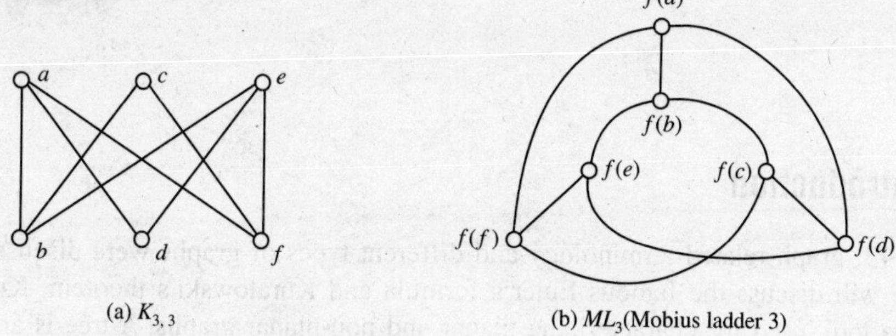

(a) $K_{3,3}$ (b) ML_3(Mobius ladder 3)

Figure 13.26 Isomorphic graphs.

Exercises

13.1 Define the following terms:

(i) Directed graph
(ii) Undirected graph
(iii) Degree
(iv) Path
(v) K-regular graph
(vi) Isomorphic undirected graphs
(vii) Graph homeomorphism
(viii) Complete bipartite graphs
(ix) Regular graphs.

13.2 Draw $K_{6,4}$ and $K_{6,6}$.

14

Planarity and Trees

14.1 Introduction

In Chapter 13, graph related terminology and different types of graphs were discussed. In this chapter, we will discuss the famous Euler's formula and Kuratowski's theorem. Kuratowski's theorem is a tool, used for characterizing planar and non-planar graphs. A tree is an important part of graph theory, and has numerous applications in science and engineering. For instance, binary tree is used as a data structure in computer science. Expression tree is used for expression evaluation. Therefore, these topics will also form a part of our discussion.

14.2 Euler's Formula

Euler gave a formula that relates number of edges, vertices, and regions of a connected graph. Let $G = (V, E)$ be a connected graph. Then,

$$v - e + r = 2 \tag{14.1}$$

where, v, e, and r are the number of vertices, edges and regions respectively, of the graph. For region, refer Section 13.13.

This formula can be proved by induction on the number of vertices and the number of edges.

Proof: Consider an arbitrary connected graph G with number of vertices v, number of edges e, and number of regions r.

(i) **Basis:** We will consider two possible cases: one for vertex and the other for edge. These are as follows:

(a) Let the graph G contains only one vertex.

Then $$v = 1,\ e = 0 \text{ and } r = 1.$$

Therefore, formula given in Eq. (14.1) is satisfied.

(b) Let graph G contains only one edge.

Then $$v = 2,\ e = 1 \text{ and } r = 1.$$

Again, Eq. (14.1) is satisfied.

(ii) **Inductive step:** Assume that the formula given in Eq. (14.1) is true for v, the number of vertices, e, the number of edges and r, the number of regions. Then, perform the induction on the two parameters: v and e, which are initialized in the two base cases. These are described below in two separate cases: one for v and another for e.

(a) **Addition of a vertex:** When a vertex is added into an existing graph, one edge has also to be added, otherwise the graph will become disconnected. By such addition, r will not change. Therefore, formula given in Eq. (14.1) remains valid.

That is, $$(v + 1) - (e + 1) + r = 2.$$

$$\Rightarrow \quad v - e + r = 2$$

(b) **Addition of an edge:** When an edge is added into the existing graph, either by joining two vertices or by creating a self-loop, it creates one additional region. Again Eq. (14.1) remains valid, because

$$v - (e + 1) + (r + 1) = 2.$$

$$\Rightarrow \quad v - e + r = 2$$

Therefore, it can be concluded from the inductive step that the formula is always valid.

The inductive step describes the construction of the graph. Euler's formula is used in finding the logical complexity of a programming language.

14.3 Kuratowski's Theorem

In 1930, K Kuratowski, a Polish mathematician gave a criteria for characterizing planar and non-planar graphs. It is known as Kuratowski's theorem. It says, "A graph is non-planar if it contains a subgraph homeomorphic to $K_{3,\,3}$ or K_5". The proof of this theorem uses the following two theorems.

(i) Graph $K_{3,\,3}$ is non-planar.

(ii) Graph K_5 is non-planar.

Geometrically, it can be seen that there is no possible way to find planar representation of $K_{3,\,3}$ and K_5, hence it can be concluded that these graphs are non-planar. However, the logical proofs of theorems 14.3(i) and 14.3(ii) are lengthy, therefore, these are omitted here.

14.4 Applications of Kuratowski's Theorem

Kuratowski's theorem can be used in determining the planarity of a graph.

EXAMPLE 14.1 Consider a graph given in Figure 14.1.

In this graph, it is easy to find a sub graph of graph G which is homeomorphic to K_5. If we delete the edges (a, b), (a, c), (a, d), (b, g), (f, g) and (e, g), the remaining graph is K_5. Therefore, according to Kuratowski's theorem, it is non-planar.

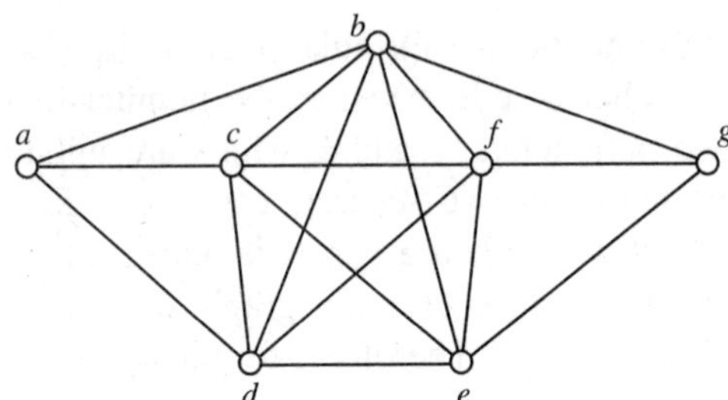

Figure 14.1 Non-planar graph.

14.5 Tree and Related Terminology

In this section, we will discuss a special type of graph called tree.

14.5.1 Tree

A **tree** is a connected graph T, such that it has no closed path or cycle.

A tree may be both directed and undirected. A tree with all its edges directed, is called directed tree. Similarly, if all the edges of the tree are undirected then it is termed undirected. A collection of trees is called a **forest**. There are different types of trees depending on the applications. Some of them are described here.

14.5.2 Rooted Tree

A tree T with one vertex designated as root is called rooted tree. All the paths of the rooted tree are directed away from the root, and thus, the rooted tree is viewed as a directed tree. In the pictorial representation of the rooted tree, the arrows on the edges are generally omitted. Each vertex of a rooted tree has only one incoming edge, but may have any number of outgoing edges. The **length** of a path in the tree is the number of edges in the path from root r to a vertex v. It is called the level or depth of v. A pictorial view of a rooted tree is given in Figure 14.2.

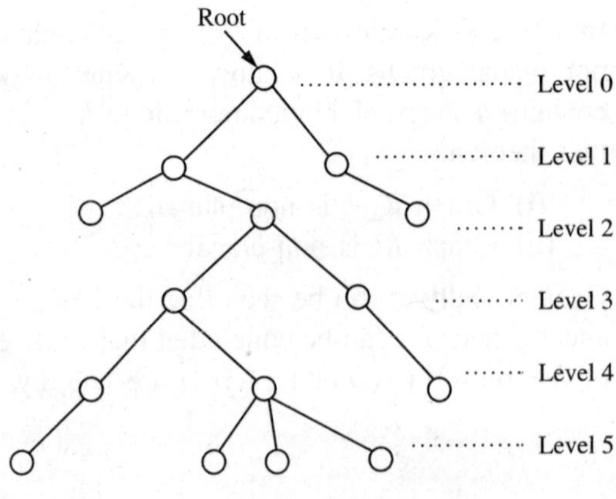

Figure 14.2 Rooted tree.

An important application of the rooted tree is to create a data structure, which is used to store data in a computer system. This data structure is called **binary tree**. In a binary tree, each vertex has at most two outgoing edges. All the vertices, which have no outgoing edges are called **leaves** of the tree. A path from a vertex to a leaf is called a **branch** of the tree. Figure 14.3(a) is a graphical representation of a binary tree in which edges are shown as undirected. An alternate form, in which edges are shown with arrows, is given in Figure 14.3(b). Both these forms represent the same binary tree without any significant difference. It is just a matter of choice among the authors to adapt a particular form.

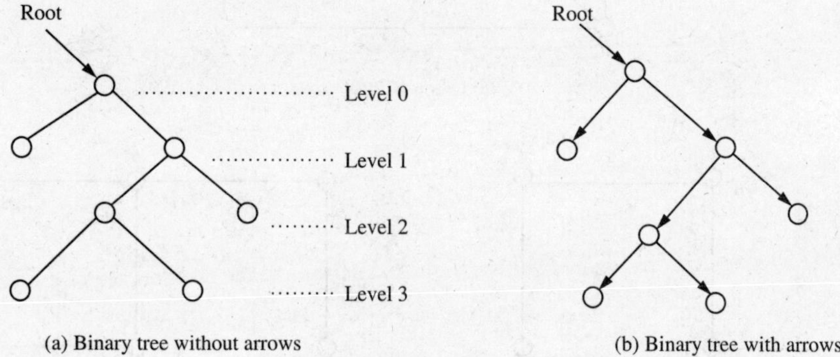

(a) Binary tree without arrows

(b) Binary tree with arrows

Figure 14.3 Binary tree.

14.5.3 Spanning Tree

A connected sub graph of a connected graph G is called a spanning tree T if it contains all the vertices of the graph, and has no loops or closed paths. A connected graph G and its two spanning trees are shown in Figure 14.4.

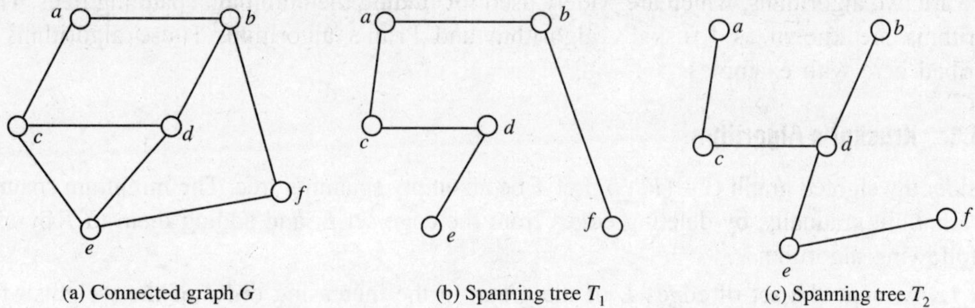

(a) Connected graph G

(b) Spanning tree T_1

(c) Spanning tree T_2

Figure 14.4 Graph and its two spanning trees.

14.5.4 Minimum Spanning Trees

The spanning trees of a weighted graph are called weighted spanning trees. The sum of the weights of all the edges of a spanning tree is called the weight of the spanning tree. A spanning tree with least weight among all the spanning trees of a weighted graph is called a **minimum spanning tree**. A minimum spanning tree is not unique, but the minimum weight is unique. If two or more edges of a connected graph G have the same weight, then G may have more than one minimum spanning tree. For instance, there are two minimum spanning trees T_1 and T_2 given in Figure 14.5(b and c), respectively of a graph G given in Figure 14.5(a).

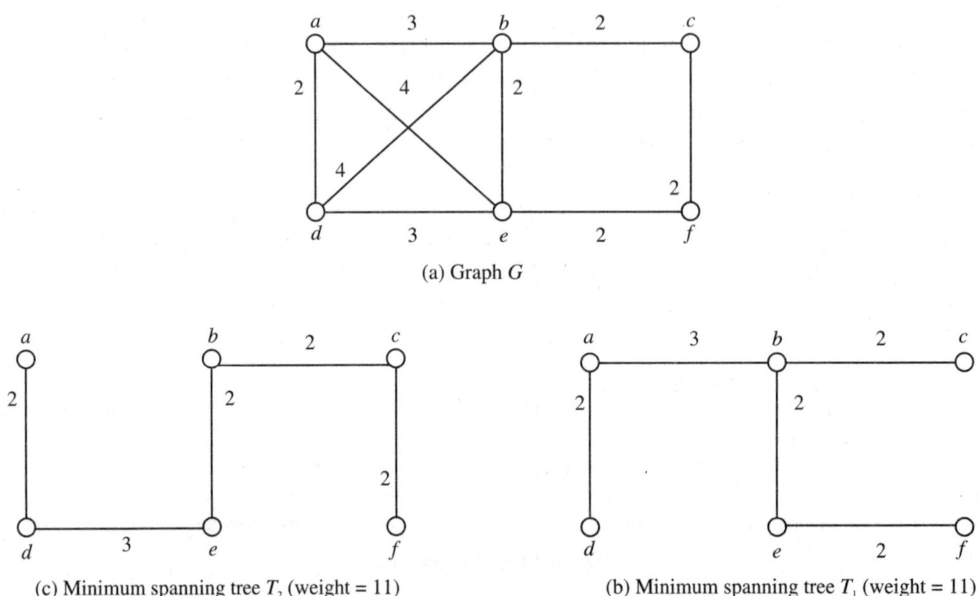

Figure 14.5 Graph with minimum spanning trees.

14.6 Finding Minimum Spanning Tree

There are two algorithms, which are widely used for finding the minimum spanning trees. These algorithms are known as Kruskal's algorithm and Prim's algorithm. These algorithms are described here with examples.

14.6.1 Kruskal's Algorithm

Consider a weighted graph $G = (V, E)$. Let T be an empty spanning tree. The minimum spanning tree T is built gradually, by deleting edges from the edge set E, and adding them to T by using the following algorithm.

Step 1. Arrange the set of edges E of the graph in the increasing order of the weights of the edges.

Step 2. Delete the first edge in the set E and add it to the tree T, if and only if no loop (closed path) is formed in the tree.

Step 3. Repeat from step 2, if E is not empty, otherwise stop.
T is the required minimum spanning tree.

EXAMPLE 14.2 Consider the weighted graph G given in Figure 14.6(a). Let us find a minimum spanning tree of this graph.

Given $G = (V, E)$,

where $V = \{a, b, c, d, e, f\}$,

and $E = \{(a, b), (a, c), (b, c), (b, d), (e, f), (e, d), (c, e), (d, f), (a, f)\}$.

Let T be an empty tree.

1. Arrange the edges in the set E in increasing order of the weights on the edges.

   $E = \{(b, d, 1), (e, f, 1), (a, b, 2), (e, d, 2), (a, c, 3), (c, e, 3), (b, c, 4), (d, f, 4), (a, f, 7)\}$.

   The third component attached with each of the element in E is the weight of the edge.

2. Delete the first edge from E and add it to T. Deletion is shown by striking through an edge in the set.

   E = { ~~$(b, d, 1)$~~, $(e, f, 1), (a, b, 2), (e, d, 2), (a, c, 3), (c, e, 3), (b, c, 4), (d, f, 4), (a, f, 7)\}$,

   $T = \{(b, d, 1)\}$.

   Since E is not empty, therefore, repeat step 2 of the algorithm.

   E = {~~$(b, d, 1)$~~, ~~$(e, f, 1)$~~, $(a, b, 2), (e, d, 2), (a, c, 3), (c, e, 3), (b, c, 4), (d, f, 4), (a, f, 7)\}$

   $T = \{(b, d, 1), (e, f, 1)\}$.

   Again, E is not empty, therefore, repeat step 2.

   E = {~~$(b, d, 1)$~~, ~~$(e, f, 1)$~~, ~~$(a, b, 2)$~~, $(e, d, 2), (a, c, 3), (c, e, 3), (b, c, 4), (d, f, 4), (a, f, 7)\}$,

   $T = \{(b, d, 1), (e, f, 1), (a, b, 2)\}$.

   Next step,

   E = {~~$(b, d, 1)$~~, ~~$(e, f, 1)$~~, ~~$(a, b, 2)$~~, ~~$(e, d, 2)$~~, $(a, c, 3), (c, e, 3), (b, c, 4), (d, f, 4), (a, f, 7)\}$,

   $T = \{(b, d, 1), (e, f, 1), (a, b, 2), (e, d, 2)\}$

   Next step,

   E = {~~$(b, d, 1)$~~, ~~$(e, f, 1)$~~, ~~$(a, b, 2)$~~, ~~$(e, d, 2)$~~, ~~$(a, c, 3)$~~, $(c, e, 3), (b, c, 4), (d, f, 4), (a, f, 7)\}$

   $T = \{(b, d, 1), (e, f, 1), (a, b, 2), (e, d, 2), (a, c, 3)\}$

   Next step,

   E = {~~$(b, d, 1)$~~, ~~$(e, f, 1)$~~, ~~$(a, b, 2)$~~, ~~$(e, d, 2)$~~, ~~$(a, c, 3)$~~, ~~$(c, e, 3)$~~, $(b, c, 4), (d, f, 4), (a, f, 7)\}$

   $T = \{(b, d, 1), (e, f, 1), (a, b, 2), (e, d, 2), (a, c, 3)\}$

   Note that in this step, though the edge $(c, e, 3)$ is deleted from the set E, but it is not added to the tree T because it will form a closed path a-b-d-e-c-a.

   The process is repeated and stopped when E is empty, i.e., all the edges in the set E are crossed. Eventually, we get

   E = {~~$(b, d, 1)$~~, ~~$(e, f, 1)$~~, ~~$(a, b, 2)$~~, ~~$(e, d, 2)$~~, ~~$(a, c, 3)$~~, ~~$(c, e, 3)$~~, ~~$(b, c, 4)$~~, ~~$(d, f, 4)$~~, ~~$(a, f, 7)$~~}.

   That is, $E = \{\ \}$ and

   $T = \{(b, d, 1), (e, f, 1), (a, b, 2), (e, d, 2), (a, c, 3)\}$.

T is the required minimum spanning tree. Its weight is $W = 9$. The solution steps are also shown pictorially from Figures 14.6(b to f). Observe that this graph has another minimum spanning tree, which is shown in Figure 14.6(g).

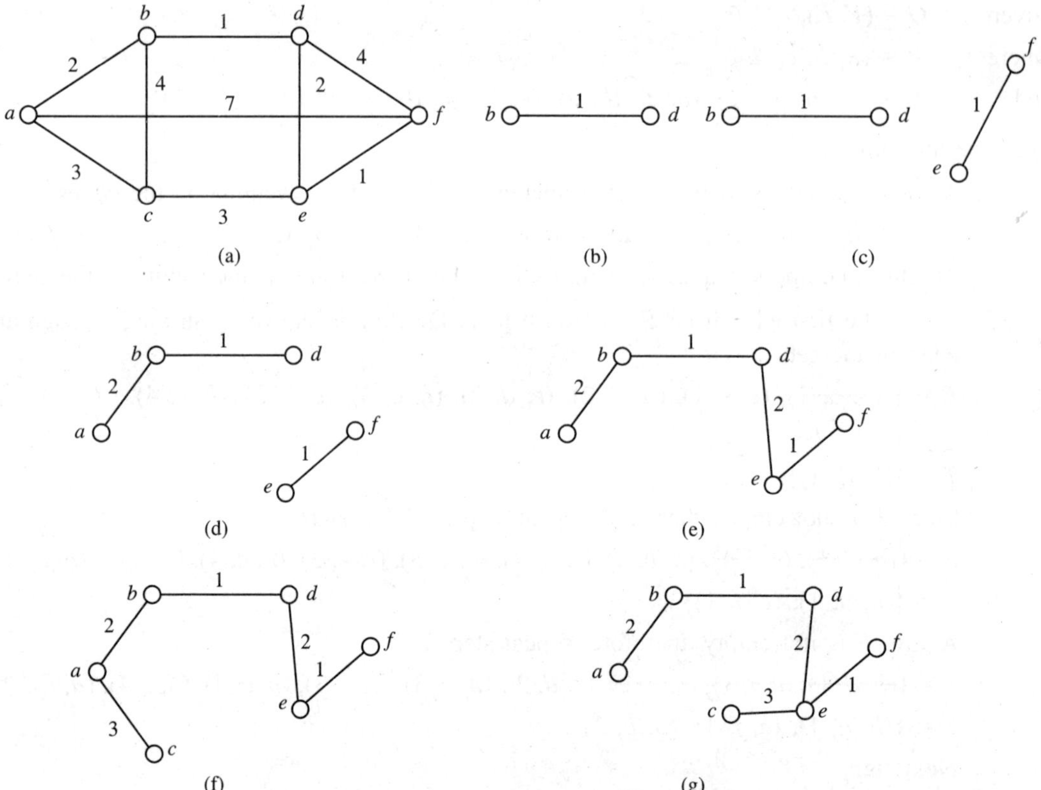

Figure 14.6 Graph G with minimum spanning tree.

14.6.2 Prim's Algorithm

In this algorithm, one can start from any vertex and build the tree by adding edges to it. The systematic procedure is given below.

Consider a weighted graph

$$G = (V, E)$$

Let T be an empty spanning tree and V', an empty vertex set. Start from any vertex. Build the minimum spanning tree by deleting edges from the edge set E gradually, one by one, and adding them to the tree T. Following are the steps of the procedure.

Step 1. Select an arbitrary vertex $v \in V$ and add it to V'.

Step 2. Select an edge (x, y) from E with minimum weight, such that one of its end vertices is in V' and the other in $V - V'$. Add this edge to the tree T only if it does not cause to form a closed path in T. If more than one edge satisfies this condition then select any one of these edges, arbitrarily. Also, add vertices u and v to V', if already not there.

Step 3. If $V' \neq V$, repeat from step 2, otherwise stop.

T is the required minimum spanning tree.

EXAMPLE 14.3 Consider the weighted graph $G = (V, E)$ given in Figure 14.7(a). Let us find the minimum spanning tree using Prim's algorithm.

In this graph:

$V = \{a, b, c, d, e, f, g\}$ and

$E = \{(a, b, 2), (a, c, 3), (b, c, 4), (b, g, 3), (b, d, 1), (d, g, 5), (e, f, 1), (e, d, 2), (c, e, 3), (d, f, 4), (g, f, 2), (a, f, 7)\}$.

where, the third component attached with each element in E is the weight of the edge.

Initially, take an empty tree T and an empty set V', i.e., $T = \{\ \}$ and $V' = \{\ \}$.

Now, apply Prim's algorithm as follows:

Step 1. Start from any vertex. Let it be g.

Therefore, $V' = \{g\}$.

Step 2. Select an edge from E with minimum weight, such that its one end vertex is in V' and the other in $V - V'$. There are three edges $(g, b, 3)$, $(g, d, 5)$ and $(g, f, 2)$, which have one end in V' and the other in $V - V'$. Select edge $(g, f, 2)$, since it has minimum weight. Add it to the tree T, because no closed path is formed by this addition.

That is, $T = \{(g, f, 2)\}$.

Also, add vertex f to V' since it is not there.

Thus, $V' = \{g, f\}$.

Step 3. Select edge $(e, f, 1)$.

Then $T = \{(g, f, 2), (e, f, 1)\}$

and $V' = \{g, f, e\}$.

Step 4. Select edge $(b, g, 3)$.

Then $T = \{(g, f, 2), (e, f, 1), (b, g, 3)\}$

and $V' = \{g, f, e, b\}$.

Step 5. Select edge $(b, d, 1)$.

Then $T = \{(g, f, 2), (e, f, 1), (b, g, 3), (b, d, 1)\}$

and $V' = \{g, f, e, b, d\}$.

Step 6. Select edge $(a, b, 2)$.

Then $T = \{(g, f, 2), (e, f, 1), (b, g, 3), (b, d, 1), (a, b, 2)\}$

and $V' = \{g, f, e, b, d, a\}$.

Step 7. Select edge $(a, c, 3)$.

Then $T = \{(g, f, 2), (e, f, 1), (b, g, 3), (b, d, 1), (a, b, 2), (a, c, 3)\}$

and $V' = \{g, f, e, b, d, a, c\}$.

Now, $V' = V$, therefore, stop.

T is the required minimum spanning tree. Its weight is

$$W = 12.$$

The solution steps are also shown pictorially from Figures 14.7(b to g).

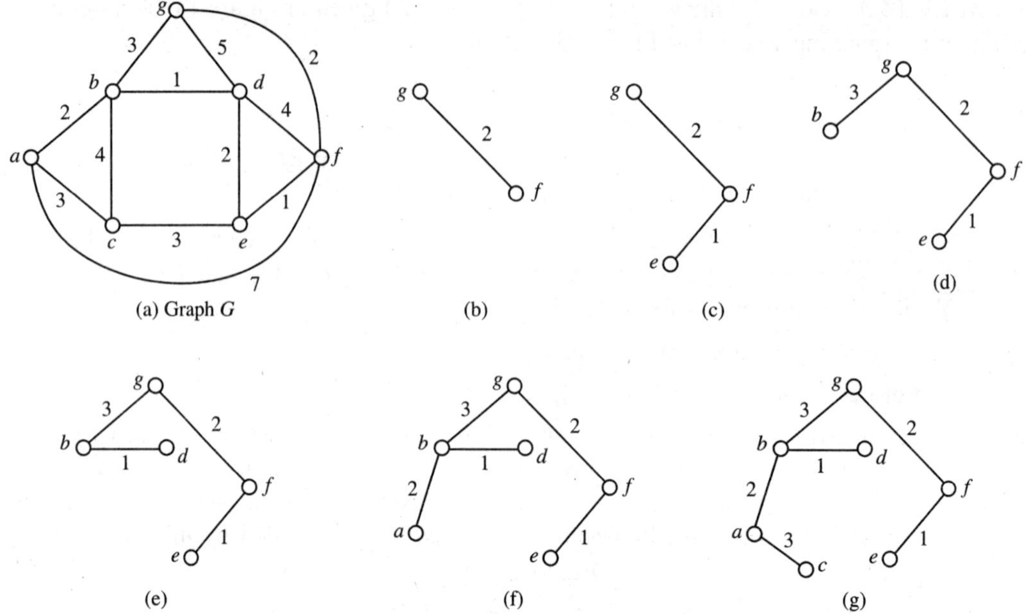

Figure 14.7 Solution steps for graph G.

14.7 Tree Traversal

A binary tree is an important data structure used in computer science. This is used to store data. Retrieval of data is very fast in a binary tree in comparison to other data structures. Tree traversal means reading the data stored in the nodes of a binary tree.

A binary tree can only be accessed through its root, i.e., the starting vertex of the traversal is the root of the tree. The reason is that the root stores the address of the tree. Further, the address of each vertex in the tree is stored in its preceding vertex. Note that the rooted tree is a linked representation. A binary tree with data stored in the nodes is given in Figure 14.8.

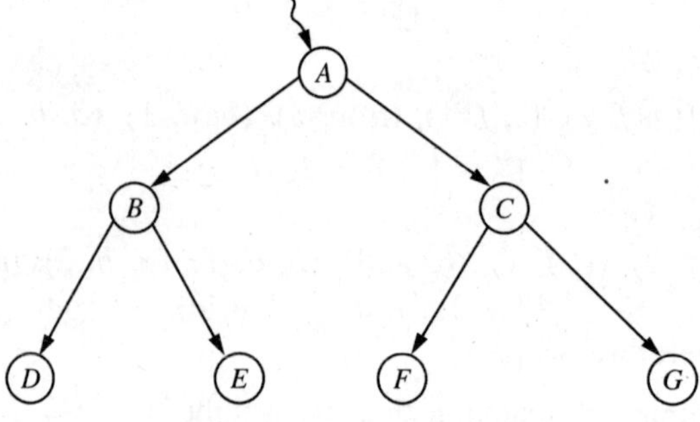

Figure 14.8 Binary tree.

In a binary tree, each node (or vertex) has at most one incoming edge and two outgoing edges. Tree on the left of a node is called its **left sub tree,** and on its right is called **right**

sub tree. Traversal is only in one direction from top to bottom, though there are some variants of the tree, which allow movement in either direction. These variants are omitted here, however, the interested readers may refer any text on data structure for more comprehension. The vertices of the tree, which have no outgoing edges, are called **leaves** of the tree. All the edges in a tree are directed, and the direction of a path is from the root to a leaf. Generally, data stored in a binary tree is such that every node contains a value larger than the values in its left sub tree, and less than the values in its right sub tree. It makes search faster.

There are three popular algorithms for a tree traversal. These algorithms are named as preorder, inorder and postorder tree traversals, and are written recursively in the following way.

14.7.1 Pre-order Tree Traversal

Consider a binary tree T with root R.

1. Process the root R.
2. Traverse the left sub tree of R in pre-order.
3. Traverse the right sub tree of R in pre-order.

Process means performing some operations like reading the data. In short, it is read as **root-left-right.**

14.7.2 In-order tree Traversal

1. Traverse the left sub tree of R in in-order.
2. Process the root R.
3. Traverse the right sub tree of R in in-order.

It is read as **left-root-right.**

14.7.3 Post-order Tree Traversal

1. Traverse the left sub tree of R in post-order.
2. Traverse the right sub tree of R in post-order.
3. Process the root R.

It is read as **left-right-root.**

Notice that the three algorithms, written above, are similar except the changes in the order of execution of the three statements.

EXAMPLE 14.4 Consider binary tree given in Figure 14.9. Let us read the data stored in this tree.

Data can be read in three ways using the three algorithms given above in Section 14.7.1.

1. Pre-order : 25 15 12 18 16 20 50 30 62 56 75
2. In-order : 12 15 16 18 20 25 30 50 56 62 75
3. Post-order : 12 16 20 18 15 30 56 75 62 50 25

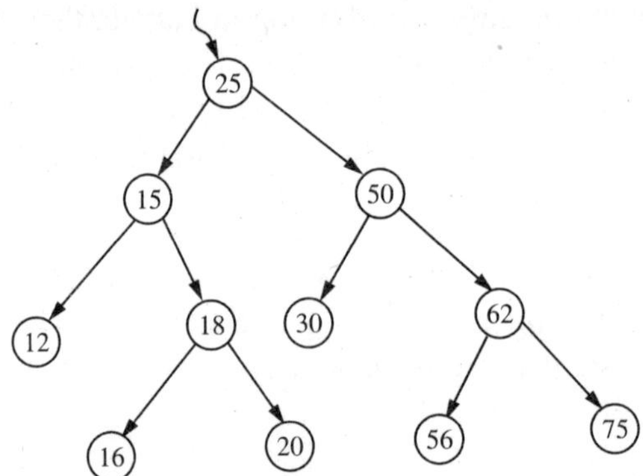

Figure 14.9 Binary tree and tree traversal.

14.8 Expression Trees and Reverse Polish Notation

An arithmetic expression can be computed using an expression tree. **Expression tree** is a binary tree that stores numeric data in the leaves of the tree. Operators of the expression are stored in the interior nodes of the tree. Computation is performed in three ways: prefix, infix and postfix that resemble the three traversal procedures given above. For instance, consider the arithmetic expression

$$x = 2 + 3 * 4 - 7.$$

This expression may be evaluated ambiguously resulting in different values if we do not have operator precedence in mind, i.e., it may be

$$(2 + 3) * 4 - 7 \text{ or } 2 + (3 * 4) - 7 \text{ etc.}$$

Therefore, expression must be parenthesized to remove the ambiguity. A fully parenthesized expression, like $((2 + 3) * 4) - 7$ has no ambiguity and results in one value. The expression tree for the expression $((2 + 3) * 4) - 7$ is given below in Figure 14.10 along with its three computational forms prefix, infix and postfix.

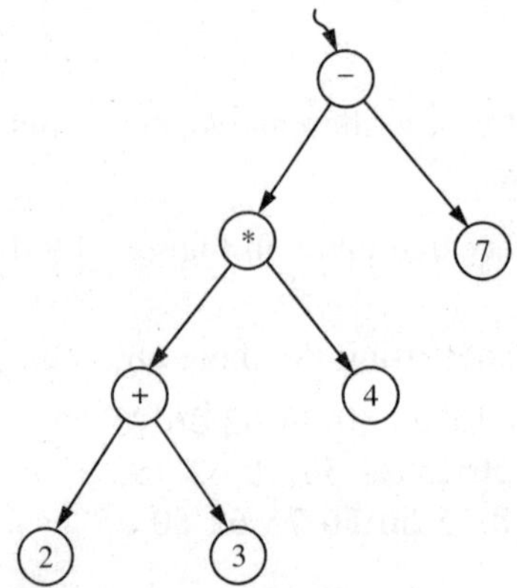

Figure 14.10 Expression tree for $((2 + 3) * 4) - 7$.

1. Prefix notation : – * + 2 3 4 7
2. Infix notation : ((2 + 3) * 4) – 7
3. Postfix notation: 2 3 + 4 * 7 –

If an expression is fully parenthesized, its expression tree will be unique. Construction of an expression tree is easy if the expression is given in infix form. The following discussion will help in understanding the construction steps of an expression tree.

Let the expression be ((2 + 5) * 4) – ((8 + 3) * (4 – 7)). This is fully parenthesized expression and hence is unambiguous. This can be divided into two sub expressions which are on the left and right of the operator '–'.

That is

1. ((2 + 5) * 4) and
2. ((8 + 3) * (4 – 7)).

Operator '–' will form the root of the tree. The first expression will constitute the left sub tree and the second right sub tree. Pictorial illustration is given below.

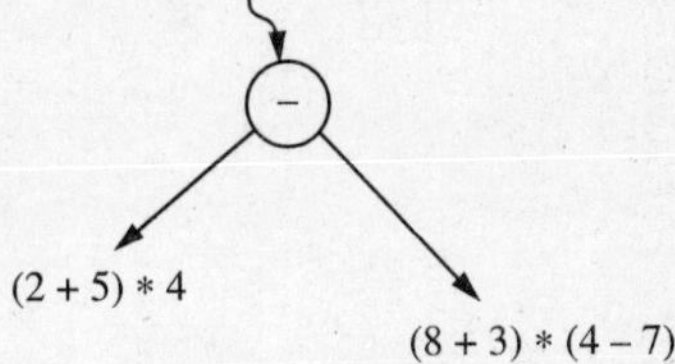

Further, construct the left and right sub trees in a similar way to complete the process.

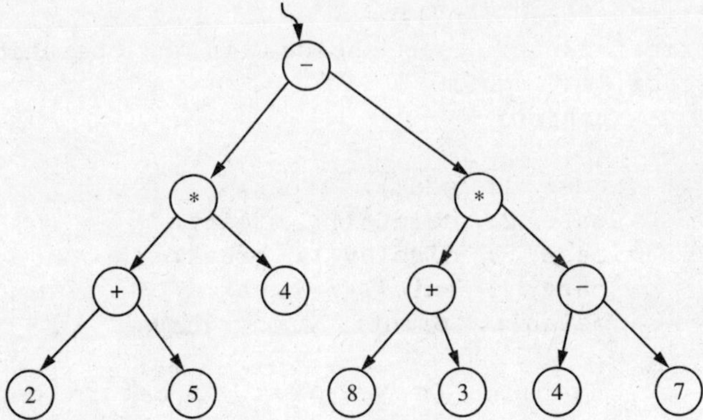

The prefix and postfix notations of this tree will be as under:

1. Prefix notation : – * + 2 5 4 * + 8 3 – 4 7
2. Postfix notation: 2 5 + 4 * 8 3 + 4 7 – * –

Postfix notation is also called **reverse polish notation** (RPN) and it is used in some calculators for arithmetic computation. An RPN can be computed easily by using a data structure, called **stack.** The stack is a data structure in which elements are added and deleted from only one end, called top of the stack. The data added last is deleted first. Hence, it is called last in first out **(LIFO)** structure. It is a user defined data structure and can easily be programmed in any general-purpose programming language. A C++ implementation of stack functions is given below. This code can further be modified to evaluate expressions.

C++ implementation of stack functions

```
/*********************************************************************/
#include<iostream.h>
#define SIZE 100
struct st
    {
    int top;
    char stack[SIZE];
    } s;
int empty(void);
int full(void);
void add(void);
void delet(void);
void display(void);
int main()
    {
       char p;
       int ch;
          do
             {
             clrscr();
             printf("1.ADD\n");
             printf("2.DELETE\n");
             printf("3.DISPLAY\n");
             printf("4.QUIT\n");
             printf("Enter your choice for an operation : ");
             scanf("%d",&ch);
                switch(ch)
                   {
                   case 1: add(); break;
                   case 2: delete(); break;
                   case 3: display(); break;
                   case 4: exit(1); break;
                   default: printf("Wrong number \n"); break;
                   }
                    printf("Do you want to return back[y/n]");
                    p = getch();
             }
                   while(p == 'y' || p == 'Y');
       return(0);
    }
int full(void)
    {
    if (s.top == SIZE)
       return(1);
       else
     return(0);
    }
```

```
int empty(void)
       {
          if (s.top == 0)
             return(1);
          else
             return(0);
       }
void add(void)
          {
          char ch, x;
             do
            {
            if(full() == 1)
             {
                printf("\nStack is full\n"); break;
             }
          else
             {
                s.top = s.top + 1;
                printf("\nEnter an element to be added: ");
                scanf("%d",&x);
                s.stack[s.top] = x;
             }
             printf("\nWould you like to add more elements?[y/n]");
             ch = getch();
            } while(ch == 'y' || ch == 'Y');
          }
void delete(void)
          {
             char ch;
                do
                   {
                      if(empty() == 1)
                         {
                            printf("\nStack is empty\n"); break;
                         }
                      else
                         {
                            printf("\n%d ", s.stack[s.top]);
                            s.top = s.top - 1;
                         }
                            printf("\nWould you like to delete
                                      another element?[y/n]");
                            ch = getch();
                   } while(ch == 'y' ||ch == 'Y');
          }
void display(void)
    {
       int i;
```

```
        clrscr();
           if (empty() == 1)
              printf("\nStack is empty.");
           else
              {
              printf("elements on the stack are:\n");
           for(i = s.top; i > 0;i--)
           printf("%d", s.stack[i]);
              }
     }
/*****************************************************************/
```

The postfix expression 2 5 + 4 * 8 3 + 4 7 – * – is evaluated using stack in the following manner.

Take two stacks s_1 and s_2. Add all the elements including operators in the stack s_1 by performing function Add( ). Delete an element from stack s_1 by using function delete( ), and add it to stack s_2. Repeat this delete-add action until an operator is encountered on top of the stack s_1. Now, delete the operator from s_1, delete two operands from s_2, and perform the specified operation. Store back the result in s_2. Continue this process until the stack s_1 becomes empty. Execution steps are shown in Figure 14.11. Here arrows point the flow of data.

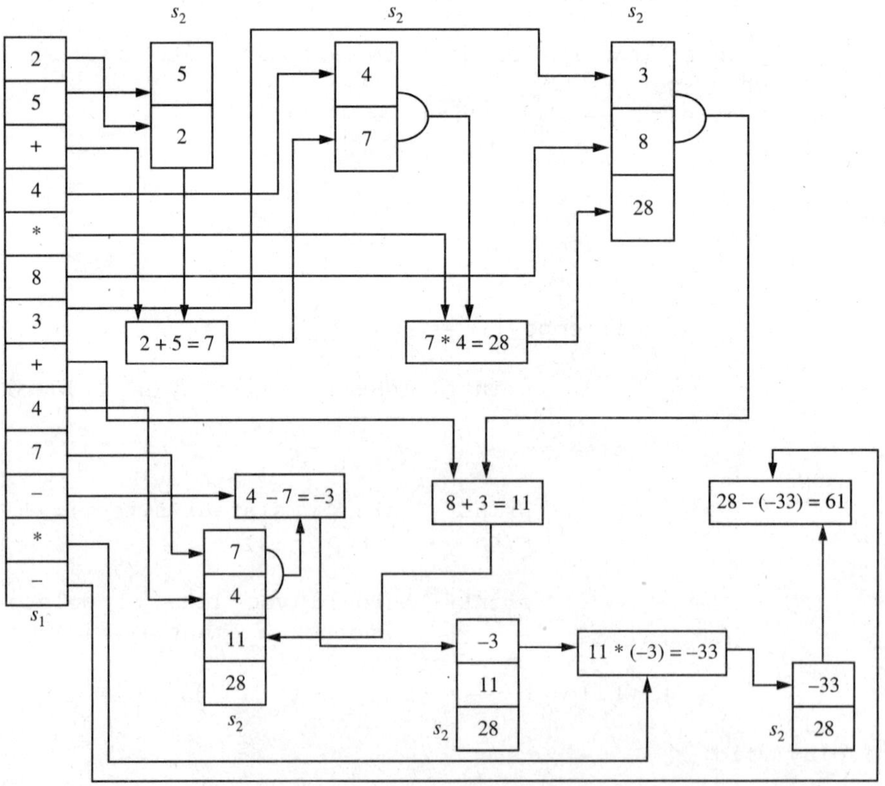

Figure 14.11 Expression evaluation using stack.

14.9 Representing Graphs in Computer System

To process large graphs with the help of computers, some data structures like matrix and linked list have been found useful. These are illustrated here.

14.9.1 Adjacency Matrix

Adjacency matrix is useful when a graph is dense. A graph G is said to be dense if it has significantly large number of edges. The adjacency matrix $M = [a_{ij}]$ of a graph G with n number of vertices $\{v_1, v_2, v_3, \ldots, v_n\}$ is an $n \times n$ matrix, such that

$$a_{ij} = \begin{cases} 1, & \text{if } v_i \text{ is adjacent to } v_j \text{ and} \\ 0, & \text{otherwise.} \end{cases}$$

Here, a_{ij} is an element in ith row and jth column. Rows and columns of the matrix are tagged with the names of the vertices. If G is a multi graph then a_{ij} may denote the number of parallel edges between the corresponding vertices (v_i and v_j). In the case of a weighted graph, a_{ij} may represent the weight of the edge (v_i, v_j). However, adjacency matrix has some limitations, as it is difficult to represent a weighted multi graph. Therefore, other tools are used. Figure 14.12 shows a graph G along with its adjacency matrix M.

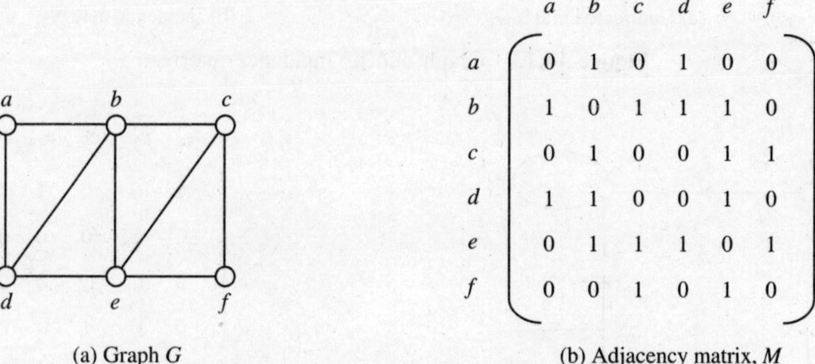

	a	b	c	d	e	f
a	0	1	0	1	0	0
b	1	0	1	1	1	0
c	0	1	0	0	1	1
d	1	1	0	0	1	0
e	0	1	1	1	0	1
f	0	0	1	0	1	0

(a) Graph G (b) Adjacency matrix, M

Figure 14.12 Graph and its adjacency matrix.

14.9.2 Incidence Matrix

A graph can be stored in a computer system by using an incidence matrix. However, an incidence matrix is possible only for those graphs whose edges have been assigned names. An incidence matrix of a graph G is the matrix M whose rows and columns are labelled by the names of the vertices and the edges, respectively. The elements of the matrix are defined as:

$$M[v, e] = \begin{vmatrix} 0 & \text{if } v \text{ is not an end point of } e, \\ 1 & \text{if } v \text{ is an end point of } e, \\ 2 & \text{if } e \text{ is a self-loop,} \end{vmatrix}$$

where, v is a vertex and e is an edge of the graph.

If G is a directed graph then M is defined as:

$$M[v, e] = \begin{cases} 0 & \text{if } v \text{ is not an end point of } e \\ 1 & \text{if } v \text{ is starting or head vertex of } e \\ -1 & \text{if } v \text{ is ending or tail vertex of } e \\ 2 & \text{if } e \text{ is a self-loop at } v \end{cases}$$

Two graphs G_1 and G_2, along with their incidence matrices are given in Figures 14.13 and 14.14.

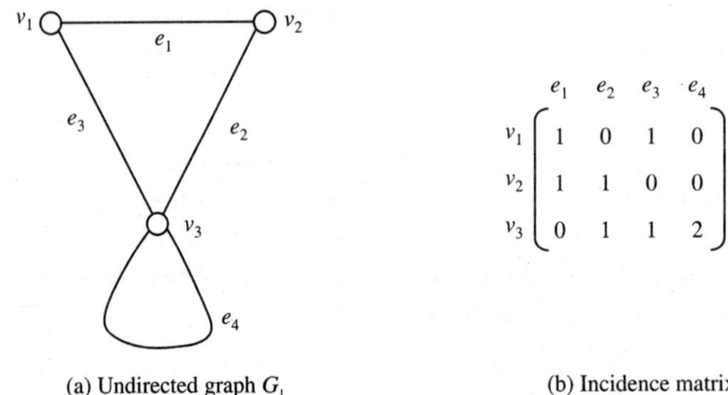

(a) Undirected graph G_1 (b) Incidence matrix

Figure 14.13 Graph and its incidence matrix.

(a) Directed graph G_2 (b) Incidence matrix

Figure 14.14 Directed graph and its incidence matrix.

14.9.3 Linked List Representation

Like adjacency matrix, incidence matrix also has some limitations. It is not suitable for weighted graphs. A Linked list representation, called adjacency list, is a better option. It is also space efficient, since adjacency and incidence matrices are mostly sparse matrices. Linked list for graphs is represented by a collection of sequential lists. A list is a collection of linked records. A linked record has two fields: one field stores vertex name, and another, address of the other record. A record in the adjacency list of a weighted graph has one additional field, which stores weight on the edge. Each list, headed by a vertex keeps record of all the vertices adjacent to the head vertex. The linked lists of two graphs undirected and directed are given in Figures 14.15 and 14.16.

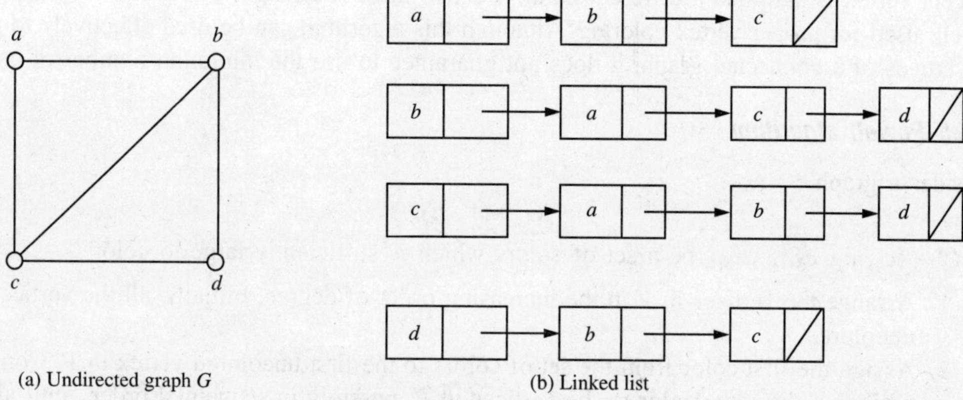

Figure 14.15 Graph and its linked list.

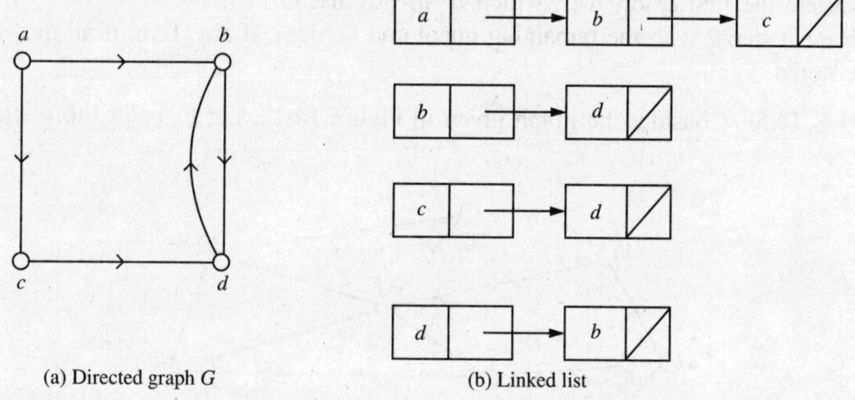

Figure 14.16 Directed graph and its linked list.

14.10 Coloring of Graph

Coloring is an important application of graph theory. Usually, graph coloring means painting the vertices of a graph. A graph can be colored in many different ways, however, if no two adjacent vertices are assigned the same color, then it is called **proper vertex coloring**. The minimum number of colors k, required to color a graph properly, is called the **chromatic number** of the graph. A graph having chromatic number k is known as k-colorable. A graph that contains only isolated vertices requires only one color for its proper coloring. Coloring is not limited to only vertex coloring but the idea is extended and region coloring, and even to edge coloring. The principles of vertex coloring are also applicable to edge and region coloring.

14.10.1 Vertex Coloring

Assigning colors to the vertices is called vertex coloring. In a **proper vertex coloring,** each adjacent vertex is assigned a different color. We will discuss an algorithm given by Welch and Powell, used for proper vertex coloring. Although this algorithm can be used effectively to paint the vertices of a connected graph, it does not guarantee to use the minimum number of colors.

Welch–Powell algorithm

Consider a graph

$$G = (V, E).$$

Let $C = \{c_1, c_2, c_3, ..., c_n\}$ be a set of colors which is sufficiently large to color G.

1. Arrange the vertices in V in the increasing order of degree. Initially, all the vertices are uncolored.
2. Assign the first color from the set of colors to the first uncolored vertex in V. Continue assigning the same color to the vertices in V, possibly in sequential order, until all the vertices in V are exhausted (with the condition that no two adjacent vertices are assigned the same color).
3. Delete the first color in C, which is already used.
4. Repeat step 2 with the remaining uncolored vertices, if any. Exit, if all the vertices are colored.

EXAMPLE 14.5 Consider the graph given in Figure 14.17. Let us color the vertices of this graph.

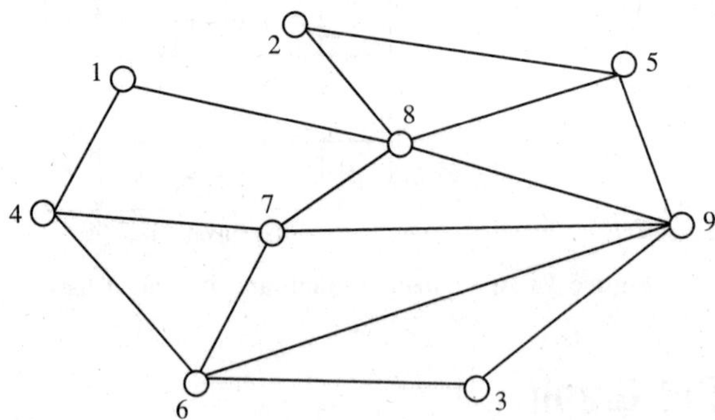

Figure 14.17 Graph G.

There are nine vertices in the graph. The vertices are named with natural numbers, which are in the increasing order of degree, i.e., vertices named 1, 2 and 3 are of degree 2, and vertices 4 and 5 are of degree 3, etc. Let $C = \{c_1, c_2, c_3, ..., c_n\}$ be the set of available colors. Assign color c_1 to vertex 1. Next, assign c_1 to vertex 2, since it is not adjacent to vertex 1. Again, assign c_1 to vertex 3, since it is also not adjacent to either vertex 1 or vertex 2. Similarly, assign c_1 to vertex 7. We are exhausted now, because no other vertex can be assigned color c_1. Delete c_1 from the set C. Now, c_2 becomes the first color in C. Repeat the process using the first color in C, i.e., c_2. All the steps are shown in Figures 14.18(a to d). Note that the chromatic number of this graph is 4.

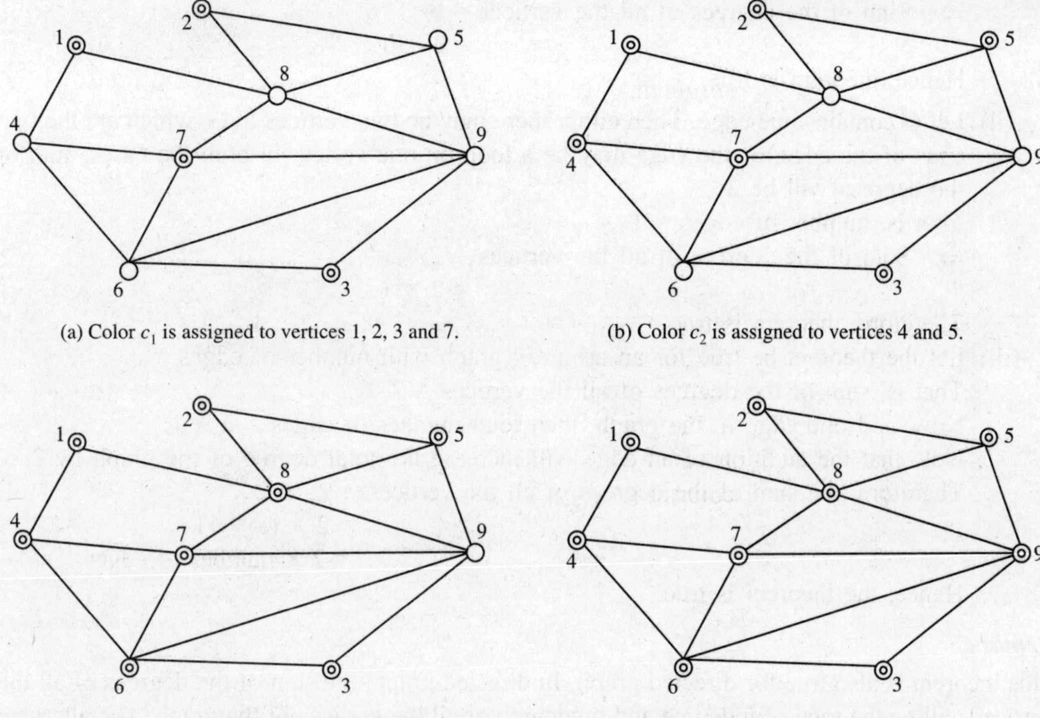

(a) Color c_1 is assigned to vertices 1, 2, 3 and 7.

(b) Color c_2 is assigned to vertices 4 and 5.

(a) Color c_3 is assigned to vertices 6 and 8.

(b) Color c_4 is assigned to vertex 9.

Figure 14.18 Vertex coloring.

14.10.2 Four Color Conjecture

In the nineteenth century, several mathematicians conjectured that every planar graph is four colorable. Finally, in 1976, Appel and Haken proved this conjecture. According to this, every planar graph is four colorable. Actually, it was tested on a large number of graphs (more than 2000 graphs) using computer system and was found to be true. Since region coloring is similar to vertex coloring, this conjecture is also valid for coloring the regions of a map.

14.11 Some Theorems

In this section, we will discuss some basic theorems of graph theory.

THEOREM 14.1 The sum of the degrees of all the vertices in a graph is equal to twice the number of edges.

Proof: Let us prove this theorem by induction. Consider a graph

$$G = (V, E)$$

(i) Let G contains no edge.
That is, number of edges = 0.

$\Rightarrow$ Sum of the degrees of all the vertices $= 0$
$= 2 \times 0$

Hence, theorem is true.

(ii) Let G contains one edge. Then either there may be two vertices in G, which are the two ends of the edge, or the edge may be a loop on one vertex. In both the cases, sum of the degrees will be 2.
That is, number of edges $= 1$.
$\Rightarrow$ Sum of the degrees of all the vertices $= 2$
$= 2 \times 1$

Therefore, theorem is true.

(iii) Let the theorem be true for an arbitrary graph with number of edges $= e$.
That is, sum of the degrees of all the vertices $= 2e$.
Now, add one edge in the graph, then total number of edges $= e + 1$.
Note that the addition of an edge will increase the total degree of the graph by 2.
Therefore, the sum of the degrees of all the vertices $= 2e + 2$,
$= 2 \times (e + 1)$,
$= 2 \times$ number of edges.

Hence, the theorem is true.

Remark

This theorem is also true for directed graph. In directed graph, the sum of the degrees of all the vertices will be the total of indegree and outdegree of all the vertices of the graph. The alternate statement of above theorem is, "the sum of the degrees of all the vertices in a graph is even".

THEOREM 14.2 In a graph G, the number of vertices of odd degree is even.

Proof: Consider a graph $G = (V, E)$, such that it contains some vertices of odd and some of even degrees. Partition the set of vertices into two categories,

(i) Vertices of odd degree and
(ii) Vertices of even degree.

Then total degree of the graph = sum of degrees of odd degree vertices
+ sum of degrees of even degree vertices.

Further,

(a) the total degree of the graph is even $= 2m$, say. (by theorem 1), and
(b) the sum of the degrees of even degree vertices will be even, naturally $= 2n$, say.

Therefore,

$$2m = \text{sum of degrees of vertices with odd degree} + 2n.$$

$$\Rightarrow \quad \text{Sum of degrees of vertices with odd degree} = 2m - 2n$$
$$= 2(m - n)$$
$$= \text{even}.$$

$\Rightarrow$ The number of vertices of odd degree will be even, because only the sum of even number of odds can be even.
Hence, theorem is proved.

THEOREM 14.3 The total number of edges in a complete graph with n vertices is $n(n - 1)/2$.

Proof: Consider a complete graph K_n.

In a complete graph with n vertices, each vertex is connected with other $n - 1$ vertices. Therefore, degree of each vertex = $n - 1$.

$\Rightarrow$ total degree = $n(n - 1)$

$\Rightarrow$ $2 \times$ number of edges in the graph = $n(n - 1)$ {using theorem 1}

Hence, total number of edges in the graph = $n(n - 1)/2$.

THEOREM 14.4 A tree T with n number of vertices has $n - 1$ edges.

Proof: Let us prove the theorem by induction. Consider a tree T.

(i) Let T contains only one vertex.

That is, $n = 1$.

$\Rightarrow$ $1 - 1 = 0$

Obviously, if there is only one vertex in the tree then number of edges will be zero. Hence, the theorem is true.

(ii) Let T contains two vertices, i.e., $n = 2$. Then there will be one edge only.

Also, $n - 1 = 2 - 1 = 1$

Hence, the theorem is true.

(iii) Let the theorem be true for an arbitrary tree with number of vertices = k.

$\Rightarrow$ Number of edges = $k - 1$

Now, add one vertex in the tree, then total number of vertices = $k + 1$.

Note that addition of one vertex in a tree will imply the addition of one edge.

Therefore, the number of edges = $k - 1 + 1$,

$= (k + 1) - 1$.

Hence, theorem is proved.

14.12 Solved Problems

PROBLEM 14.1 Verify Euler's formula $v - e + r = 2$ for the graph given in Figure 14.19.

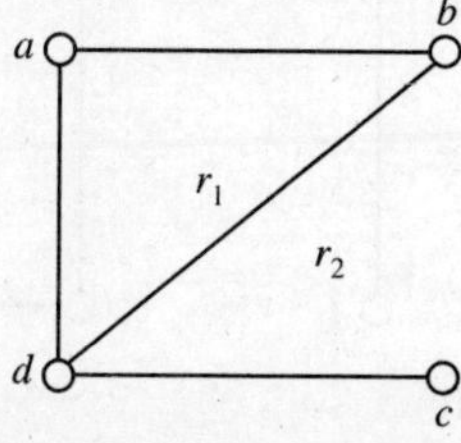

Figure 14.19 Graph of Problem 14.1.

Solution

Consider the graph given in Figure 14.19.

Here, $v = 4$

$e = 4$

$r = 2 \{r_1 \text{ and } r_2\}$.

Thus, $v - e + r = 2$.

PROBLEM 14.2 Consider a graph G, given in Figure 14.20. Find the minimum spanning trees.

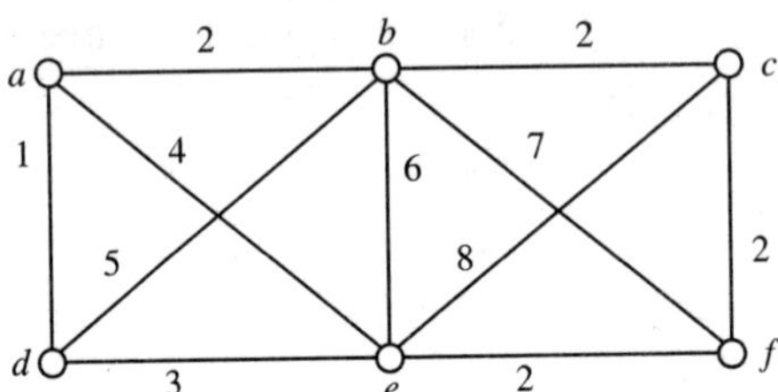

Figure 14.20 Graph G of Problem 14.2.

Solution

We can find minimum spanning trees by using any one of the two algorithms discussed earlier in Section 14.6. There are four minimum spanning trees. These trees are given in Figures 14.21(a, b, c and d).

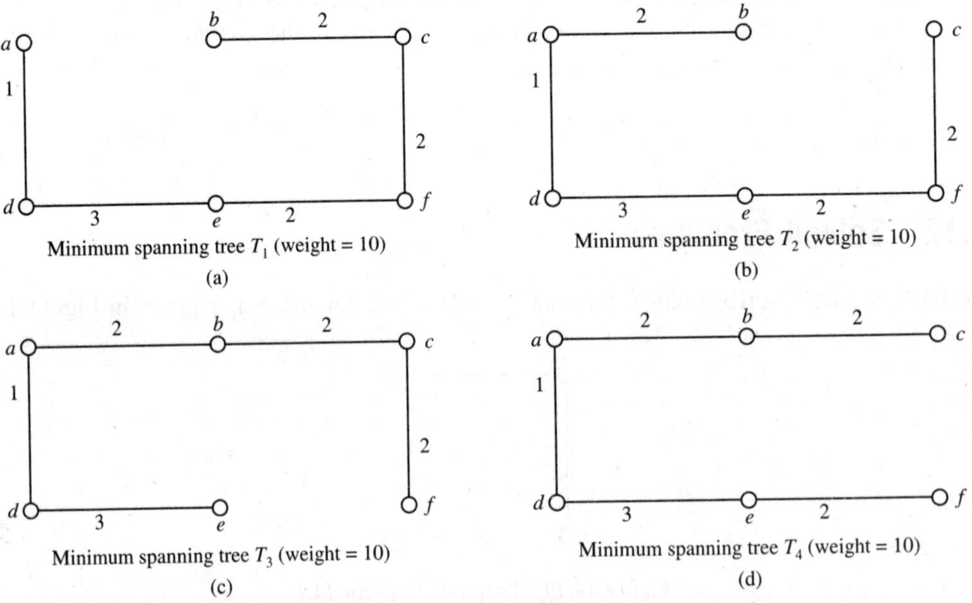

Figure 14.21 Minimum spanning trees for Problem 14.2.

PROBLEM 14.3 Color the vertices of the graph G given in Figure 14.22.

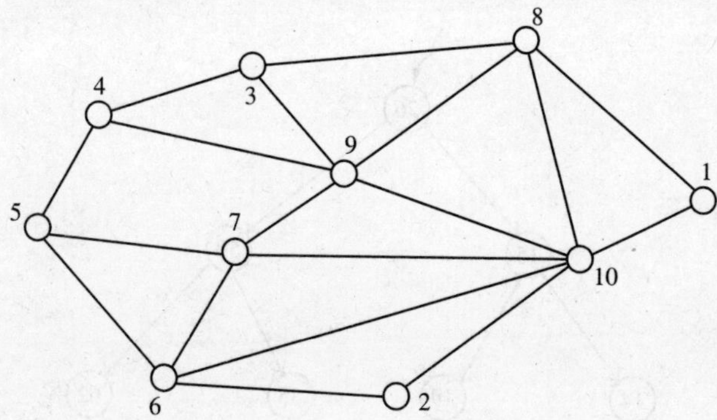

Figure 14.22 Graph G of Problem 14.3.

Solution

There are ten vertices in the graph. The vertices are named with natural numbers, which are in the increasing order of degree, i.e., vertices 1 and 2 are of degree 2, and vertices 3, 4 and 5 are of degree 3, etc. Let $C = \{c_1, c_2, c_3, ..., c_n\}$. Assign color c_1 to vertex 1. Next, assign c_1 to vertex 2, since it is not adjacent to vertex 1. Again, assign c_1 to vertex 3 since it is not adjacent to both the vertices, vertex 1 and vertex 2. Similarly, assign c_1 to vertex 5. We are exhausted now, since no other vertex can be assigned color c_1. Delete c_1 from the set C. Now, c_2 becomes the first color in C. Repeat the process using the first color in C, i.e., c_2. All the steps are shown in Figures 14.23(a to d). The chromatic number of this graph is 4.

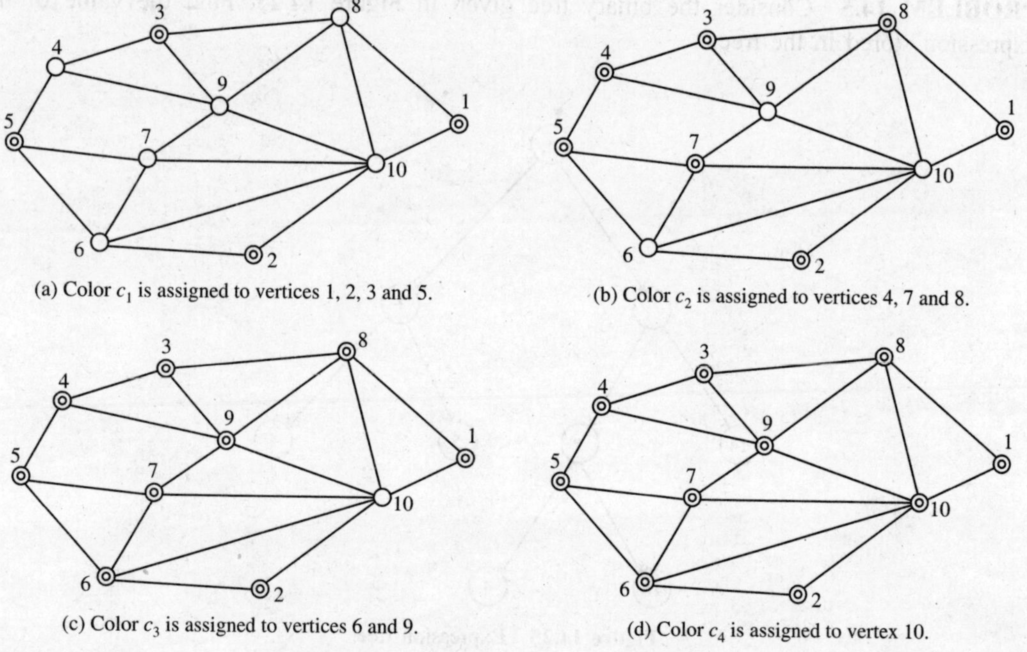

(a) Color c_1 is assigned to vertices 1, 2, 3 and 5.

(b) Color c_2 is assigned to vertices 4, 7 and 8.

(c) Color c_3 is assigned to vertices 6 and 9.

(d) Color c_4 is assigned to vertex 10.

Figure 14.23 Vertex coloring of Problem 14.3.

PROBLEM 14.4 Consider the binary tree given in Figure 14.24. Traverse the tree in pre-order, in-order and post-order.

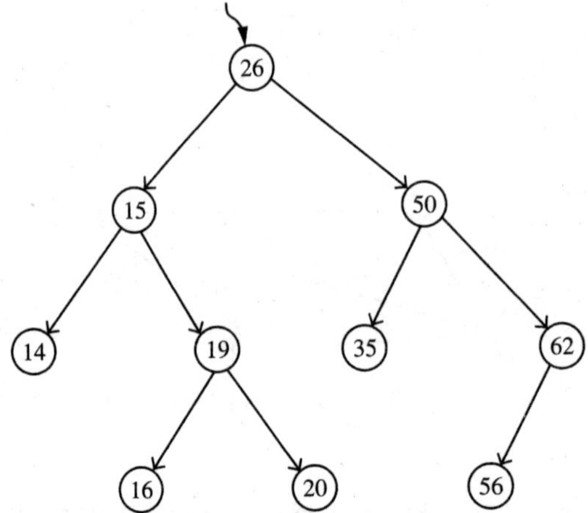

Figure 14.24 Binary tree of Problem 14.4.

Solution

Data in a binary tree can be read in three ways using three algorithms given in Section 14.7.

1. Pre-order : 26 15 14 19 16 20 50 35 62 56
2. In-order : 14 15 16 19 20 26 35 50 56 62
3. Post-order : 14 16 20 19 15 35 56 62 50 26

PROBLEM 14.5 Consider the binary tree given in Figure 14.25. Find the value of the expression stored in the tree.

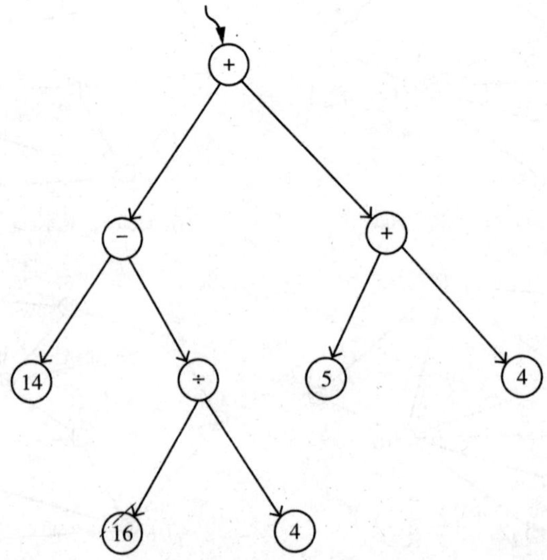

Figure 14.25 Expression tree.

Solution

Data stored in this binary tree can be read in three ways using three traversal algorithms as follows:

1. Pre-order : + – 14 ÷ 16 4 + 5 4
2. In-order : (14 – (16 ÷ 4)) + (5 + 4)
3. Post-order : 14 16 4 ÷ 5 4 + +

Value of the expression is 19.

Exercises

14.1 Draw the following graphs:

(i) $K_{3,5}$ and (ii) K_8.

Also, discuss the planarity of these graphs.

14.2 Are the following graphs planar? If so, draw their planar representation.

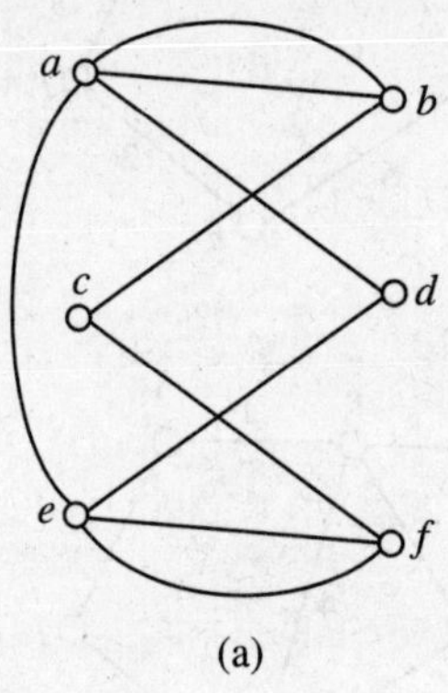

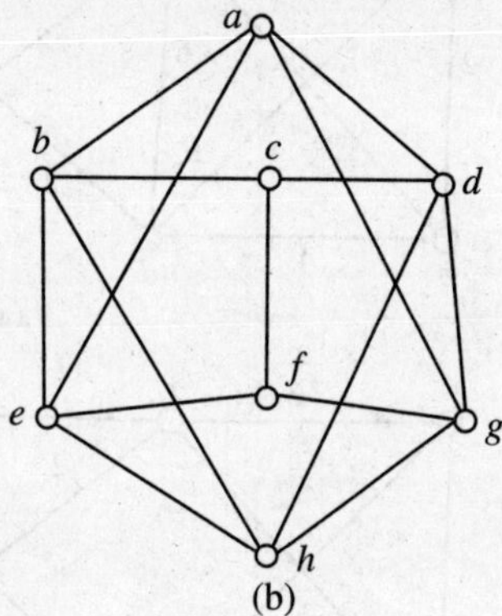

Figure 14.26

14.3 Find all 2-regular sub graphs of the following graph.

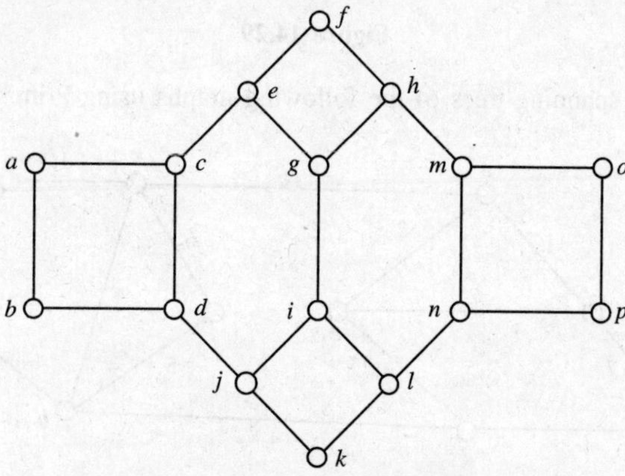

Figure 14.27

14.4 Find any two spanning trees of the following graph.

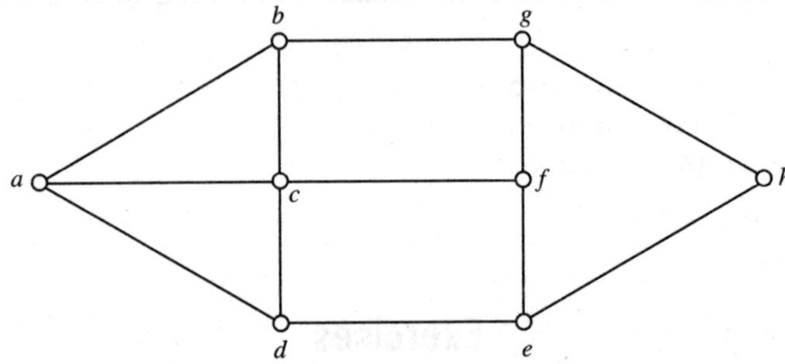

Figure 14.28

14.5 Find the minimum spanning trees of the following graphs using Kruskal's algorithm.

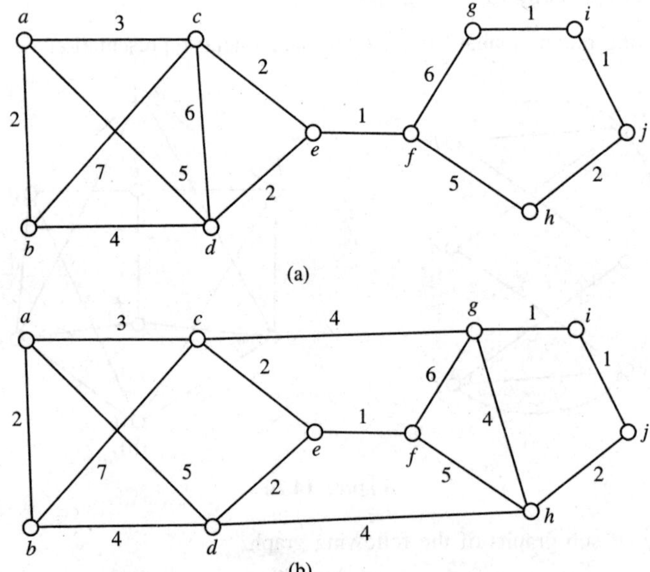

Figure 14.29

14.6 Find the minimum spanning trees of the following graphs using Prim's algorithm.

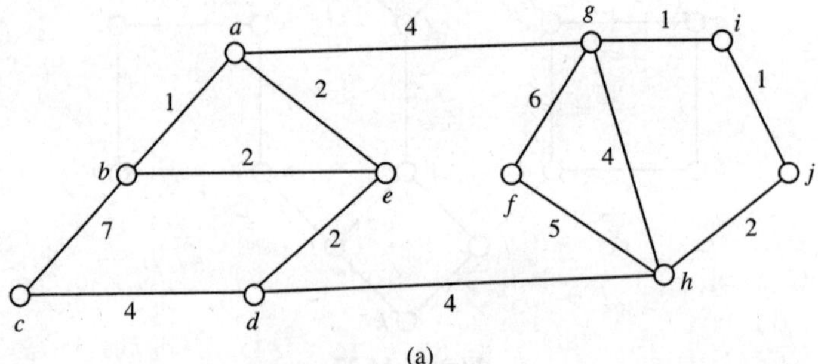

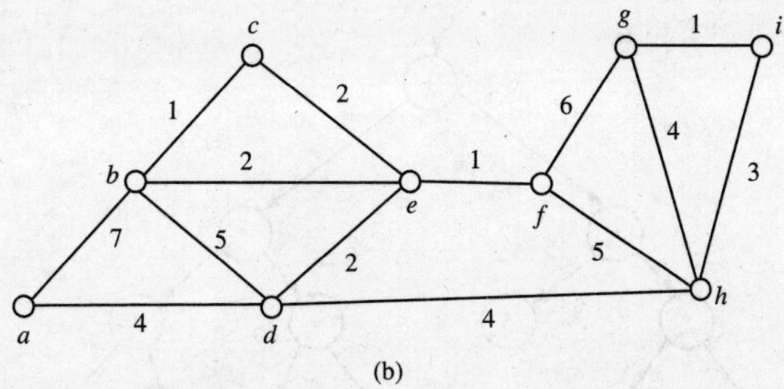

(b)

Figure 14.30

14.7 Traverse the following trees into pre-order, in-order and post-order.

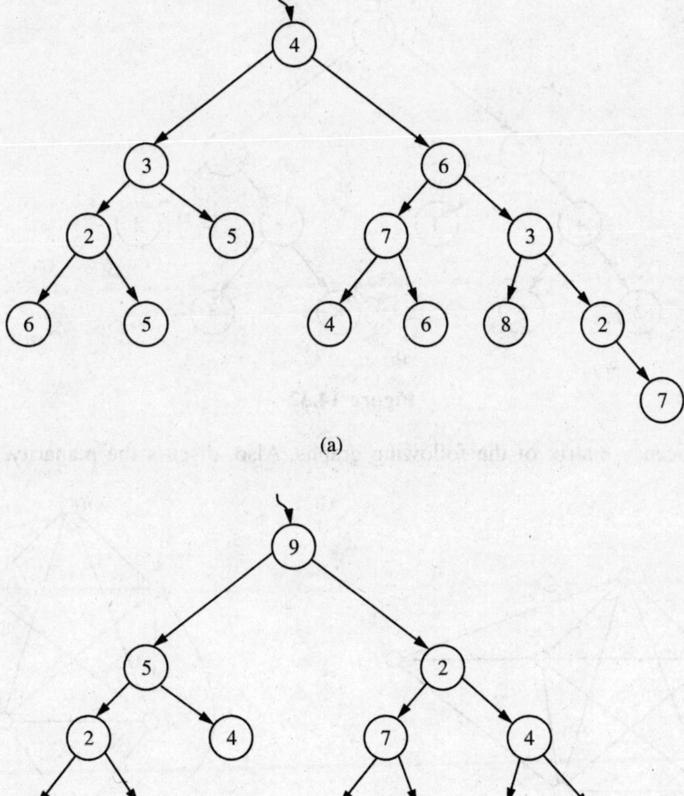

(a)

(b)

Figure 14.31

14.8 Find the prefix, infix and postfix notations of the following binary trees. Also, compute the value in each by using postfix notation.

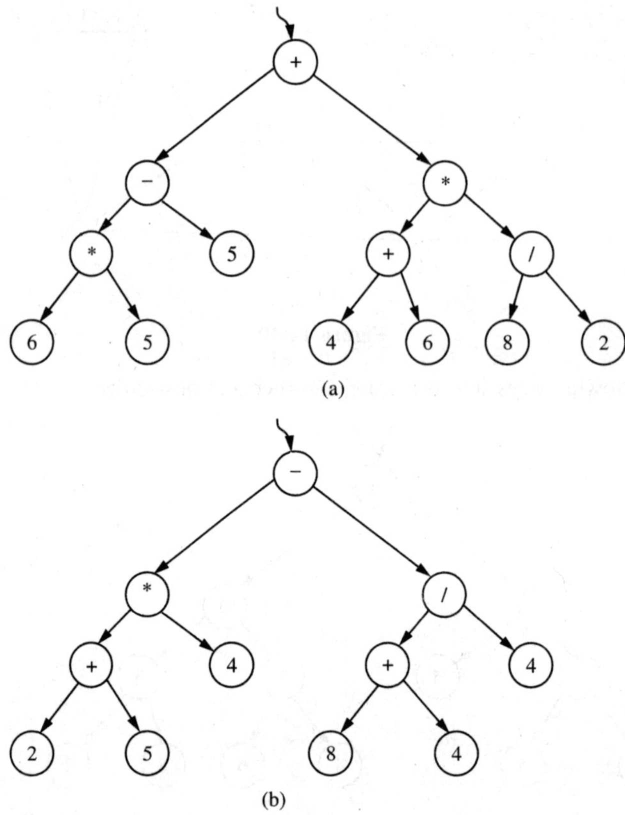

Figure 14.32

14.9 Find the adjacency matrix of the following graphs. Also, discuss the planarity.

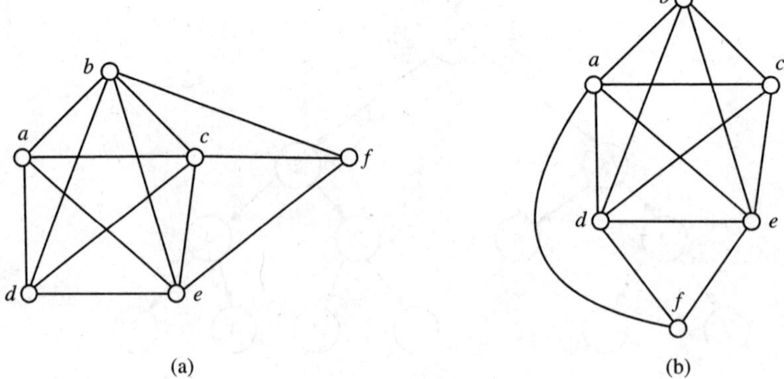

Figure 14.33

14.10 Find the incidence matrix and adjacency list of each of the following graphs.

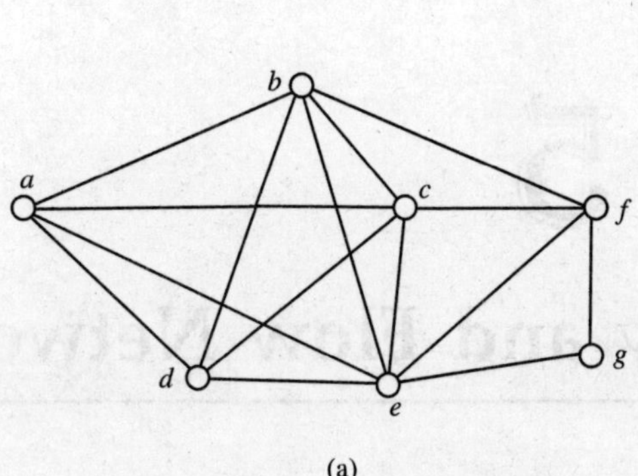

(a) (b)

Figure 14.34

14.11 Find the chromatic number of the following graphs. Also, color their vertices.

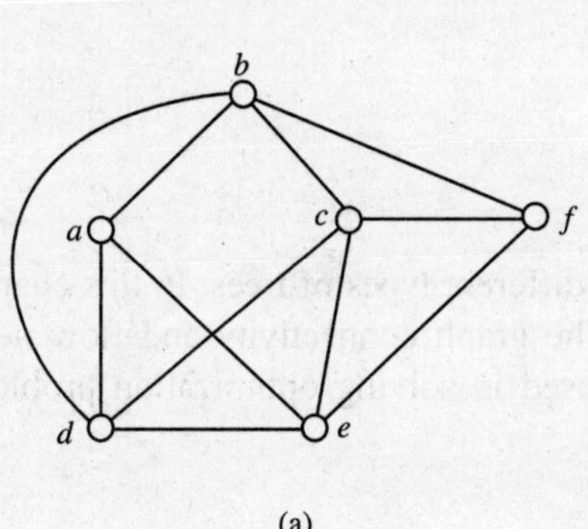

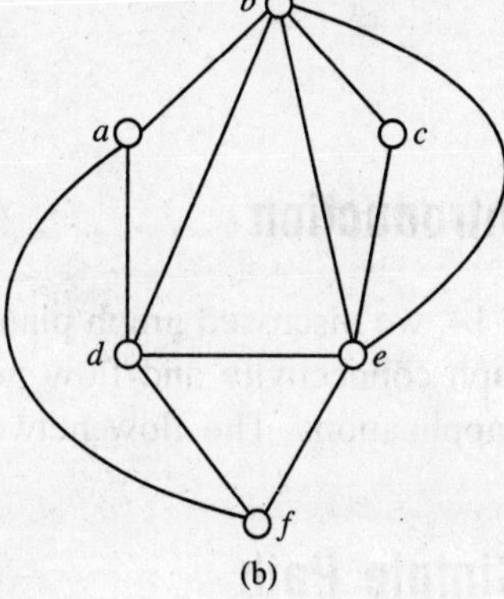

(a) (b)

Figure 14.35

15

Graph Connectivity and Flow Network

15.1 Introduction

In Chapter 14, we discussed graph planarity and different types of trees. In this chapter, we will discuss graph connectivity and flow network. The graph connectivity and flow network have important applications. The flow networks are used in solving optimization problems.

15.2 Simple Path

A path in the connected graph that has an alternating sequence of vertices and edges is called a simple path. All the edges in a simple path are different, i.e., no edge is repeated in a simple path. However, a vertex may appear more than once. For instance, in Figure 15.1, *a-b-e-f* and *a-c-d-f* are simple paths of graph G from vertex a to f. In some texts, simple path is also termed **walk, edge train, trail** or a **chain.**

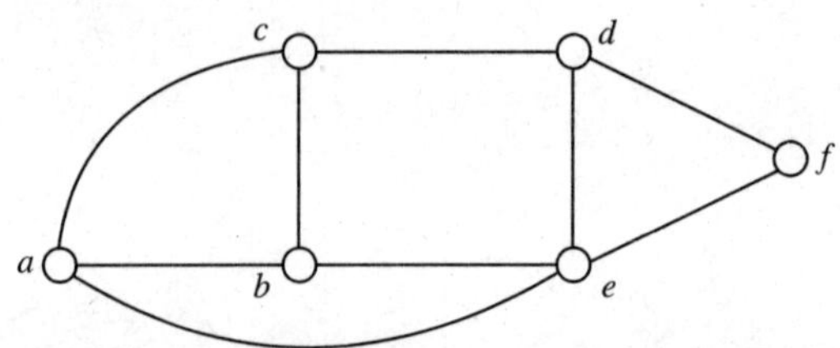

Figure 15.1 Graph G.

15.3 Königsberg Bridge Problem

Königsberg bridge problem was introduced in Chapter 13. The eighteenth century Königsberg town (formerly in East Prussia but now in Russia and known as Kaliningrad) was situated on the two banks and two islands of the river Pregel, which were connected to one another by seven bridges as shown in Figure 15.2(a). The problem was to find a walk that crossed each of the seven bridges exactly once, before returning to the starting point. The Prussian emperor, Frederick the Great brought this problem to the attention of the famous Swiss mathematician L Euler[$]. Euler represented the town Königsberg by another diagram [Figure 15.2(b)] consisting of points or nodes which were joined by arcs called edges. The islands and the two sides of the river were represented by nodes, and the bridges were replaced by edges. After analysis, Euler concluded that no such walk was possible. The conclusion is a direct consequence of Euler theorem given in the following section.

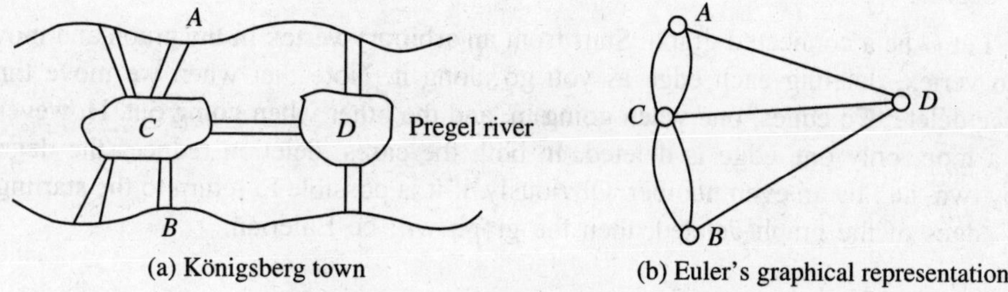

(a) Königsberg town (b) Euler's graphical representation

Figure 15.2 Königsberg town and its graph.

15.4 Eulerian Path and Eulerian Circuit

A simple path that contains all the edges of a graph G is called the **Eulerian path**. Further, if this path starts and ends at the same vertex then it is called an **Eulerian circuit**. In Figure 15.3(a), *b-c-d-e-f-a-b-f* is an Eulerian path, and in Figure15.3(b), *a-b-c-d-e-f-g-b-d-g-a* is an Eulerian Circuit.

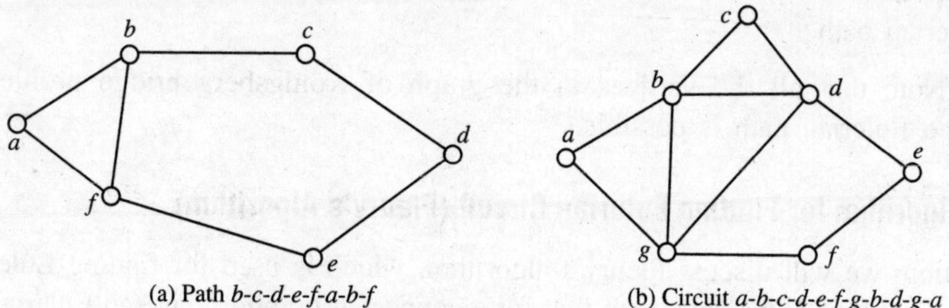

(a) Path *b-c-d-e-f-a-b-f* (b) Circuit *a-b-c-d-e-f-g-b-d-g-a*

Figure 15.3 Euler's path and circuit.

[$] **Leonhard Euler** (1707 –1783) was a pioneering Swiss mathematician and physicist. He made important discoveries in fields like infinitesimal calculus, graph theory, mathematical analysis, mechanics, fluid dynamics, optics, and astronomy.

A graph that contains an Eulerian circuit is called the **Eulerian graph**. The Königsberg bridge problem can now be visualized as finding the Eulerian circuit in the graphical representation of Königsberg town. In other words, we have to determine whether the graphical representation of Königsberg town is the Eulerian graph or not. The theorem given below will give the criteria for determining a graph to be the Eulerian graph.

15.4.1 Euler Theorem

The Euler theorem is an important result that categorizes Eulerian and non-Eulerian circuits. It solves the Königsberg bridge problem completely. The theorem is stated and proved below.

THEOREM 15.1 (Euler's theorem): A finite connected graph is Eulerian if and only if each vertex has even degree.

In other words, a connected graph that has an Eulerian circuit must have all its vertices of even degree.

Proof: Let G be a connected graph. Start from an arbitrary vertex in the graph and move from vertex to vertex, deleting each edge as you go along it. Note that when we move through a vertex, we delete two edges, one when going in, and the other when going out. However, in the case of a loop, only one edge is deleted. In both the cases, deletion reduces the degree of a vertex by two, i.e., by an even number. Obviously, if it is possible to return to the starting vertex with all edges of the graph deleted, then the graph will be Eulerian.

15.4.2 Corollary

A connected graph has an Eulerian path if and only if it has either zero or two vertices of odd degree.

Proof: This is an extension of the Euler theorem since each Eulerian circuit, is itself an Eulerian path. In an Eulerian circuit, each vertex is of even degree, i.e., there are zero vertices of odd degree. Further, if the last edge connecting the starting vertex to the last is missing or deleted, then it is not possible to return to the first vertex. This implies that these two vertices will be of odd degree. Nevertheless, if it is still possible to delete all the edges by moving from vertex to vertex in the similar way, as in Euler theorem (refer Section 15.4.1), then the graph has an Eulerian path.

Remark Note that all the vertices in the graph of Königsberg bridge problem are odd, therefore, no Eulerian path is possible.

15.4.3 Algorithm for Finding Eulerian Circuit (Fleury's Algorithm)

In this section, we will discuss Fleury's algorithm, which is used for finding Eulerian circuit.

In Section 15.4.1, we have seen that for a connected graph to have an Eulerian circuit, all its vertices must be even. Let $G = (V, E)$ be a connected graph with all its vertices even. Then, by executing the following algorithm, we will find an Eulerian circuit in it.

1. Choose an arbitrary starting vertex $v \in V$ in the graph and consider it as current vertex c.

2. Start from c and move to an adjacent vertex p, which is connected to it by an edge (which is not a bridge). Consider a bridge, if it is the only option.
3. Delete the edge or bridge after moving to p.
4. Make p the current vertex c
5. Continue from step 2 and stop when all the edges in the graph are deleted, and starting vertex is reached.
6. Note down the sequence of vertices traversed. This is the Eulerian circuit.

EXAMPLE 15.1 Consider the graph G given in Figure 15.4. We will execute Fleury's algorithm described above and find Eulerian circuit in it.

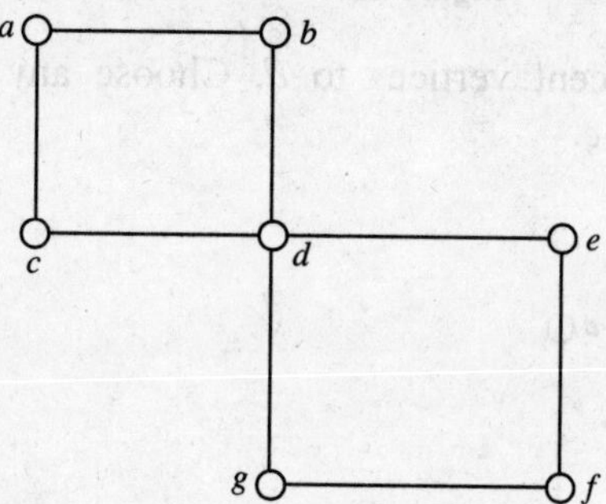

Figure 15.4 Graph *G*.

Since all the vertices in the given graph are even, therefore, it is an Eulerian graph. Steps to find the Eulerian circuit are stated below along with sub graphs shown in Figures 15.5 to 15.12.

Choose any vertex to begin with. Let it be '*a*'.

1. Choose vertex *a*.
2. Move to vertex *b*, other choice is vertex *c*.
3. Delete edge *a*-*b*.

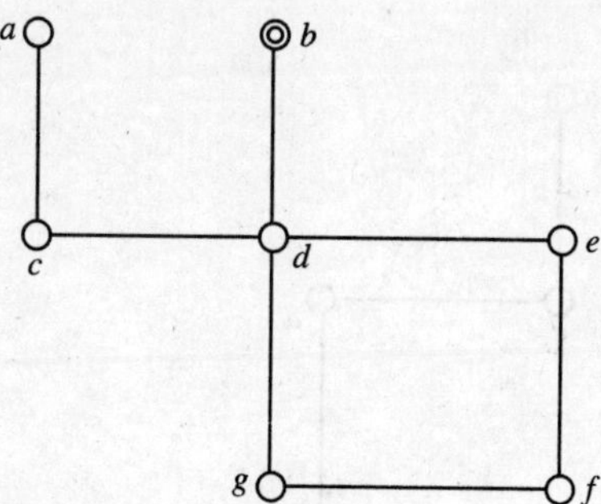

Figure 15.5 Vertex *a* chosen.

4. *b*-*d* is the only edge, which is also a bridge.
5. Move to vertex *d*.
6. Delete *b*-*d*.

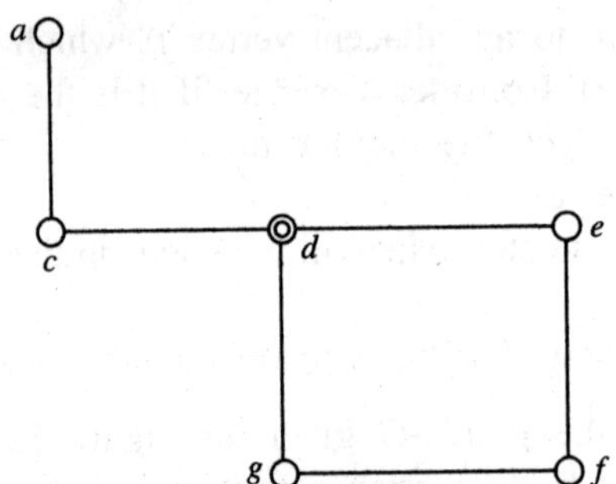

Figure 15.6 *b-d* deleted.

7. There are three adjacent vertices to *d*. Choose any of these, except *c*, to move arbitrarily, say vertex *e*.
8. Move to vertex *e*.
9. Delete *d-e*.

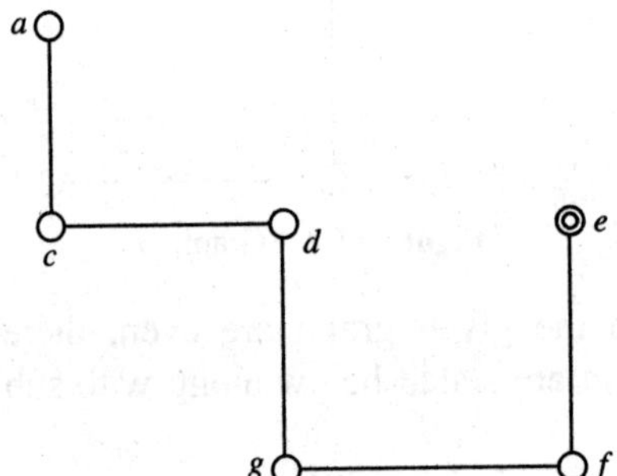

Figure 15.7 Vertex *e* chosen.

10. *e-f* is the only edge, which is also a bridge.
11. Move to vertex *f*.
12. Delete *e-f*.

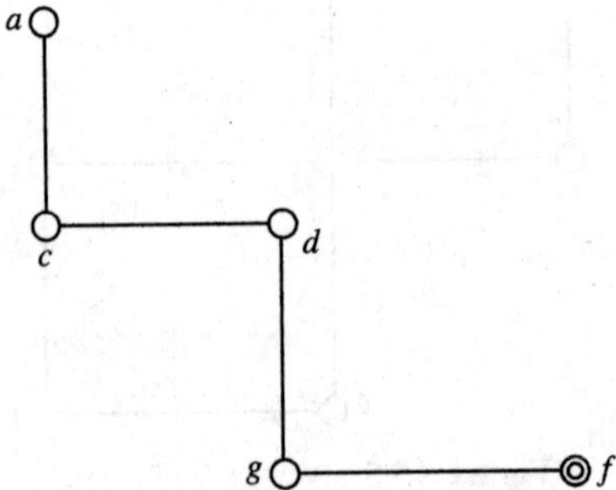

Figure 15.8 *e-f* deleted.

13. *f-g* is the only edge, which is also a bridge.
14. Move to vertex *g*.
15. Delete *f-g*.

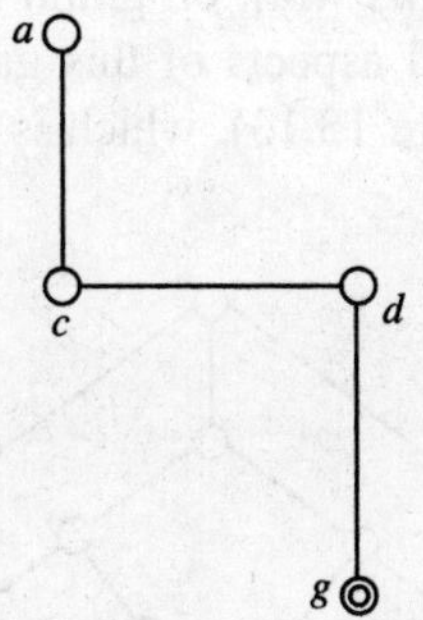

Figure 15.9 *f-g* deleted.

16. *g-d* is the only edge, which is also a bridge.
17. Move to vertex *d*.
18. Delete *g-d*.

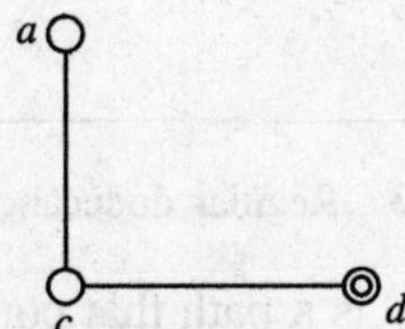

Figure 15.10 *g-d* deleted.

19. *d-c* is the only edge, which is also a bridge.
20. Move to vertex *c*.
21. Delete *d-c*.

Figure 15.11 *d-c* deleted.

22. *c-a* is the only edge which is also a bridge.
23. Move to vertex *a*.
24. Delete *c-a*.

Figure 15.12 *c-a* deleted.

No edge is left in the graph, and the starting vertex '*a*' is reached, therefore, we stop. The graph has an Eulerian circuit. Note down the sequence of vertices at steps 1, 2, 5, 8, 11, 14, 17, 20 and 23. The Eulerian circuit is *a-b-d-e-f-g-d-c-a*.

15.5 Hamiltonian Path and Hamiltonian Circuit

Sir William Rowan Hamilton (1805-1865) was a great Irish mathematician. He worked in several areas of mathematics. During his work on graph theory, he invented a puzzle known as the **Icozian Game**. One of the several aspects of this game was to find a Hamiltonian circuit in a regular dodecahedral graph (Figure 15.13), which is the planar version of the solid puzzle. Try it*.

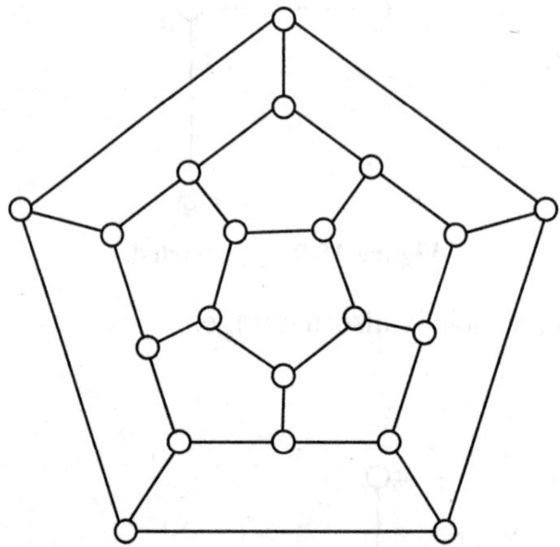

Figure 15.13 Regular dodecahedral graph.

A **Hamiltonian path** in a graph G is a path that contains each vertex once and only once. Therefore, an edge may appear more than once in the Hamiltonian path but not a vertex. The **Hamiltonian circuit** is a Hamiltonian path that starts and ends at the same vertex. A graph that contains a Hamiltonian circuit is called a **Hamiltonian graph**.

EXAMPLE 15.2 Graph given in Figure 15.14(a) has a Hamiltonian circuit, *a-b-c-d-a* but there is no Hamiltonian circuit in Figure 15.14(b).

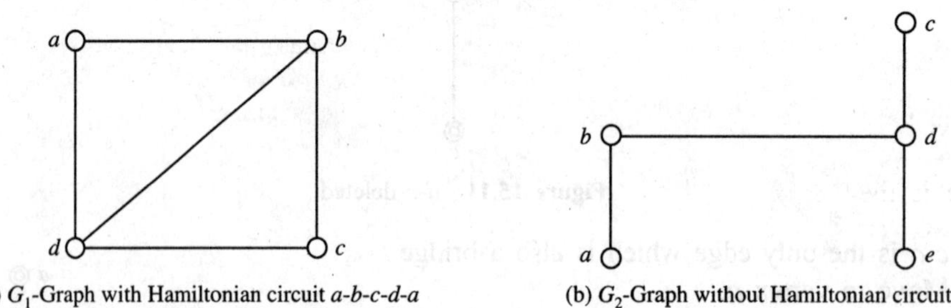

(a) G_1-Graph with Hamiltonian circuit *a-b-c-d-a* (b) G_2-Graph without Hamiltonian circuit

Figure 15.14 Graphs G_1 and G_2.

Finding a Hamiltonian Circuit

Contrary to Eulerian circuit, there is no straightforward procedure to find a Hamiltonian circuit in a graph. However, a theorem by G.A. Dirac helps in determining whether a given graph is Hamiltonian. It is discussed as follows.

Note: Every complete graph is a Hamiltonian graph. In complete graph, starting from any vertex, one can visit all other vertices sequentially in a desired order.

THEOREM 15.2 (by G A Dirac): Let G be a connected graph with n vertices. Then G is Hamiltonian if $n > 2$ and degree of each vertex is at least $n/2$.

The proof of this theorem is lengthy and hence is omitted here. The graph given in Figure 15.15(a) is Hamiltonian, since $n = 6$ and degree of each vertex is 3, i.e., conditions given in the theorem are satisfied. Also, the graph given in Figure 15.15(b) is Hamiltonian, though the conditions of the theorem are not satisfied (since it has all its vertices of degree $2 < 5/2$). Further, the graph given in Figure 15.15(c) is not Hamiltonian, since it does not satisfy the conditions. From Figure 15.15(b), we observe that a graph may be Hamiltonian even though the conditions may not be satisfied. Thus, these conditions are merely sufficient but not necessary, and hence, the existence and nonexistence of Hamiltonian circuit cannot be determined if a graph does not satisfy the said conditions.

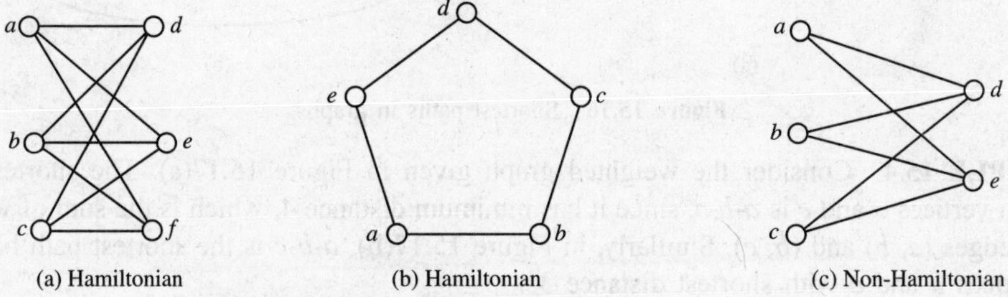

Figure 15.15 Hamiltonian and non-Hamiltonian graphs.

15.6 Shortest Path Problem

Visualize the map of a country in which cities are connected by roads, railways, etc. A salesperson has to travel from one city to another for his/her professional duties. It is obvious that the salesperson would like to optimize his/her journey for financial reasons. Therefore, finding the shortest distance between two cities is a real life natural problem. Graph theory can be used to find the solution of this problem. In a graph, the **shortest path** between two vertices is a simple path that has **minimum number of edges**. The **shortest distance** or **distance** between two vertices is the number of edges in the shortest path. In a weighted graph, **weight of a path** is the sum of weights of the edges in the path. Therefore, the shortest distance between two vertices in a weighted graph is the weight of the simple path that has minimum weight. **Path length** is the number of edges in a path.

The definitions given above remain unchanged even if the graph is **directed,** i.e., the shortest path between two vertices in a directed graph is the simple path that has minimum number of connected directed edges. If directed graph is weighted, then the shortest path between two vertices is the directed simple path that has minimum weight. Also, the shortest distance between two vertices in a directed weighted graph is the weight of the directed simple path that has minimum weight.

EXAMPLE 15.3 Consider the graph given in Figure 15.16(a). The shortest path between vertices *a* and *c* is *a-b-c*, since it has minimum number of edges which is two. Other possible paths are *a-d-b-c*, *a-d-e-c* and *a-b-d-e-c*. Further, there are two shortest paths, *a-c-f* and *a-d-f*, each with 2 edges between the vertices *a* and *f* in Figure 15.16(b). Thus, there may be more than one shortest path between a pair of vertices.

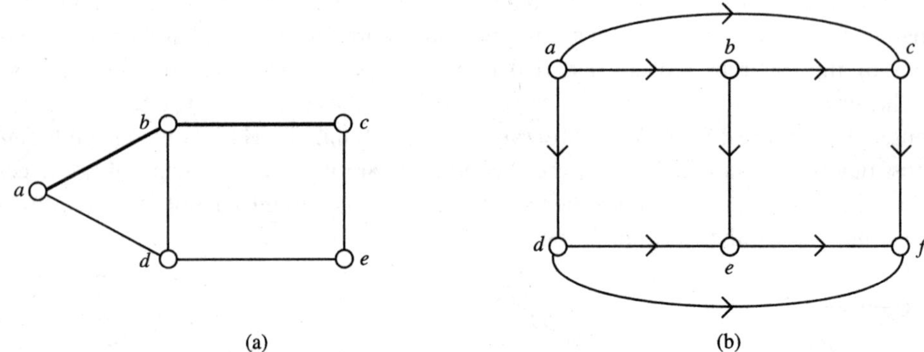

Figure 15.16 Shortest paths in graphs.

EXAMPLE 15.4 Consider the weighted graph given in Figure 15.17(a). The shortest path between vertices *a* and *c* is *a-b-c*, since it has minimum distance 4, which is the sum of weights on the edges (*a*, *b*) and (*b*, *c*). Similarly, in Figure 15.17(b), *a-b-c* is the shortest path between the vertices *a* and *c* with shortest distance 3.

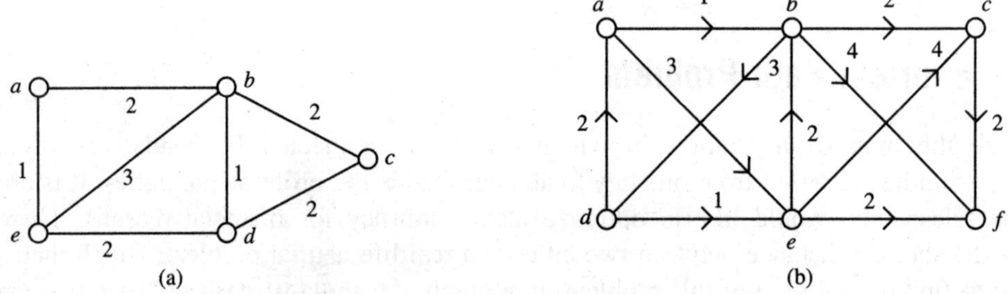

Figure 15.17 Shortest paths in weighted graphs.

15.6.1 Dijkstra's Shortest Path Algorithm

There are several algorithms for finding the shortest path between a pair of vertices in a connected graph. However, the algorithm given by **E.W. Dijkstra**** is efficient and suitable for coding in a programming language, and therefore is discussed here.

** **Edsger Wybe Dijkstra** (1930–2002) a Dutch computer scientist, was one of the early pioneers in the field of distributed computing. He received the 1972 **Turing Award** for his contributions to developing programming languages. He was born in Rotterdam, and studied **theoretical physics** at Leiden University; but he was more interested in computer science. He worked at the University of Texas at Austin, in the United States. Among

(Contd.)

Consider a connected graph $G = (V, E)$. Let a and z be two vertices. The aim is to find the shortest path between a and z.

Partition the set V of vertices into two sets, P and T, such that P contains a, and

$$T = V - P$$

1. Initially $P := \{a\}$; mark vertex a visited by enclosing it in a circle.

   $T := V - \{a\}$.

   For every vertex r in T

   $i(r) := w(a, r)$,

   where $w(a, r) = \infty$, if a and r are not adjacent,

   $=$ weight of the edge (a, r), otherwise.

   Call $i(r)$, the index of vertex r.

   Also attach path, the sequence of vertices with $i(r)$ whose weight is $w(a, r)$, if $w(a, r) \neq \infty$. Leave blank, otherwise. Blank is shown by a dash symbol(–).

   Note: Here operator ' := ' is an assignment operator used in Pascal programming language. It assigns the value of right hand operand to the left hand operand. On the other hand, operator ' = ' is used in the logical sense, 'equal to', i.e., comparing the two operands.
2. Select a vertex x in T which has the minimum index; mark x visited by enclosing it in a circle.

   If $x = z$, stop and exit; shortest path is the sequence of vertices attached with $i(x)$, and distance is $i(x)$.

   Otherwise, move to step 3.
3. $P' := P \cup \{x\}$,

   $T' := T - \{x\}$,

   for every vertex y in T' and adjacent to x, modify its index by the formula

   $i(y) := \min\{i(y), i(x) + w(x, y)\}$.

   Also $p := p'$,

   $T := T'$.

   Modify the path (or sequence of vertices) attached with each vertex y in T according to the new index $i(y)$, i.e., the sequence of the vertices which gives the value $i(y)$.
4. Go to step 2.

(*Contd.*)

his contributions to computer science are the **shortest path-algorithm**, Reverse Polish Notation and related Shunting yard algorithm. **Dijkstra** was against the use of the GOTO statement in computer programming. He argued for effective replacement of GOTO by structured control constructs, such as the while loop. Dijkstra was a fan of ALGOL 60, and he worked with the team that implemented the first compiler for it. Dijkstra was known for his habit of carefully composing manuscripts with his fountain pen. Despite having invented much of the technology of software, Dijkstra rarely used computers in his own work. He used computer only for internet surfing.

EXAMPLE 15.5 Consider the weighted graph $G = (V, E)$ given in Figure 15.18. Let us find the shortest path and the shortest distance between the vertices a and h.

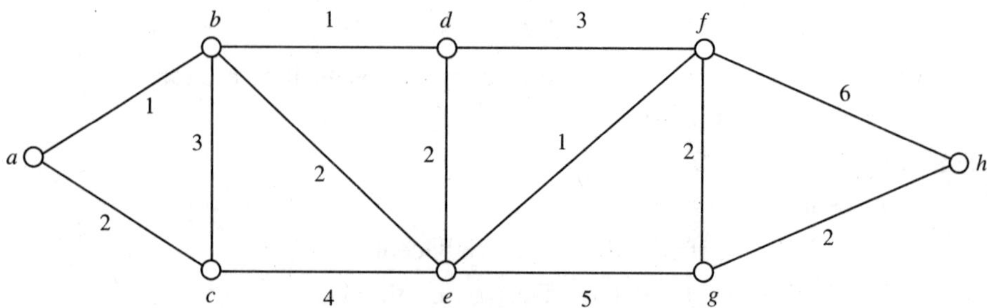

Figure 15.18 Shortest path between *a*-*h*.

Start from the vertex a and mark it visited by enclosing it in a circle. There is no change in the symbolic representation of other vertices.

Here $$P = \{a\},$$

and $$T = \{b, c, d, e, f, g, h\}$$

1. Calculate the index of each vertex in T. These are given below in capital brackets along with their associated path. These are also shown attached with each vertex in Figure 15.19.

   $i(b) = [1, a\text{-}b]$
   $i(c) = [2, a\text{-}c]$
   $i(d) = [\infty, -]$
   $i(e) = [\infty, -]$
   $i(f) = [\infty, -]$
   $i(g) = [\infty, -]$
   $i(h) = [\infty, -]$

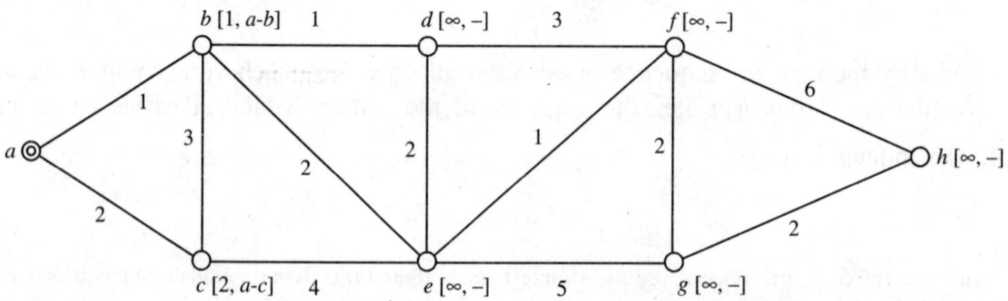

Figure 15.19 Starting at vertex *a*.

2. Now, minimum index is

$$i(b) = [1, a\text{-}b]$$

Mark b visited by enclosing it in a circle.
Since $b \neq h$, we have to continue.

Now $P' = \{a, b\}$,

and $T' = \{c, d, e, f, g, h\}$.

Modify the index of each vertex in T', which is adjacent to b.

$$i(c) = \text{minimum } \{2, 1 + 3 = 4\} = [2, a\text{-}b]$$
$$i(d) = \text{minimum } \{\infty, 1 + 1 = 2\} = [2, a\text{-}b\text{-}d]$$
$$i(e) = \text{minimum } \{\infty, 1 + 2 = 3\} = [3, a\text{-}b\text{-}e]$$

The indices of other vertices will remain unchanged.
These are shown attached with each vertex in Figure 15.20.

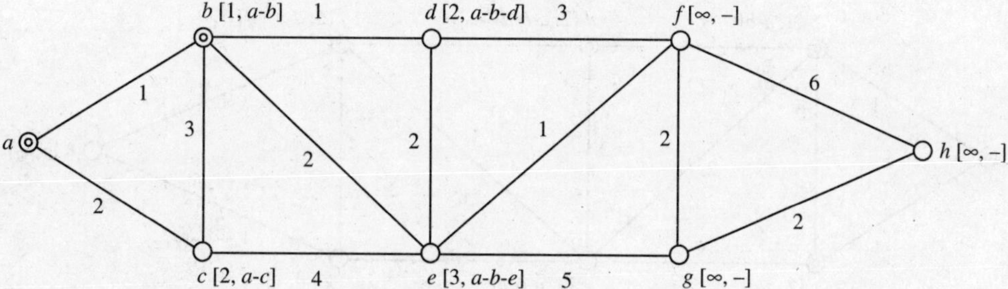

Figure 15.20 Move to vertex b.

3. Now, minimum index is

$$i(c) = [2, a\text{-}c]$$

and

$$i(d) = [2, a\text{-}b\text{-}d]$$

Choose any one of these arbitrarily. Say $i(c)$.
Mark c visited by enclosing it in a circle.
Since $c \neq h$, we have to continue.

Now, $P' = \{a, b, c\}$

and $T' = \{d, e, f, g, h\}$

Modify the index of each vertex in T', which is adjacent to c.

$$i(e) = \text{minimum } \{3, 2 + 4 = 6\} = [3, a\text{-}b\text{-}e]$$

The indices of other vertices will remain unchanged. See Figure 15.21.

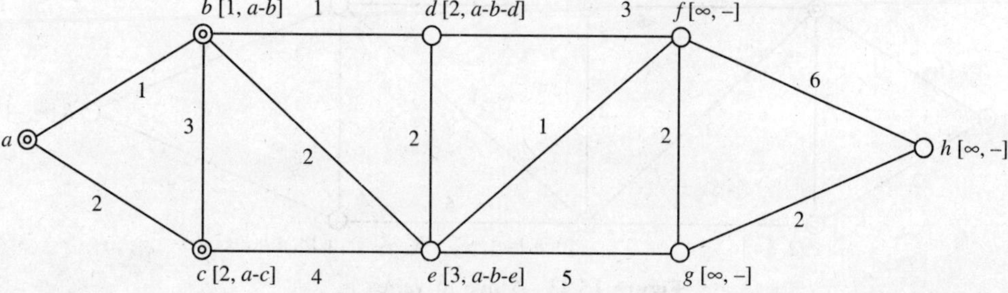

Figure 15.21 Move to vertex c.

4. Now, minimum index is

$$i(d) = [2, a\text{-}b]$$

Mark d visited by enclosing it in a circle.
Since $d \neq h$, we have to continue.

Now, $P' = \{a, b, c, d\}$

and $T' = \{e, f, g, h\}$

Modify the index of each vertex in T', which is adjacent to d.

$$i(e) = \text{minimum } \{3, 2 + 2 = 4\} = [3, a\text{-}b\text{-}e]$$

$$i(f) = \text{minimum } \{\infty, 2 + 3 = 5\} = [5, a\text{-}b\text{-}d\text{-}f]$$

The indices of other vertices will remain unchanged. See Figure 15.22.

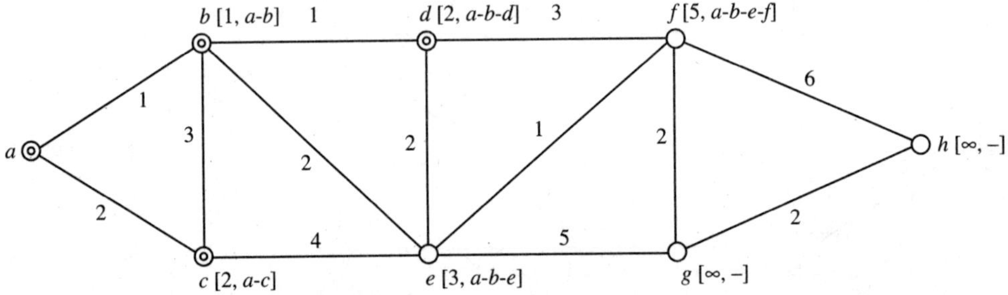

Figure 15.22 Move to vertex d.

5. Now, minimum index is

$$i(e) = [3, a\text{-}b\text{-}e]$$

Mark e visited by enclosing it in a circle.
Since $e \neq h$, we have to continue.

Now, $P' = \{a, b, c, d, e\}$

and $T' = \{f, g, h\}$

Modify the index of each vertex in T', which is adjacent to e.

$$i(f) = \text{minimum } \{5, 3 + 1 = 4\} = [4, a\text{-}b\text{-}e\text{-}f]$$

$$i(g) = \text{minimum } \{\infty, 3 + 5 = 8\} = [8, a\text{-}b\text{-}e\text{-}g]$$

See Figure 15.23.

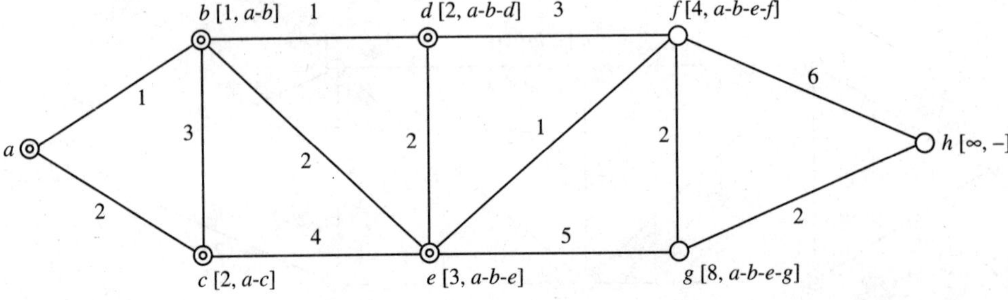

Figure 15.23 Move to vertex e.

6. Now, minimum index is

$$i(f) = [4, a\text{-}b\text{-}e\text{-}f]$$

Mark f visited by enclosing it in a circle.
Since $f \neq h$, we have to continue.

Now, $P' = \{a, b, c, d, e, f\}$

and $T' = \{g, h\}$

Modify the index of each vertex in T', which is adjacent to f.

$$i(g) = \text{minimum } \{8, 4 + 2 = 6\} = [6, a\text{-}b\text{-}e\text{-}f\text{-}g]$$

$$i(h) = \text{minimum } \{\infty, 4 + 6 = 10\} = [10, a\text{-}b\text{-}e\text{-}f\text{-}h]$$

See Figure 15.24.

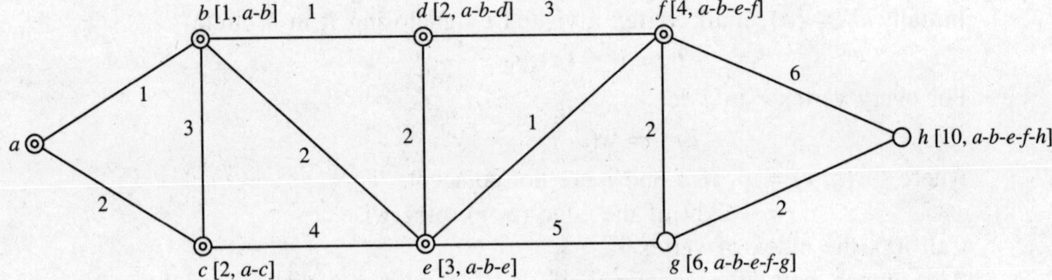

Figure 15.24 Move to vertex f.

7. Now minimum index is

$$i(g) = [6, a\text{-}b\text{-}e\text{-}f\text{-}g]$$

Mark g visited by enclosing it in a circle.
Since $g \neq h$, we have to continue.

Now, $P = \{a, b, c, d, e, f, g\}$

and $T = \{h\}$.

Modify the index of each vertex in T', which is adjacent to g. It is now h only.

$$i(h) = \text{minimum } \{10, 6 + 2 = 8\} = [8, a\text{-}b\text{-}e\text{-}f\text{-}g\text{-}h]$$

See Figure 15.25.

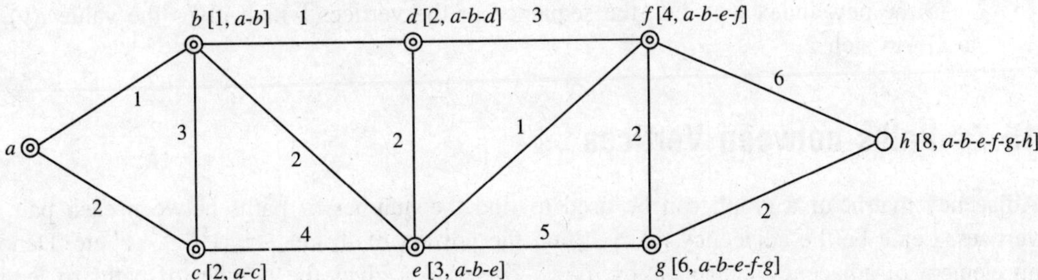

Figure 15.25 Move to vertex g.

8. Now, minimum index is $i(h)$ and it is at vertex h, therefore, stop.

The shortest path = a-b-e-f-g-h,

Shortest distance or distance = 8.

15.6.2 Shortest Paths to All Vertices

Dijkstra's shortest path algorithm can be modified to find the shortest path from a vertex to all other vertices of the graph.

Algorithm: Consider a connected graph $G = (V, E)$. The aim is to find the shortest path from a vertex a to all other vertices of the graph.

Partition the set V of vertices into two sets, P and T, such that P contains a and

$$T = V - P$$

1. Initially $P := \{a\}$; mark vertex a visited by enclosing it in a circle.

   $$T := V - \{a\},$$

   For every vertex r in T let

   $$i(r) := w(a, r),$$

   where $w(a, r) = \infty$, if a and r are not adjacent,

   $\qquad\qquad$ = weight of the edge (a, r), otherwise.

   Call $i(r)$, the index of vertex r.

   Also attach path, the sequence of vertices with $i(r)$ whose weight is $w(a, r)$ if $w(a, r) \neq \infty$. Leave blank, otherwise. Blank is shown by a dash symbol (–).
2. If T is empty then stop and exit. The shortest path of each vertex r from the vertex a is the sequence of vertices attached with $i(r)$, and the distance is $i(r)$.

   Else, select a vertex x in T that has the minimum index; mark x visited by enclosing it within a circle.
3. $P' := P \cup \{x\}$

   $T' := T - \{x\}$, for every vertex y adjacent to x in T', modify its index by the formula

   $$i(y) := \min\{i(y), i(x) + w(x, y)\}$$

   Also, $\quad p := p'$

   $$T := T'.$$

   Modify the path (or sequence of vertices) attached with each vertex y in T according to the new index $i(y)$, i.e., the sequence of the vertices which gives the value $i(y)$.
4. Go to step 2.

15.7 Paths between Vertices

Adjacency matrix of a graph can be used to find the number of paths between each pair of vertices. Let A be the adjacency matrix. Find the powers of A, i.e., A^2, A^3, ..., A^n etc. Denote an element of adjacency matrix A^k by $a_{k(i,\, j)}$. Then $a_{k(i,\, j)}$ gives the number of paths of length k from vertex v_i to vertex v_j.

Also, let $B_r = A + A^2 + A^3 + \cdots + A^r$. Then b_{ij}, element in ith row and jth column gives the number of paths of length r or less from vertex v_i to vertex v_j.

EXAMPLE 15.6 Consider the directed graph and its adjacency matrix given in Figure 15.26(a and b). Matrices A^2 and A^3 have also been computed and shown in Figure 15.26(c and d).

Notice that there is only one element in the matrix A^2. It shows that there is only one path of length 2 in the graph. This path is *a*-*b*-*c*. Matrix A^3 is null and, therefore, it verifies that there is no path of length 3 in the graph.

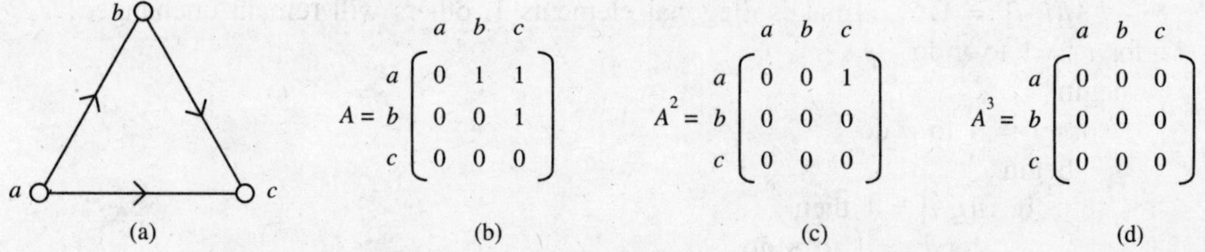

Figure 15.26 Graph G with adjacency matrix A and matrices A^2 and A^3.

15.8 Path Matrix

Let G be a graph with n vertices. Then a square matrix P of order $n \times n$ is called the path matrix of the graph G, such that the element

$p_{ij} = 1$, if there is a path from vertex v_i to v_j

$= 0$, otherwise.

Let A be the adjacency matrix of the graph G with n vertices. Compute A^2 , A^3, ..., A^n. Also, find

$$B_n = A + A^2 + A^3 + \cdots + A^n.$$

Form a matrix P by replacing the non zero elements of matrix B_n by 1. This P will be the path matrix of the graph G. If all the elements in the path matrix of a graph are 1, then it is called strongly connected.

EXAMPLE 15.7 Consider the graph given in Figure 15.26. The matrix B_n can be computed by adding matrices A, A^2 and A^3. Form path matrix P by replacing all non-zero elements of matrix B_n by 1. The matrix B_n and path matrix P is given in Figure 15.27.

$$B_n = \begin{array}{c} \\ a \\ b \\ c \end{array} \begin{array}{c} \begin{matrix} a & b & c \end{matrix} \\ \begin{bmatrix} 0 & 1 & 2 \\ 0 & 0 & 1 \\ 0 & 0 & 0 \end{bmatrix} \end{array} \qquad P = \begin{array}{c} \\ a \\ b \\ c \end{array} \begin{array}{c} \begin{matrix} a & b & c \end{matrix} \\ \begin{bmatrix} 0 & 1 & 1 \\ 0 & 0 & 1 \\ 0 & 0 & 0 \end{bmatrix} \end{array}$$

(a) Matrix B_n (b) Path matrix

Figure 15.27 Path matrix.

15.8.1 Warshall's Algorithm

There are several algorithms to find the path matrix of a graph G. We will discuss a popular algorithm given by **Warshall**#. A path matrix is also called **transitive closure** of the graph. Here, the algorithm is written in pseudo Pascal code. Consider a directed graph $G = (V, E)$ and let M be its adjacency matrix. The matrix M is considered Boolean consisting of elements 0 or 1. Path matrix P is computed by execution of this algorithm.

```
begin
    for i := 1 to n do
        M[i, i] := 1;     {makes diagonal elements 1, others will remain unchanged}
    for i := 1 to n do
     begin
        for j := 1 to n do
         begin
            if M[j, i] = 1 then
               for k := 1 to n do
                  M[j, i] := M[j, i] ∨ M[i, k];     {matrix M is modified at each iteration}
         end                                          {end of inner loop}
        end                                           {end of outer loop}
    for i := 1 to n do
        begin
            for j := 1 to n do
               P[j, i] := M[j, i];          {stores values of matrix M into path matrix P}
        end
end;                                        {end of algorithm}
```

Explanation

$M[i, i]$ is an element in the matrix M at ith row and jth column. Statement, for $i := 1$ to n do $M[i, i] := 1$, makes $M[i, i]$ to be executed n times with starting value of i as 1. Value of i is increased by 1 each time. In the statement $M[j, i] := M[j, i] \vee M[i, k]$, $\vee$ is a logical OR operator already discussed in logic. Sentences in the curly brackets are just for explanation and do not form part of the algorithm.

Stephen Warshall (1935–2006) was born in New York City. During his career, Warshall carried out research and development in operating systems, compiler design, language design, and operations research. Warshall received a bachelor's degree in mathematics in 1956 from Harvard University. He never went for an advanced degree since he did not find courses in his areas of interest in any of the universities. After graduating from Harvard, Warshall worked at ORO (Operation Research Office), a program set up by Johns Hopkins to do research and development for the United States Army. In 1961, he founded Massachusetts Computer Associates. Later, this company became part of Applied Data Research (ADR). There is an interesting anecdote about his proof on Warshall's algorithm. He and one of his colleagues at Technical Operations bet on who first could determine whether this algorithm always works. Warshall came up with a proof overnight, and won the bet, which however was symbolic. Since, Warshall did not like sitting at a desk, therefore, he did much of his work in unconventional places such as on a sailboat in the Indian Ocean or in a Greek lemon orchard. Warshall died on December 11, 2006 of cancer at his home in Gloucester, MA.

EXAMPLE 15.8 Consider the graph G given in Figure 15.28. Its adjacency matrix is given in Figure 15.29.

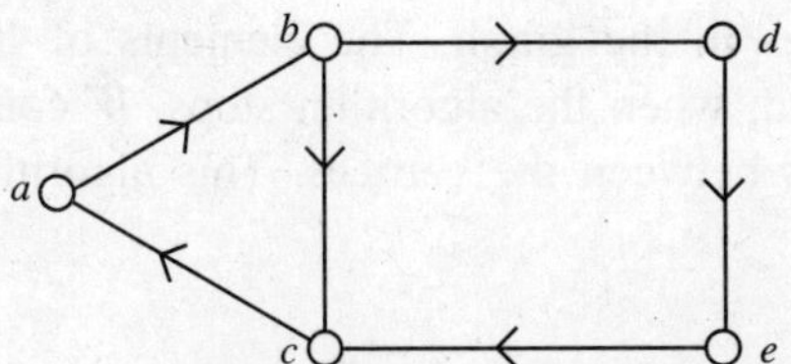

Figure 15.28 Graph G.

$$\begin{array}{c} \\ a \\ b \\ c \\ d \\ e \end{array} \begin{array}{c} \begin{array}{ccccc} a & b & c & d & e \end{array} \\ \begin{pmatrix} 0 & 1 & 0 & 0 & 0 \\ 0 & 0 & 1 & 1 & 0 \\ 1 & 0 & 0 & 0 & 0 \\ 0 & 0 & 0 & 0 & 1 \\ 0 & 0 & 1 & 0 & 0 \end{pmatrix} \end{array}$$

Figure 15.29 Adjacency matrix.

After execution of statement $M[i, i] := 1$, matrix $M[i, i]$ is transformed into the matrix given in Figure 15.30.

$$\begin{array}{c} \\ a \\ b \\ c \\ d \\ e \end{array} \begin{array}{c} \begin{array}{ccccc} a & b & c & d & e \end{array} \\ \begin{pmatrix} 1 & 1 & 0 & 0 & 0 \\ 0 & 1 & 1 & 1 & 0 \\ 1 & 0 & 1 & 0 & 0 \\ 0 & 0 & 0 & 1 & 1 \\ 0 & 0 & 1 & 0 & 1 \end{pmatrix} \end{array}$$

Figure 15.30 $M[i, i] := 1$.

$$P = \begin{array}{c} \\ a \\ b \\ c \\ d \\ e \end{array} \begin{array}{c} \begin{array}{ccccc} a & b & c & d & e \end{array} \\ \begin{pmatrix} 1 & 1 & 1 & 1 & 1 \\ 1 & 1 & 1 & 1 & 1 \\ 1 & 1 & 1 & 1 & 1 \\ 1 & 1 & 1 & 1 & 1 \\ 1 & 1 & 1 & 1 & 1 \end{pmatrix} \end{array}$$

Figure 15.31 Path matrix.

When algorithm is executed completely, the path matrix P is obtained. Since all the elements of P are 1, therefore, the graph G is strongly connected.

15.8.2 Alternative form of Warshall's Algorithm

Another form of Warshall's algorithm is given below.

```
begin
  for k := 1 to n do
    begin
      for i := 1 to n do
        begin
          for j := 1 to n do
          M[i, j] := M[i, j] ∨ (M[i, k] ∧ M[k, j]); {matrix M is modified at each iteration}
        end                                        {end of inner loop}
      end                                          {end of outer loop}
    for i := 1 to n do
      begin
        for j := 1 to n do
          P[j, i] := M[j, i];          {stores values of matrix M into path matrix P}
      end
end;                                   {end of algorithm}
```

15.8.3 Shortest Path between Each Pair of Vertices

Warshall's algorithm can be modified to find the shortest path between each pair of vertices in a directed weighted graph. Consider a directed graph $G = (V, E)$ and let W be its adjacency matrix with its elements as the weights on the edges in the graph. The elements of W are modified repeatedly in each scan in a loop. At the end, when the algorithm stops, W contains elements, which are the weights of the shortest paths between the vertices. This algorithm is similar to Dijkstra's shortest path algorithm.

```
begin
   for k := 1 to n do
     begin
       for i := 1 to n do
         begin
           for j := 1 to n do
             begin
             W[i, j] := min(W[i, j], W[i, k] + W[k, j]);   {computes minimum weight}
             end                                            {end of inner loop of variable j}
         end                                                {end of middle loop}
     end                                                    {end of outer loop of variable k}
end;                                                        {end of algorithm}
```

Remark: A program in C++ language that computes a path matrix is given below. It also computes shortest path between each pair of vertices.

A program in C++ language that computes a path matrix

```
/**********************************************************************/

//Program to find path matrix and shortest path by Warshall's algorithm:
#include<iostream.h>
#define MAX 20
#define infinity 9999
  int main()
    {
      int display(int matrix[MAX][MAX],int n)  // function prototype
      int minimum(int a, int b)                // function prototype
      int i,j,k,n;
      int w_adj[MAX][MAX], adj[MAX][MAX], path[MAX][MAX],s_path[MAX][MAX];
      printf("Enter the number of vertices : ");
      scanf("%d", &n);
      printf("Enter weighted adjacency matrix of the graph :\n");
    for(i=0; i<n; i++)
      for(j=0; j<n; j++)
        scanf("%d", &w_adj[i][j]);
        printf("The weighted adjacency matrix is :\n");
        display(w_adj, n);
    // Make boolean adjacency matrix
    for(i=0; i<n; i++)
      for(j=0; j<n; j++)
        if (w_adj[i][j]==0)
```

```
            adj[i][j]=0;
          else
            adj[i][j]=1;
            printf("The adjacency matrix is :\n");
            display(adj, n);
        for(i=0; i<n; i++)
          for(j=0; j<n; j++)
          path[i][j]=adj[i][j];
        for(k=0; k<n; k++)
          {
            for(i=0; i<n; i++)
              for(j=0; j<n; j++)
                path[i][j]=( path[i][j] || (path[i][k] && path[k][j]) );
          }
            printf("Path matrix of the given graph is :\n");
            display(path,n);
          for(i=0; i<n; i++)
            for(j=0; j<n; j++)
              if (w_adj[i][j]==0)
                s_path[i][j]=infinity;
              else
                s_path[i][j]=w_adj[i][j];
            for(k=0; k<n; k++)
              {
                for(i=0; i<n; i++)
                  for(j=0; j<n; j++)
                   s_path[i][j]=minimum(s_path[i][j],
                                              s_path[i][k]+s_path[k][j] );
              }
          printf("Shortest path matrix is :\n");
          display(s_path, n);
        }
    int display(int matrix[MAX][MAX], int n)
      {
        int i,j;
          for(i=0; i<n; i++)
            {
              for(j=0; j<n; j++)
                printf("%3d",matrix[i][j]);
                printf("\n");
            }
          return(0);
        }/*End of display()*/
int minimum(int a, int b)
   {
      if(a<=b)
        return a;
      else
        return b;
      return(0);
    }/*End of minimum()*/
/**********************************************************************/
```

15.9 Cutpoint, Bridge, Cutset and Connectivity

In this section, we will discuss cutpoint, bridge, cutset and connectivity of a graph.

15.9.1 Cutpoint

A vertex in a graph G is called a cutpoint if its deletion makes the graph disconnected. In the graph given in Figure 15.32, vertices b, e and f are the three cutpoints.

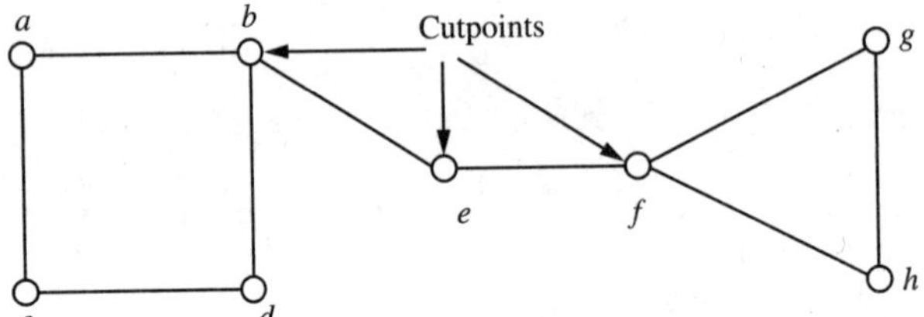

Figure 15.32 Cutpoints.

15.9.2 Bridge

An edge of a graph is called a bridge if its deletion makes the graph disconnected. In the graph given in Figure 15.33, edges (e, f) and (b, e) are bridges.

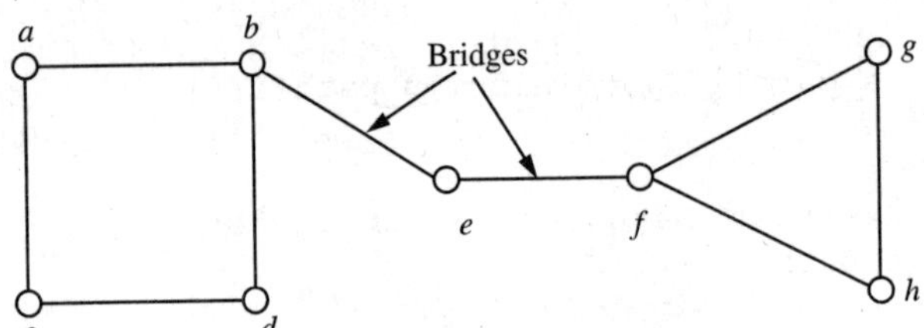

Figure 15.33 Bridges.

15.9.3 Edge-cutset

An edge-cutset of a graph is the collection of edges whose removal makes the graph disconnected. In the graph given in Figure 15.34, set $\{(b, e), (c, f), (d, g)\}$ is one of the edge-cutsets.

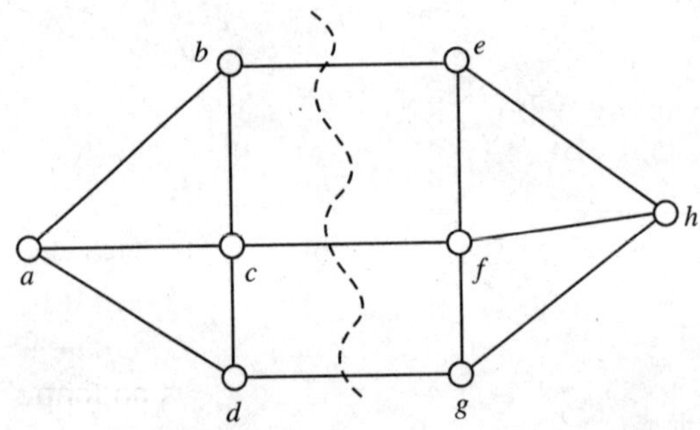

Figure 15.34 Edge-cutset $\{(b, e), (c, f), (d, g)\}$.

15.9.4 Vertex-cutset

A vertex-cutset of a graph is the collection of vertices whose removal makes the graph disconnected. In the graph given in Figure 15.35, set $\{b, c, d\}$ is one of the vertex-cutsets.

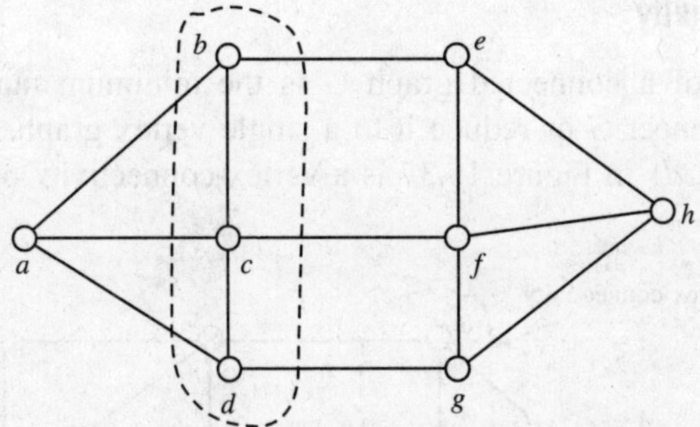

Figure 15.35 Vertex-cutset $\{b, c, d\}$.

15.9.5 Partition-cutset

Let V_1 and V_2 be two sets of vertices of graph G, such that V_1 and V_2 form a partition of G. Then the collection of all edges of G having one end point in V_1 and the other end point in V_2 is called a partition-cutset. Removal of partition-cutset from the graph makes the graph disconnected. A partition-cutset is denoted as $< V_1, V_2 >$. In the graph given in Figure 15.36, let

$$V_1 = \{a, b, c, d\}$$

and

$$V_2 = \{e, f, g, h\}$$

Then partition-cutset

$$< V_1, V_2 > = \{(b, e), (c, f), (d, g)\}$$

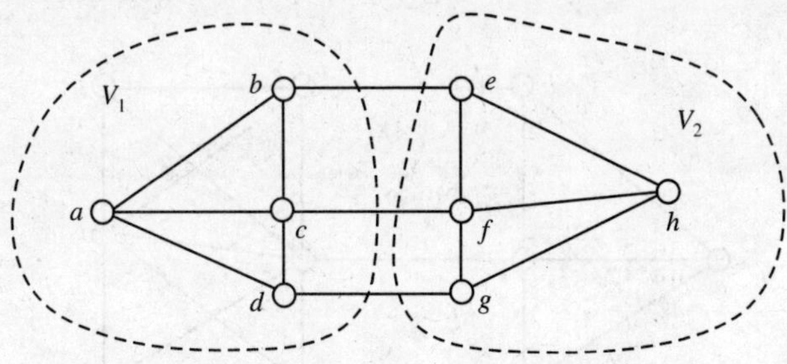

Figure 15.36 Partition-cutset $\{(b, e), (c, f), (d, g)\}$.

THEOREM 15.3 Every cutset in a connected graph G contains at least one edge of every spanning tree of G.

Proof: Consider a graph G and let S be a cut set of it. Also, let T be a spanning tree of it.

Let the cut set do not contain an edge of the tree T. Then it is obvious that the tree T, and hence, G will remain connected. Note that a spanning tree contains all the vertices of a graph, which are connected by the edges. Further, in fact, there is no loop or closed path in a spanning tree, and therefore, all the edges are essential to keep the spanning tree a connected graph. Clearly, to disconnect the tree, at least one edge has to be deleted. Therefore, every cut set in a connected graph G contains at least one edge of every spanning tree of G.

15.9.6 Vertex-connectivity

The vertex-connectivity of a connected graph G is the minimum number of vertices whose deletion can either disconnect G or reduce it to a single vertex graph. It is denoted as K_v. For instance, vertex set $\{b, c, d\}$ in Figure 15.37 is a vertex-connectivity of the given graph. Here, $K_v = 3$.

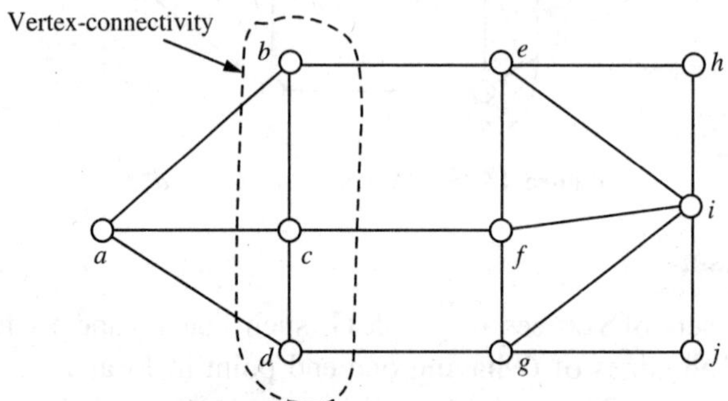

Figure 15.37 Vertex-connectivity $\{b, c, d\}$.

15.9.7 Edge-connectivity

The Edge-connectivity of a connected graph G is the minimum number of edges whose deletion can disconnect G. It is denoted as K_e. For instance, edge set $\{(b, e), (c, f), (d, g)\}$ in Figure 15.38 is a vertex-connectivity of the given graph. Here, $K_e = 3$.

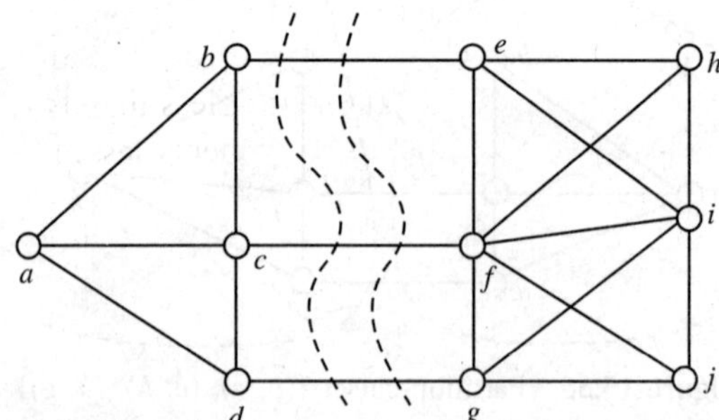

Figure 15.38 Edge-connectivity $K_e = 3$ with edge-cutset $\{(b, e), (c, f), (d, g)\}$.

15.9.8 Internally Disjoint Paths

Suppose, G be a graph. Let a and b be any two of its vertices. Let us call a path of G, ***a–b* path** if a is initial vertex and b is final vertex of the path. A vertex v is called the **internal vertex** of a–b path if it is neither the initial vertex nor the final vertex of the path. A set of ***a–b* paths** in G is said to be **internally disjoint** if no two paths in the set has an internal vertex in common.

A set of three internally disjoint *a-h* paths in the graph given in Figure 15.38 is {*a-b-e-h*, *a-c-f-h*, *a-d-g-h*}. Another set of internally disjoint *a-h* paths in this graph is {*a-b-c-f-e-h*, *a-d-g-h*}. Readers may observe that there are some more such paths in the graph.

Three internally disjoint *a-h* paths are shown by three different dotted paths in Figure 15.39.

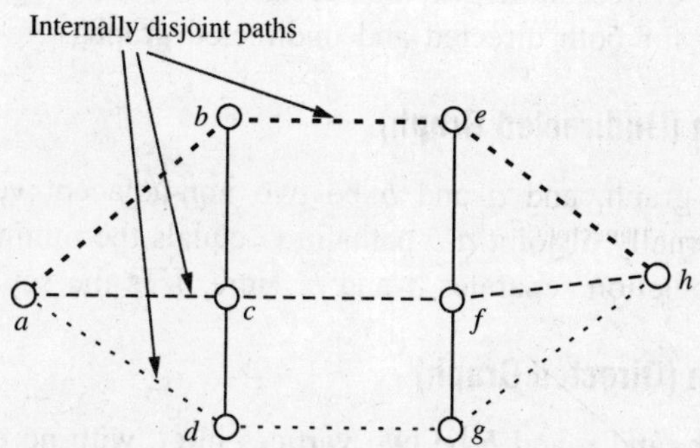

Figure 15.39 Three internally disjoint paths.

15.9.9 *k*-vertex-connected Graph

Let *G* be a graph. Then *G* is said to be *k*-vertex-connected, if *G* is connected and $K_v \geq k$. If *G* has non-adjacent vertices, then *G* is *k*-vertex-connected if every vertex-cutset has at least *k* vertices.

15.9.10 *k*-edge-connected Graph

Let *G* be a graph. Then *G* is said to be *k*-edge-connected if *G* is connected and $K_e \geq k$.

EXAMPLE 15.9 Consider a graph *G* given in Figure 15.40. In this graph, there is no cutpoint. However, there are three vertex-cutsets: {*b*, *d*}, {*e*, *g*}, and {*d*, *e*}, each consisting of two elements. Obviously, there is no vertex-cutset having less than two elements. Thus, $K_v = 2$. Similarly, there is no edge-cutset with a number of elements less than 3. Hence, $K_e = 3$.

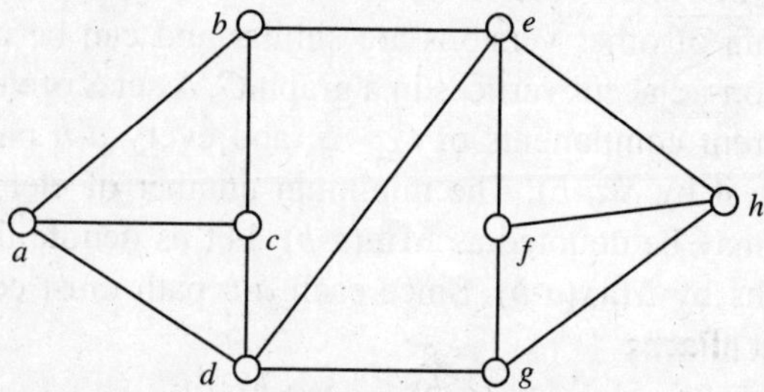

Figure 15.40 Graph *G* with $K_v = 2$ and $K_e = 3$.

15.10 Menger's Theorem

Menger's theorem is a basic result for connectivity in graphs. It was stated and proved by Karl Menger in 1927. Originally, it was stated for undirected graphs, but it applies to directed graphs too. There are several analogues of Menger's theorem, which resulted due to research work carried out by a number of researchers in this area. Vertex and edge versions of Menger's theorem are stated below for both directed and undirected graphs.

15.10.1 Vertex Version (Undirected Graph)

Let G be an undirected graph, and a and b be two non-adjacent vertices in G. Then, the maximum number of internally disjoint a-b paths in G equals the minimum number of vertices from $V - \{a, b\}$ whose deletion separates a and b. Here, V is the set of all vertices of G.

15.10.2 Vertex Version (Directed Graph)

Let D be a directed graph, and a and b be two vertices in G, with no edge from a to b. Then, the maximum number of internally disjoint directed a-b paths in D equals the minimum number of vertices from $V - \{a, b\}$ whose deletion removes all directed paths from a to b.

15.10.3 Edge Version (Undirected Graph)

Let G be an undirected graph, and a and b be two non-adjacent vertices in G. Then, the maximum number of edge-disjoint a-b paths in G equals the minimum number of edges from E whose deletion separates a and b. Here, E is the set of all edges of G.

15.10.4 Edge Version (Directed Graph)

Let D be a directed graph, and a and b be two vertices in D, with no edge from a to b. Then, the maximum number of edge-disjoint directed a-b-paths in D equals the minimum number of edges in E whose deletion removes all directed paths from a to b.

15.10.5 Proof of Menger's Theorem

We will discuss the proof of vertex version (Undirected graph) of Menger's theorem stated in Section 15.10.1. The proofs of other versions are similar and can be constructed analogously.

Let a and b be two non-adjacent vertices in a graph G. A set S of vertices is ***a-b* separating** set if a and b lie in different components of $G - S$, and every a-b path contains at least one vertex of S. Let us denote it by $S(a\text{-}b)$. The minimum number of elements in $S(a\text{-}b)$ is called the ***a-b* connectivity** and may be denoted as **Min(*a-b*)**. Let us denote the maximum number of internally disjoint a-b paths by **Max(*a-b*)**. Since each a-b path must contain a different vertex from $S(a\text{-}b)$ set, it is clear that

$$\text{Max}\,(a\text{-}b) \leq |\, S(a\text{-}b)\,|.$$

We now show that

$$\text{Max}(a\text{-}b) = \text{Min}(a\text{-}b).$$

Let

$$k = \text{Max}(a\text{-}b).$$

Then no set having less than k vertices separates a and b. Let us show, by induction on k, that if

$$\text{Max}(a\text{-}b) \geq k, \text{ then } S(a\text{-}b) \geq k$$

Thus in particular $\text{Max}(a\text{-}b) = k$ implies that $\text{Min}(a\text{-}b) = k$, which is the desired result.

Basis: Let $k = 1$. Obviously, there is one a-b path. Figure 15.41 shows $k = 1$. At least one vertex, p in this case, has to be removed to disconnect a and b.

Thus, $$\text{Max}(a\text{-}b) = \text{Min}(a\text{-}b)$$

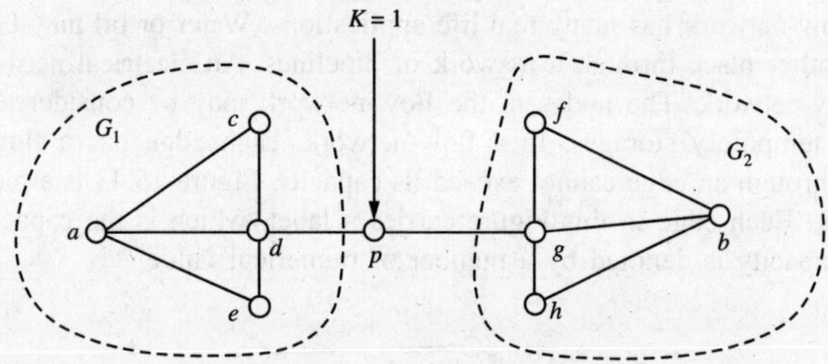

Figure 15.41 $S = \{p\}$ and $\min(a\text{-}b) = 1$.

Induction step: Let $k \geq 2$ and that theorem is true for k, i.e.,

$$\text{Min}(a\text{-}b) = \text{Max}(a\text{-}b) = k$$

Assume further that a and b are non-adjacent vertices in G with $\text{Max}(a\text{-}b) \geq k + 1$. By the induction hypothesis, there are k internally disjoint a-b paths $P_1, P_2, \ldots, P_k$. Now if

$$\text{Max}(a\text{-}b) = k + 1$$

then it is obvious that at least one additional vertex, which lie on this path needs to be deleted to separate a and b. Thus, $\text{Min}(a\text{-}b)$ will become $k + 1$ in place of k. It shows that theorem is true for $k + 1$. Induction case is shown in Figure 15.42.

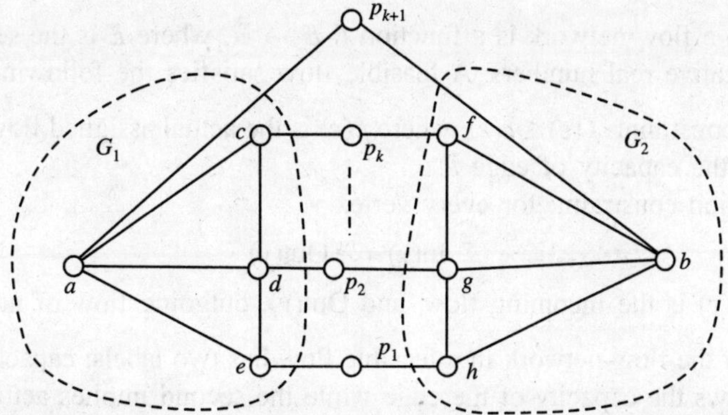

Figure 15.42 $S = \{p_1, p_2, \ldots, p_{k+1}\}$ and $\min(a\text{-}b) = \max(a\text{-}b)$.

15.11 Flow-network

A flow-network, $G = (V, E)$ is a directed graph with V, the set of vertices (or **nodes**), and E, the set of edges (or **arcs**), such that each edge $e \in E$ has a **capacity** $c(e) \geq 0$. In the flow-network, if a node has only outgoing edges it is called **source**. In contrary, if a node has only incoming edges, it is called **sink.** In graph, source node is designated with s, and sink as t.

Other names of flow-network that appear in other texts are **transport network** and **capacited-network.** However, we will keep on using the word flow-network or network in our discussion. A flow-network has many real life applications. Water or oil may be supplied from one place to another place through a network of pipelines. An electrical network can also be treated as a flow-network. The nodes in the flow-network may be considered as distribution points, or even temporary storages. In a flow-network, each edge has a flow capacity, and therefore, flow through an edge cannot exceed its capacity. Figure 15.43 is a pictorial model of the flow-network. Each edge in this Figure carries a label, which is the capacity of the edge. Generally, the capacity is denoted by a number or numerical value.

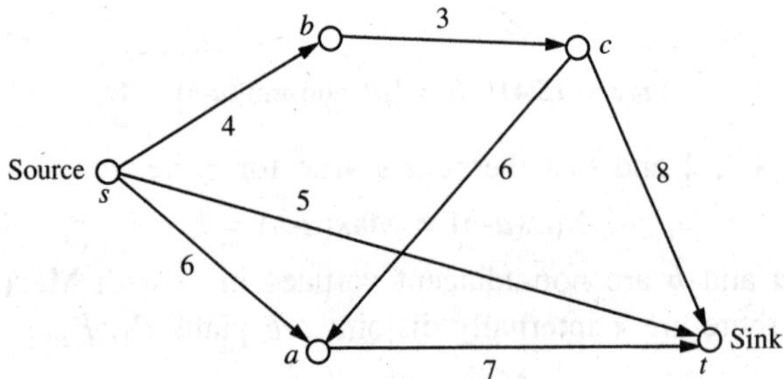

Figure 15.43 Flow-network.

15.12 Feasible Flow

A **feasible flow** in a flow-network is a function $f: E \to \mathbb{R}$, where E is the set of edges, and $\mathbb{R}$, the set of non-negative real numbers. A feasible flow satisfies the following constraints:

1. Capacity constraint: $f(e) \leq c(e)$, where $f(e)$ is the actual assigned flow value to an edge, and $c(e)$, the capacity of edge e.
2. Conservation constraint: for every vertex v,

$$\Sigma \text{ In}(v) = \Sigma\text{Out}(v)$$

where, In(v) is the incoming flow, and Out(v), outgoing flow of node v.

Each edge of the flow-network in a feasible flow has two labels: capacity and actual flow. The first digit shows the capacity of the edge while the second implies actual flow through it. A feasible flow is shown in Figure 15.44.

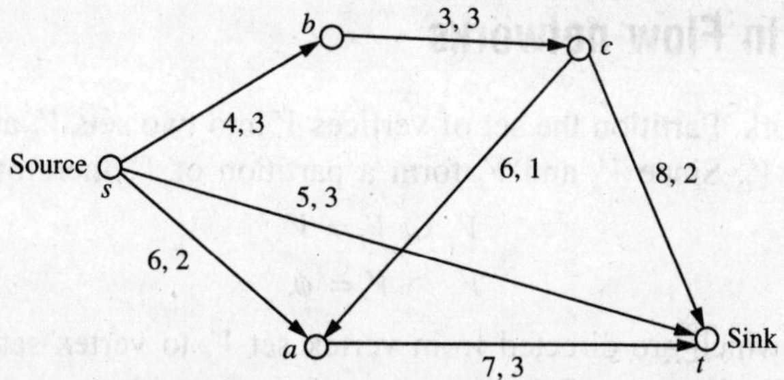

Figure 15.44 A feasible flow in a flow-network.

15.13 Multiple Sources/Sinks

A flow-network may contain more than one source, and more than one sink nodes. However, for all theoretical purposes, such network can always be converted into a single source and single sink network by adding all sources of the network to a single source called **super source**, and all sinks to a single sink, called **super sink**. A single source and single sink flow-network is called an *s-t* network Figure 15.45 shows a multiple sources/sinks flow-network. This network is modified into a single source and single sink network and is given in Figure 15.46.

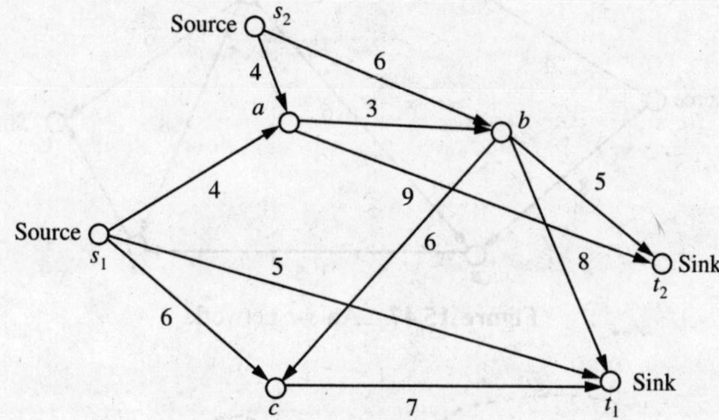

Figure 15.45 Multiple sources/sinks.

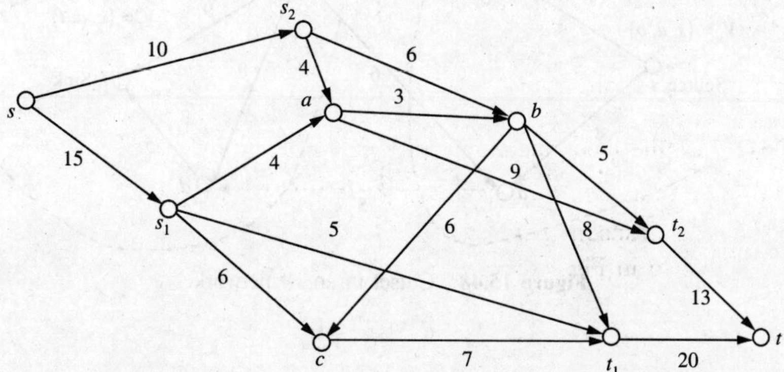

Figure 15.46 Super-source and super-sink *s-t* network.

15.14 Cutsets in Flow-networks

Let F be a flow-network. Partition the set of vertices V into two sets V_s and V_t, such that source $s \in V_s$, and sink $t \in V_t$. Since V_s and V_t form a partition of V, therefore, it is obvious that

$$V_s \cup V_t = V$$

and
$$V_s \cap V_t = \phi.$$

The set of all edges, which are directed from vertex set V_s to vertex set V_t is called an s-t cut and is denoted as $< V_s, V_t >$. A **minimum *s-t* cut** is a cut with minimum capacity.

EXAMPLE 15.10 Consider the flow-network given in Figure 15.47.

Let
$$V_s = \{s, a, b\}$$

and
$$V_t = \{c, d, t\}.$$

Then, the cutset of this flow-network is $\{(b, c), (a, d)\}$. It is also a **minimum cutset** with capacity 3 + 5 = 8. The edges of the cutset are shown by dotted arrows in Figure 15.48. Note that edge (c, a) is not in the cutset of this network, since it is directed from vertex c, which is in V_t, to the vertex a in V_s.

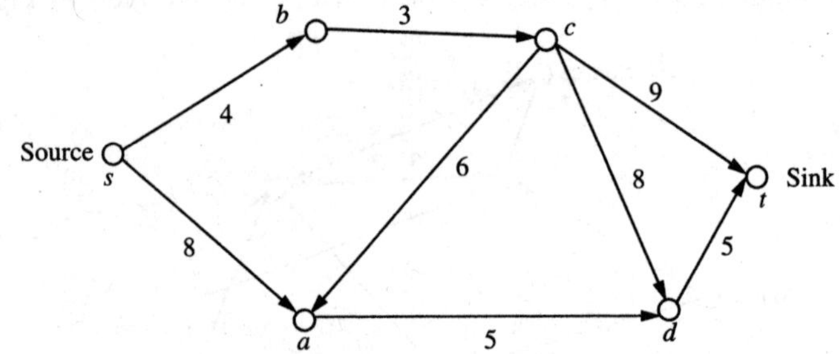

Figure 15.47 An s-t network.

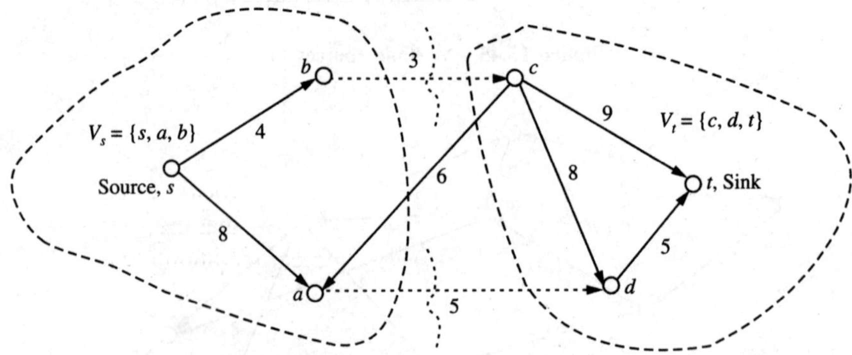

Figure 15.48 Cutset in an s-t network.

15.15 Relation between Flows and Cuts

There is a direct relationship between a flow and a cut. In fact, the value of any flow equals the total flow across the edges of the cutset $< \{s\}, V-\{s\} >$. Here, V is the set of vertices,

Further, $$V_s = \{s\}$$

and $$V_t = V - \{s\}.$$

Let **Val(*f*)** denotes the **net flow** leaving the source s, then

$$\text{Val}(f) = \sum_{e \in \text{Out}(s)} f(e) - \sum_{e \in \text{In}(s)} f(e)$$

In other words,

$$\text{Val}(f) = \sum_{e \in <\{s\}, V-\{s\}>} f(e) - \sum_{e \in <V-\{s\}, \{s\}>} f(e).$$

EXAMPLE 15.11 Consider the flow-network given in Figure 15.49. In this network, Val(f) = 8.

$$\begin{aligned}\text{Val}(f) &= \sum_{e \subset <\{s,a,b\}, \{c,d,t\}>} f(e) - \sum_{e \in <V-\{c,d,t\},\ \{s,a,b\}>} f(e) \\ &= 9 - 1 \\ &= 8\end{aligned}$$

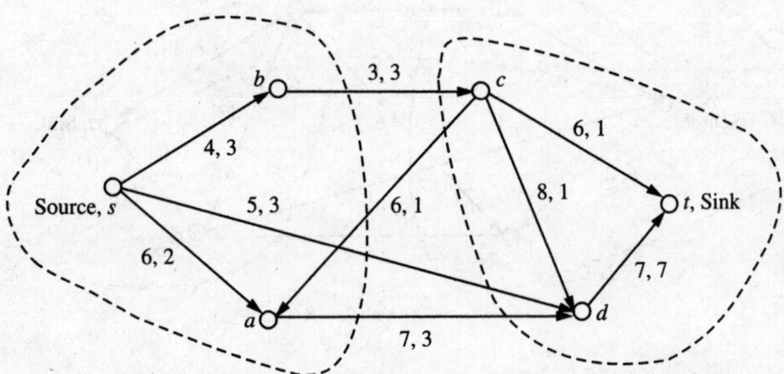

Figure 15.49 Value of flow Val(f) = 8.

15.16 Maximum Flow Problem

Finding maximum flow in a network is a problem of great significance. It is an **assignment of the flow value $f(e)$** to each edge of the network in such a way that the net flow of the network is maximum, such that the two constraints: capacity and conservation, given in Section 15.12, are maintained. The maximum flow in an s-t network can be found by augmenting the flow iteratively until no further augmentation is possible. It may be seen that an assignment of a maximum flow is not unique, however, the **maximum flow value** is unique. Let f be a flow in a network G (V, E), and let there be a directed path

$$p = s\text{-}e_1\text{-}v_1\text{-}e_2\text{-}v_2\text{-}\cdots\text{-}e_k\text{-}t$$

such that

$$f(e_i) < c(e_i) \quad \text{for} \quad i = 1 \text{ to } k$$

where $\qquad s, v_1, v_2, \ldots, t \in V$

and $\qquad e_1, e_2, \ldots, e_k \in E.$

Now, considering the capacity on each edge, the flow can be augmented by as much as

$$\Delta_p = c(e_i) - f(e_i).$$

However, to maintain the conservation of flow at each node v_i, increase on all the edges of the path must be equal.

Thus, the largest possible value of Δ_p = minimum $\{c(e_i) - f(e_i)\}$.

EXAMPLE 15.12 Consider the flow-network given in Figure 15.50(a). In this network, Val(f) = 8. Further, consider the path *s-b-c-t*. On this path, we can increase the flow by $\Delta_p = 1$, since this is the maximum possibility on the edge *s-b*. The network after augmenting the flow is given in Figure 15.50(b).

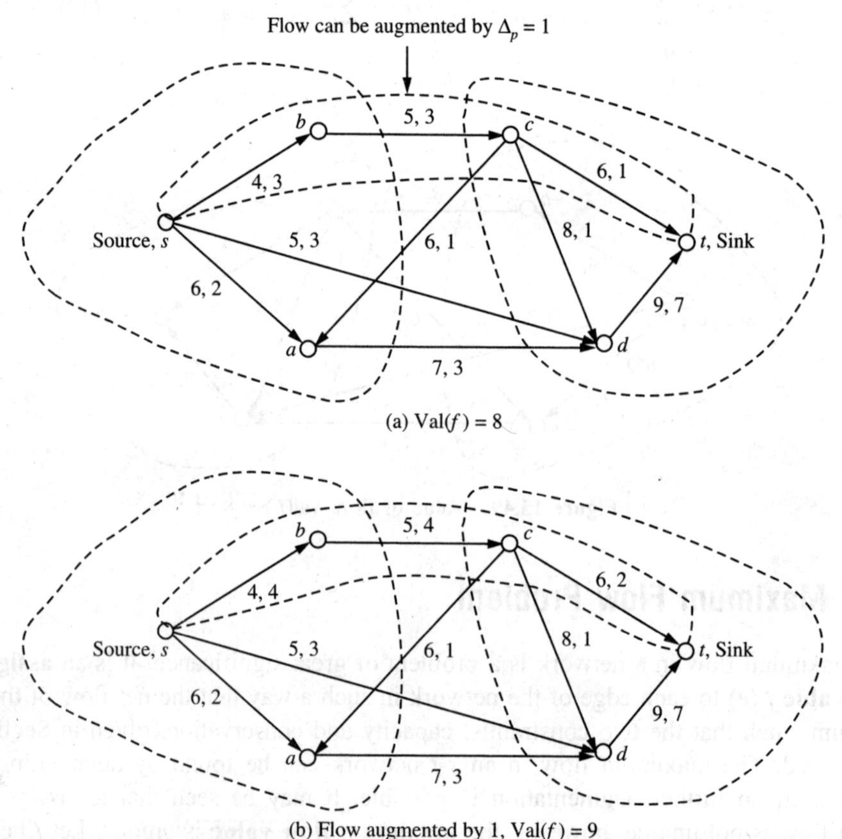

Figure 15.50 Flow augmenting in a network.

15.16.1 *f*-augmenting Path

A semi *s-t* path is a sequence of vertices and edges

$$p = s\text{-}e_1\text{-}v_1\text{-}e_2\text{-}v_2\text{-}\dots\text{-}e_k\text{-}t$$

such that each vertex except t, is followed by an edge, which may be either a **forward edge** or a **backward edge**. An edge e_i is called a forward edge if it is directed from vertex v_{i-1} to vertex v_i, and edge e_i is called backward edge if it is directed from v_i to v_{i-1}. For instance, edge e_2 is backward edge in semi *s-t* path given in Figure 15.51(a). Rest of the edges e_1, e_3 and e_4 are forward edges. A **directed *s-t* path** is a semi *s-t* path whose edges are all forward. A directed *s-t* path is given in Figure 15.51(b).

An *f*-augmenting path is a semi *s-t* path such that the flow on each forward edge can be increased, and the flow on each backward edge can be decreased. Further, for each edge e on an *f*-augmenting path:

$f(e) < c(e)$, if e is a forward edge,
$f(e) > 0$, if e is a backward edge.

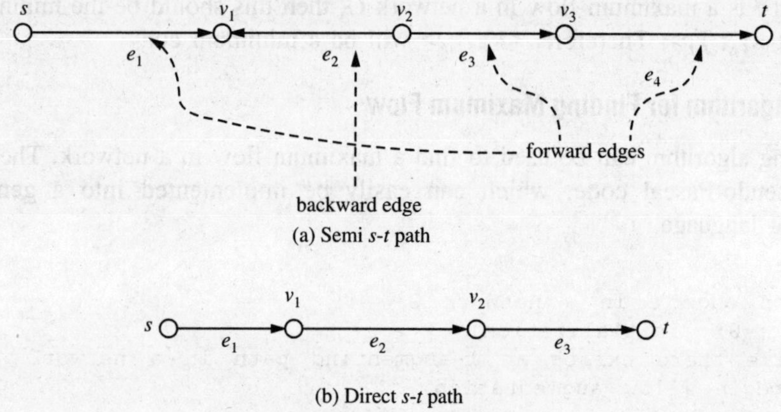

(a) Semi *s-t* path

(b) Direct *s-t* path

Figure 15.51 *s-t* paths.

For each edge e on an *f*-augmenting path, the quantity Δ_e is called the **slack** on edge e, and is given by,

$$\Delta_e = \begin{cases} c(e) - f(e), & \text{if } e \text{ is a forward edge,} \\ f(e), & \text{if } e \text{ is a backward edge.} \end{cases}$$

Let p be an *f*-augmenting path with minimum slack Δ_p on its edges. Then, a feasible augmented flow is given by

$$\hat{f}(e) = \begin{cases} f(e) + \Delta_p, & \text{if } e \text{ is a forward edge of } p \\ f(e) - \Delta_p, & \text{if } e \text{ is a forward edge of } p, \\ f(e), & \text{otherwise.} \end{cases}$$

In this case, $\text{Val}(\hat{f}) = \text{Val}(f) + \Delta_p$.

15.16.2 Max-flow Min-cut Theorem

The theorem states that for a given network, the value of a maximum flow is equal to the capacity of a minimum cut.

Proof: We have used induction method to prove Menger's theorem in Section 15.10. A similar approach may work for this theorem too. It is left as an exercise for the students. However, the logic given below will help in establishing the validity of the theorem.

Let $< V_s, V_t>$ be an *s-t* cut in a network *G*, and let *f* be a flow assigned to the network such that

$$f(e) = \begin{cases} c(e), & \text{if } e \in < V_s, V_t > \\ 0, & \text{if } e \in < V_t, V_s >. \end{cases}$$

Then, $$\text{Val}(f) = \sum_{e \in <V_s, V_t>} f(e).$$

Therefore, it is obvious that *f* is a maximum flow.

Now, if *f* is a maximum flow in a network *G*, then this should be the minimum capacity of the cutset $<V_s, V_t>$. Therefore, $<V_s, V_t>$ will be a minimum cut.

15.16.3 Algorithm for Finding Maximum Flow

The following algorithm can be used to find a maximum flow in a network. The algorithm is given in pseudo-Pascal code, which can easily be implemented into a general-purpose programming language.

```
begin
   for each edge e in a network G
     f(e) :=0; {initialization}
       while there exists an f-augmenting path in a network
         begin {flow augmentation}
           find an f-augmenting path p;
             let Δp = minimum{Δe}, where e ∈ p;
               for each edge e ∈ p
                 begin
                   if e is a forward edge then
                     f(e) := f(e) + Δp
                   else { e is a backward edge}
                     f(e) := f(e) - Δp;
                 end;
         end;
end;
```

15.17 Matching in a Graph

A **matching** in a graph *G* is a set of edges, which have no endpoints in common. It is also called a **matching set** and is denoted by *M*. Let $e \in M$ be an edge of graph *G* and the vertices *x* and *y* be the endpoints of edge e, then, it is said that vertex x **is matched with vertex** y. A **maximum matching** is a matching with maximum number of edges. A matching is called **maximal,** if it is not a proper subset of any other matching. **A maximum matching is always maximal, but a maximal may not be a maximum.** A **perfect matching** is a matching, which matches all the vertices of the graph, that is, every vertex of the graph is the end-vertex of an edge of the matching. Note that if a graph has an odd number of vertices, then it cannot have a perfect matching. Every perfect matching is maximum, and thus maximal. A **near-perfect matching** is one in which exactly one vertex is unmatched. This may occur when a graph has an odd number of vertices, and the matching defined is maximum.

EXAMPLE 15.13 Consider the graph given in Figure 15.52. The singleton matching sets are: $\{e_1\}$, $\{e_2\}$, $\{e_3\}$ and $\{e_4\}$. The edge sets $\{e_1, e_4\}$ and $\{e_2, e_3\}$ are both, maximum as well as maximal matchings. These are also perfect matchings. However, edge set $\{e_5\}$ is maximal but not maximum matching.

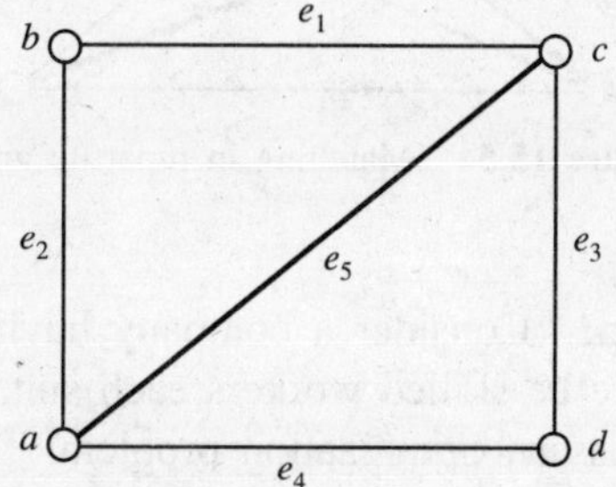

Figure 15.52 Matching $\{e_5\}$ is maximal but not maximum.

EXAMPLE 15.14 Consider the graph given in Figure 15.53. A matching $M = \{(b, c), (d, e)\}$ is an example of a near-perfect matching.

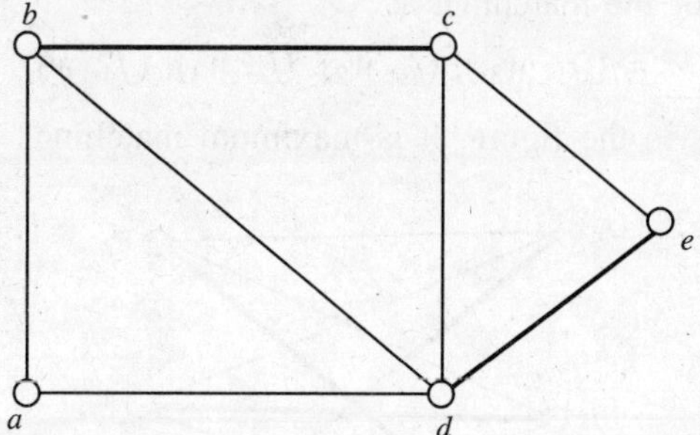

Figure 15.53 Near perfect matching.

15.17.1 Matching in a Bipartite Graph

We know that in a bipartite graph, a vertex bipartition exists. Let G be a bipartite graph, such that $< X, Y >$ be a bipartition, then the set of edges having one end in X and other in Y without common endpoints is a matching. The bipartite matching has a number of applications. Personnel-assignment and groom-bride matching are the potential matching problems.

EXAMPLE 15.15 Consider a bipartite graph $K_{3,3}$ given in Figure 15.54.

Here $$X = \{x_1, x_2\ x_3\}$$

and $$Y = \{y_1, y_2, y_3\}.$$

In this graph, edge sets $\{e_1, e_5, e_9\}$, $\{e_3, e_4, e_8\}$ and $\{e_2, e_6, e_7\}$ are all, maximum as well as maximal matchings.

One of these matchings, $\{e_2, e_6, e_7\}$ is shown by bold edges in the Figure 15.54.

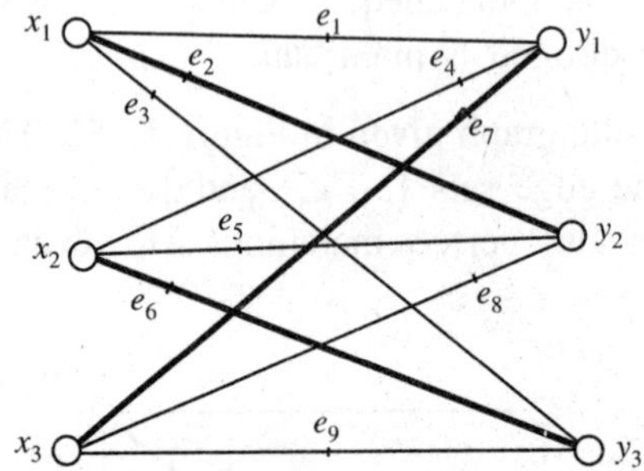

Figure 15.54 Matching in bipartite graph.

EXAMPLE 15.16

Personnel assignment problem: Consider a company having some different types of jobs that need special skills. Let there be skilled workers each suitable for some specific job. Then assigning a job to each worker is an optimization problem. This can be solved by maximum matching in the graph. Consider a graph given in Figure 15.55.

Here $$X = \{j_1, j_2, j_3, j_4\}$$

is a set of jobs,

and $$Y = \{w_1, w_2, w_3, w_4\}$$

is a set of workers. One of the matchings is

$$M = \{(j_1, w_3), (j_2, w_2), (j_3, w_1), (j_4, w_4)\}.$$

It is shown by bold edges in the figure. It is maximum matching.

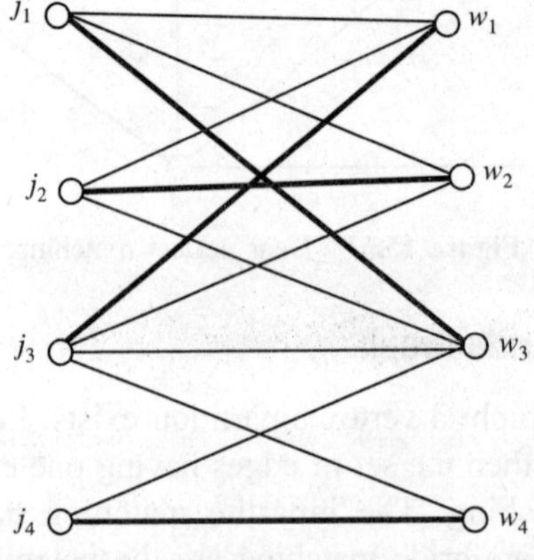

Figure 15.55 Personnel-assignment matching.

EXAMPLE 15.17

Brides-grooms matching problem: Consider matching among n grooms and n brides. The data on which couples will or will not be happy together is given. The aim is to create as many happy couples as possible. The compatibilities among brides and grooms may be represented in terms of a graph as depicted in Figure 15.56. The problem of finding a matching that maximizes the number of happy couples is known as the matching problem.

Let
$$X = \{b_1, b_2, \ldots, b_n\}$$
is a set of brides,

and
$$Y = \{g_1, g_2, \ldots, g_n\}$$
is a set of grooms. There may be many matching solutions depending on the choices of brides and grooms.

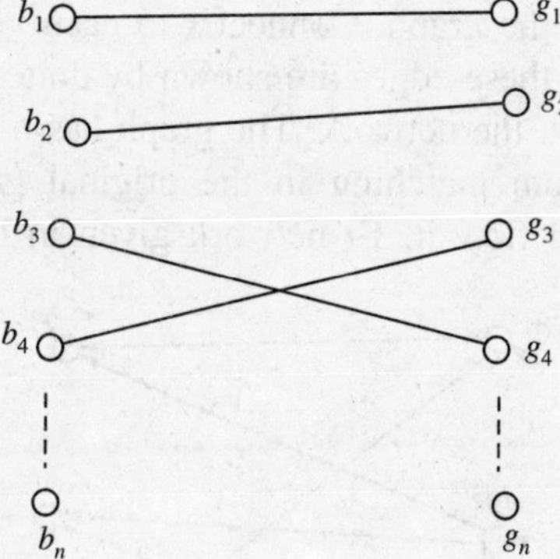

Figure 15.56 Brides-grooms matching.

15.17.2 Converting a Matching Problem into a Flow Problem

The problem of finding a maximum matching in a graph g can be transformed into a maximum flow problem. Let $G(V, E)$ be a graph with vertex partition $\{X, Y\}$. Let us construct an s-t network for this graph using the procedure outlined in the following steps:

1. Add a source vertex s and a sink vertex t to the graph.
2. Add a direction to each edge in G, such that it points from a vertex in X to a vertex in Y.
3. Join s to each vertex $x \in X$ by taking a forward edge pointing towards x.
4. Join each vertex $y \in Y$ to t by taking a forward edge pointing towards t.
5. Assign each edge a capacity $c(e) = 1$.

Once the problem is transformed into an s-t network, it can then be solved for finding a maximum flow, which is also a maximum matching. The procedure is explained with the help of an example given blow.

EXAMPLE 15.18 Consider a bipartite graph given in Figure 15.57.

Here
$$X = \{x_1, x_2, x_3\}$$
and
$$Y = \{y_1, y_2, y_3\}.$$

To find a maximum matching, we will transform it into an *s*-*t* network.

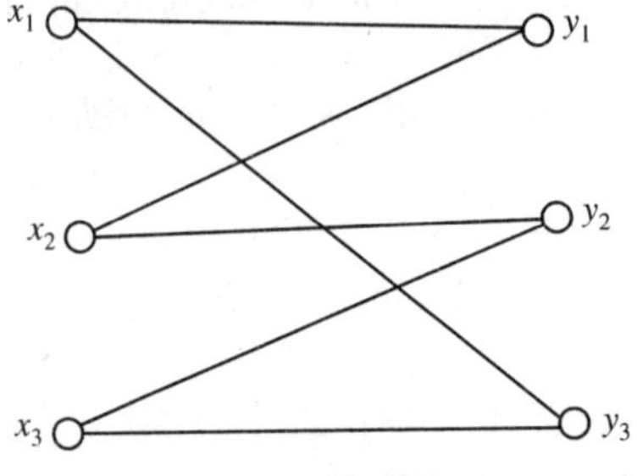

Figure 15.57 Graph *G*.

Make each edge directed such that it points from $x \in X$ to $y \in Y$. Further, add a source vertex *s*, and a sink vertex *t* to the graph. Connect *s* to each vertex $x \in X$. Similarly, connect each vertex $y \in Y$ to sink *t*. All these edges are shown by dotted arrows in Figure 15.58. Now, assign capacity 1 to each edge in the network. The graph has now been transformed into an *s*-*t* network. Now, finding maximum matching in the original graph given in Figure 15.57 is nothing but finding a maximum flow in *s*-*t* network given in Figure 15.58.

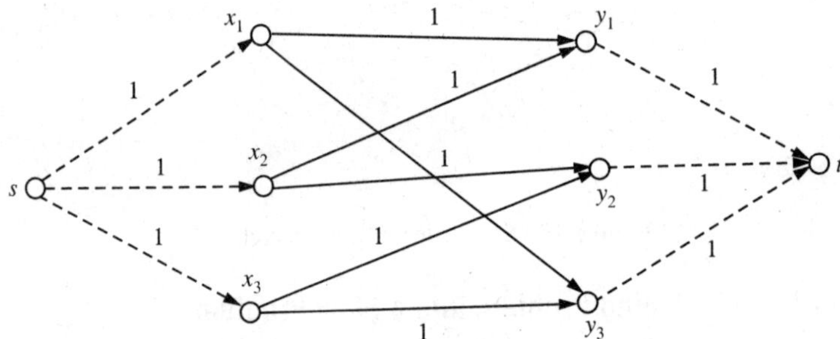

Figure 15.58 An *s*-*t* network for graph *G* given in Figure 15.57.

15.18 Covering in a Graph

In this section, we will discuss two types of covering problems of graph theory. One is vertex-covering and the other edge-covering. Both are discussed below.

15.18.1 Vertex-cover

Let *G*(*V*, *E*) be a graph. Let *C* be a subset of vertex set *V*. Then the subset *C* is called a **vertex-cover** of the graph *G* if every edge of *G* is incident on at least one vertex in *C*. Obviously, the set of all vertices, *V* is a vertex-cover. A vertex-cover is called **minimum vertex-cover** if it has the least number of vertices. Further, the endpoints of any maximal matching form a vertex-cover. Also, note that the complete bipartite graph $K_{m,\,n}$ has a minimum vertex-cover, such that its size is equal to min(*m*, *n*). A minimum vertex-cover of a graph can be obtained by finding all possible vertex-covers exhaustively, and then selecting the minimum one. However, the minimum may not be unique.

Some facts

Suppose, $G(V, E)$ is a graph. Let M be a matching in G, and C be a vertex-cover of it.

- Then $|M| \leq |C|$
- If $|M| = |C|$, then M is a maximum matching and C, a minimum vertex-cover. The converse may not be true in general. However, it does hold if G is bipartite.
- If G is bipartite then the number of edges in a maximum matching in G is equal to the number of vertices in a minimum vertex-cover of G.

EXAMPLE 15.19 Consider a graph G given in Figure 15.59(a). A **minimum vertex-cover** of this graph is shown by solid black vertices in Figure 15.59(b). In this case, the cardinality of minimum vertex-cover is equal to the cardinality of maximum matching, which is equal to 3. The maximum matching is shown by bold edges in Figure 15.59(b).

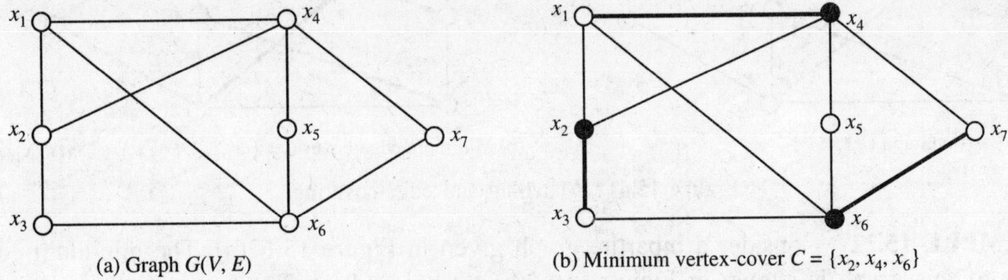

Figure 15.59 Matching and vertex-covering.

EXAMPLE 15.20 Consider a bipartite graph given in Figure 15.60(a). The graph is redrawn in Figure 15.60(b). In this Figure, the maximum matching is shown by bold edges, and **minimum vertex-cover** by solid black vertices. Note that the cardinality of both, maximum matching and minimum vertex-cover is equal, and it is 4 in this case.

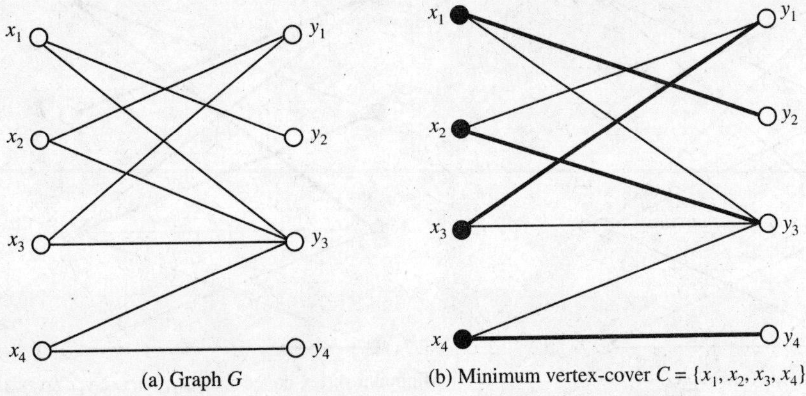

Figure 15.60 Matching and vertex-covering of bipartite graph.

15.18.2 Edge-cover

An edge-cover of a graph $G(V, E)$ is a subset C of E, such that each vertex of G is incident with at least one edge in C. A **minimum-edge covering** is an edge-covering of the least size. A perfect matching is always a minimum edge-covering. The set of all edges in a connected graph is an edge-cover. Further, the complete bipartite graph $K_{m, n}$ has a minimum edge-cover with size of equal to $\max(m, n)$. A minimum edge-cover of a graph can be obtained by finding all possible edge-covers exhaustively, and then selecting the minimum one. However, the minimum may not be unique.

EXAMPLE 15.21 Consider a graph G given in Figure 15.61(a). A minimum edge-cover of this graph is shown by bold edges in Figure 15.61(b). The cardinality of minimum edge-cover is equal to 4, in this case.

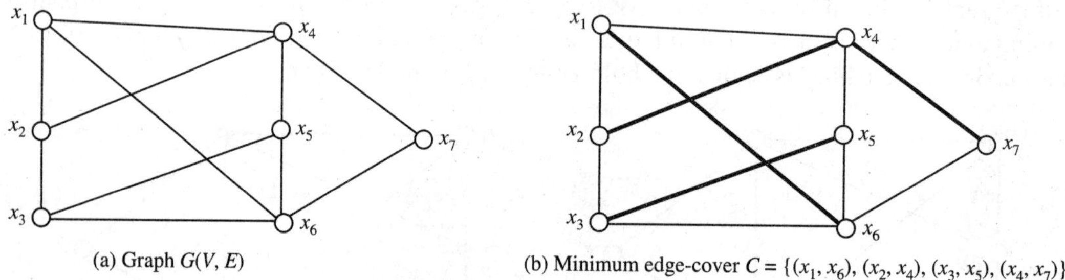

(a) Graph $G(V, E)$ (b) Minimum edge-cover $C = \{(x_1, x_6), (x_2, x_4), (x_3, x_5), (x_4, x_7)\}$

Figure 15.61 Matching and edge-covering.

EXAMPLE 15.22 Consider a bipartite graph given in Figure 15.62(a). The minimum edge-cover of this graph is shown in Figure 15.62(b) by bold edges. The cardinality of minimum edge-cover is equal to 4. A maximum matching in the graph is

$$M = \{(x_1, y_2), (x_2, y_1), (x_3, y_3), (x_4, y_4)\}$$

Note that the cardinality of both, maximum matching and minimum edge-cover is equal, in this case. Moreover, the cardinality of minimum vertex cover (shown by solid black vertices) is also 4.

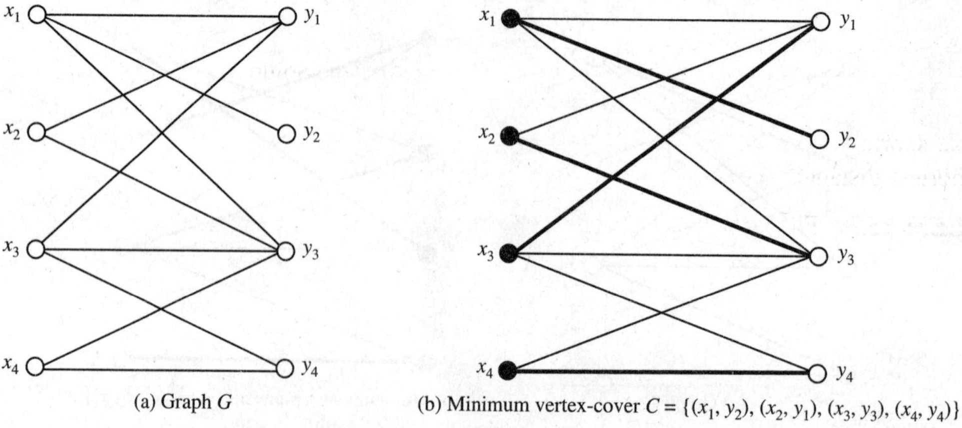

(a) Graph G (b) Minimum vertex-cover $C = \{(x_1, y_2), (x_2, y_1), (x_3, y_3), (x_4, y_4)\}$

Figure 15.62 Matching and edge covering of bipartite graph.

15.19 Solved Problems

PROBLEM 15.1 Consider the graph G given in Figure 15.63. Find Eulerian circuit in it using Fleury's algorithm.

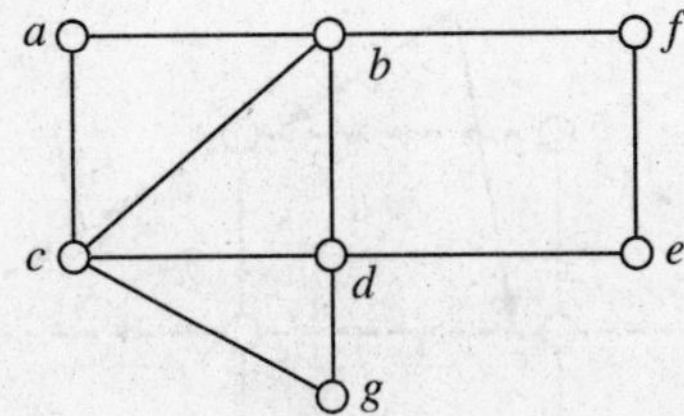

Figure 15.63 Graph G of Problem 15.1.

Solution

Since all vertices in the given graph are even, therefore, Eulerian circuit exists. We can find it using Fleury's algorithm. The solution steps are left for the readers to work-out. One Eulerian circuit is *a-b-d-e-f-b-c-g-d-c-a*.

PROBLEM 15.2 Consider the weighted graph $G = (V, E)$ given in Figure 15.64. Find the shortest path and shortest distance between the vertices a and h.

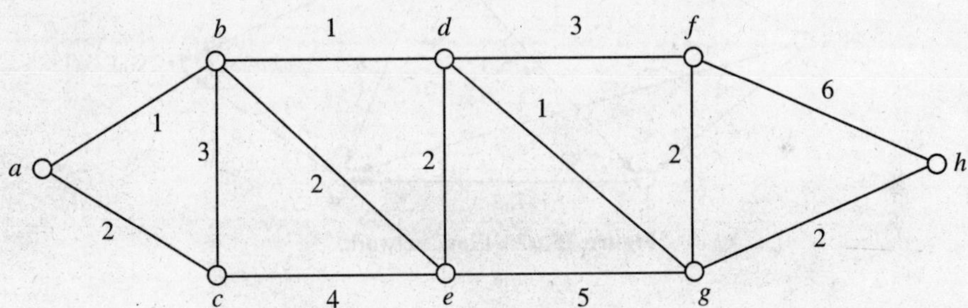

Figure 15.64 Shortest path between *a-h*.

Solution

The shortest path can be obtained by using Dijkstra's algorithm. Solution steps are left for the readers to work-out.

The shortest path is *a-b-d-g-h*.

Shortest distance = 5.

PROBLEM 15.3 Find the internally disjoint paths in the graph given in Figure 15.65.

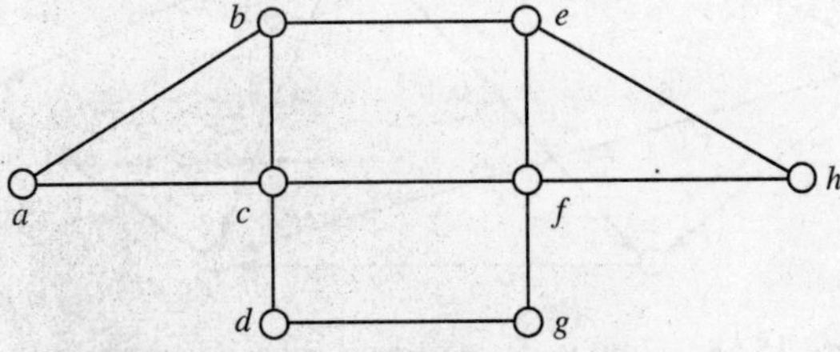

Figure 15.65 Internally disjoint paths.

Solution

Internally disjoint paths are those paths, which do not have a common vertex except s and t. A pair of two paths is shown in Figure 15.66.

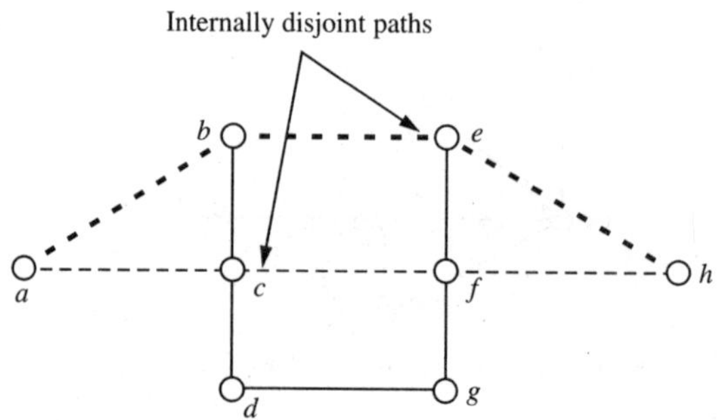

Figure 15.66 Three internally disjoint paths.

PROBLEM 15.4 Consider the flow-network given in Figure 15.67. Find maximum flow and the maximum flow value.

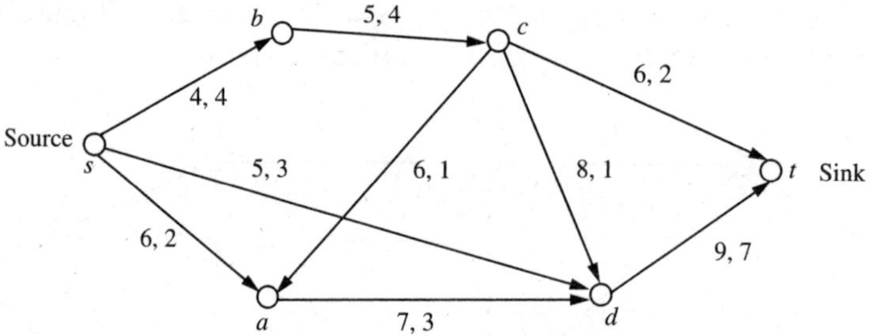

Figure 15.67 Flow network.

Solution

In this network, the flow can only be augmented across the edges (s, d) and (s, a). Edge (s, b) is assigned a flow equal to its capacity, i.e., $|c(s, b)| = |f(s, b)|$. Therefore, flow cannot be augmented on this edge. However, the flow can be decreased across the edges (c, a) and (c, d) by 1 unit, each. This flow can be added to edge (c, t).

Thus, the net flow across the edge $(c, t) = 2 + 2 = 4$. The modified flow-network is given in Figure 15.68.

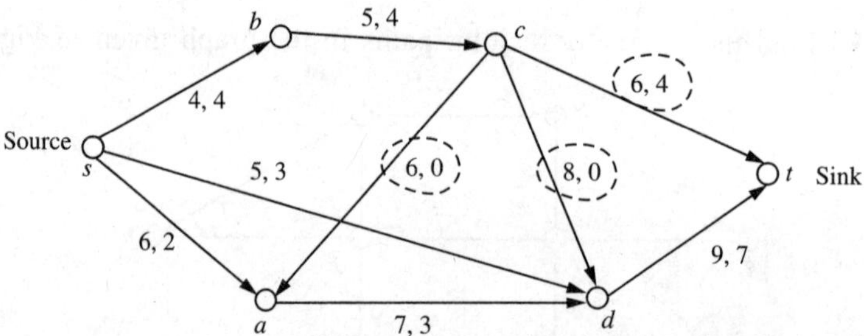

Figure 15.68 Flow is decreased on edges (c, a) and (c, d).

Further, to maintain the conservation constraint at the vertices a and d, augment the flows across edges (s, a) and (s, d), each by 1 unit. The flow-network, thus, modified is given in Figure 15.69.

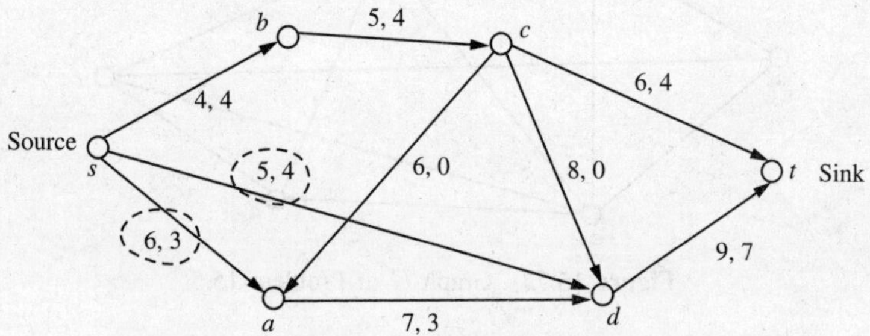

Figure 15.69 Flow after augmentation on (s, a) and (s, d).

Now, since the capacity of the edge (d, t) is 9, and the assigned flow is 7, therefore, the flow on this edge can be increased by 2. There are two possibilities:

(i) Augment the flow on edge (s, a) by 2 units.
(ii) Augment the flow on edge (s, a) and edge (s, d), each by 1 unit.

In both the cases, flow is maximum. The two solutions, thus, obtained are given in Figures 15.70 and 15.71. Maximum flow value = 13.

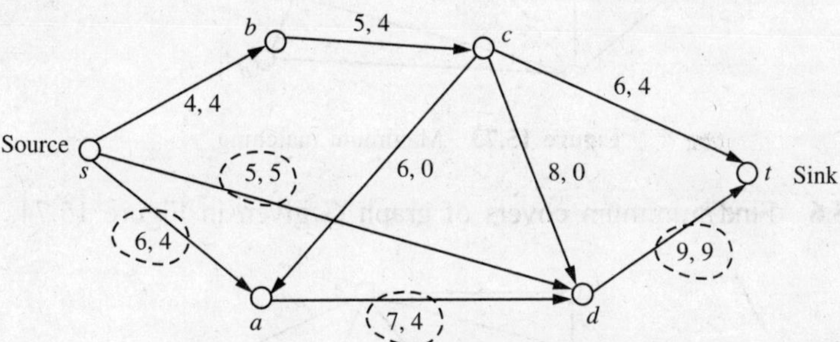

Figure 15.70 Maximum flow (i), value = 13.

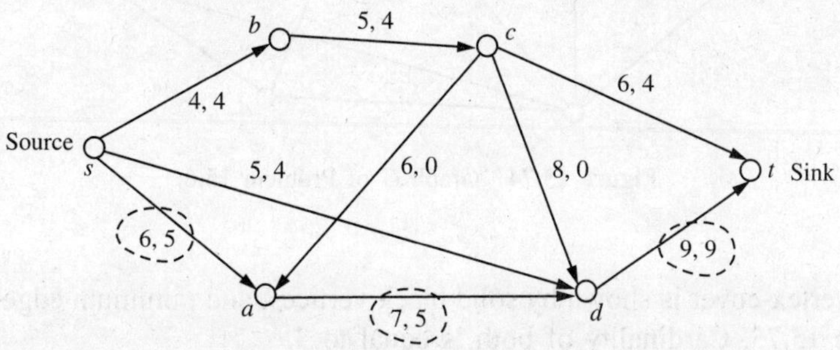

Figure 15.71 Maximum flow (ii), value = 13.

PROBLEM 15.5 Find maximum matching in graph G, given in Figure 15.72.

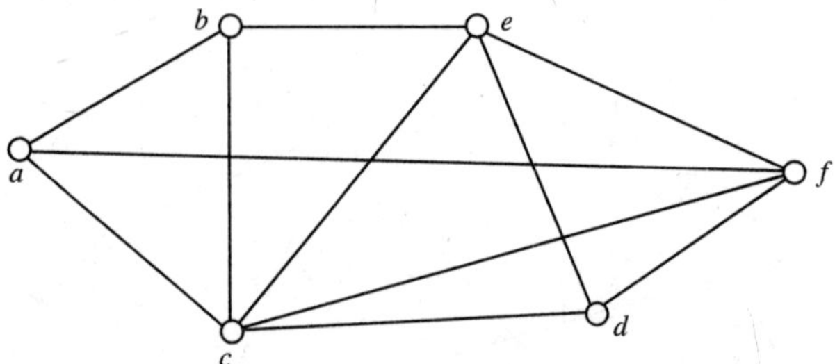

Figure 15.72 Graph G of Problem 15.5.

Solution

One of the maximum matching solutions is given in Figure 15.73. The dotted edges form the set of edges in maximum matching.

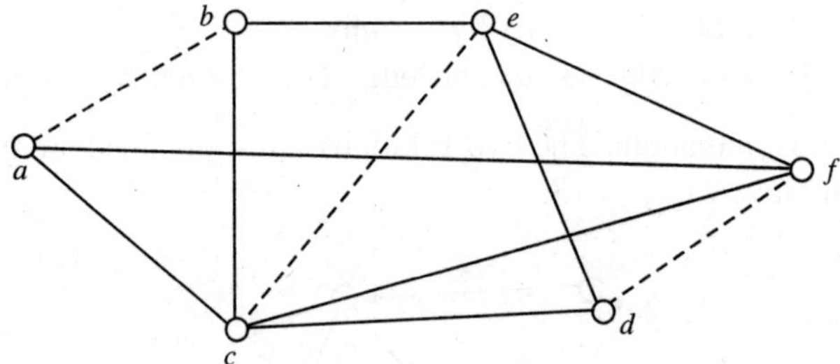

Figure 15.73 Maximum matching.

PROBLEM 15.6 Find minimum covers of graph G given in Figure 15.74.

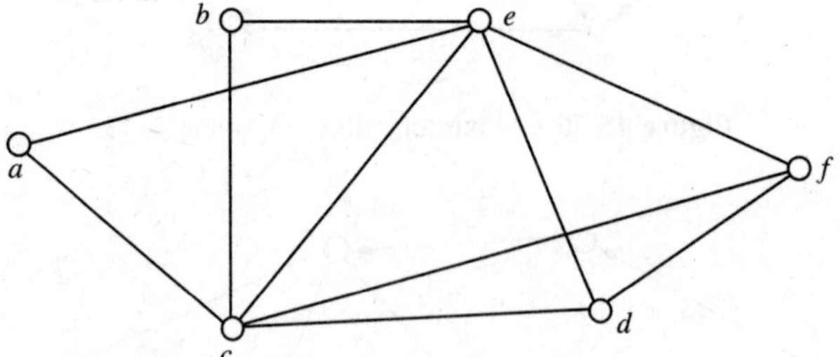

Figure 15.74 Graph G of Problem 15.6.

Solution

The minimum vertex-cover is shown by solid black vertices, and minimum edge-cover by dotted edges in Figure 15.75. Cardinality of both is equal to 3.

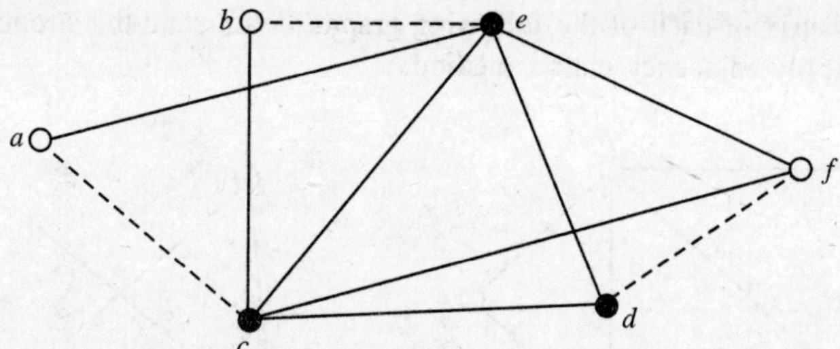

Figure 15.75 Minimum covers.

Exercises

15.1 Find the Eulerian circuit in the following graphs.

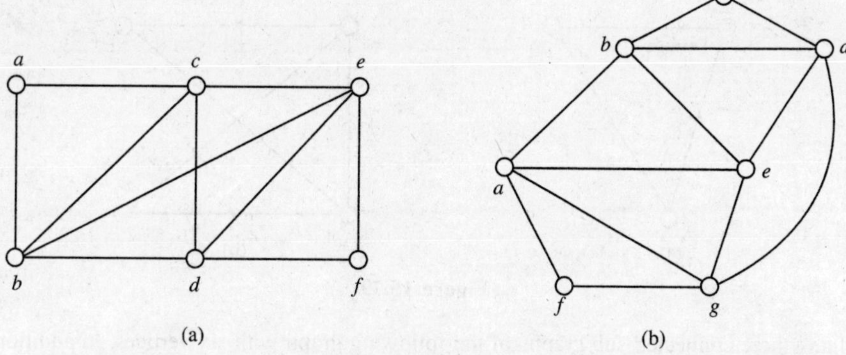

Figure 15.76

15.2 Find the shortest path from a to f in the following graphs. Use Dijkstra's shortest path algorithm.

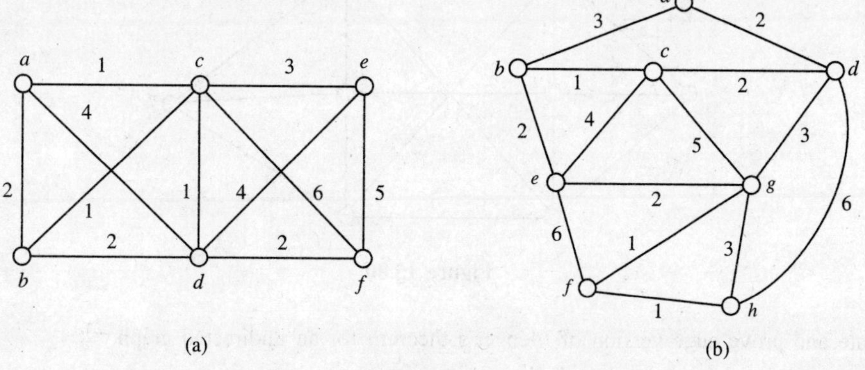

Figure 15.77

15.3 Find the path matrix of each of the following graphs. Also, state the strong or weak connectivity in each case. Apply adjacency matrix method.

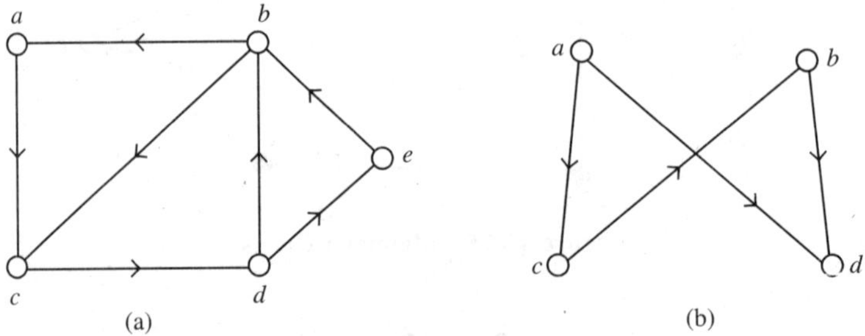

Figure 15.78

15.4 Find the path matrix of each of the following graphs. Also, discuss the connectivity in each case. Use Warshall's algorithm.

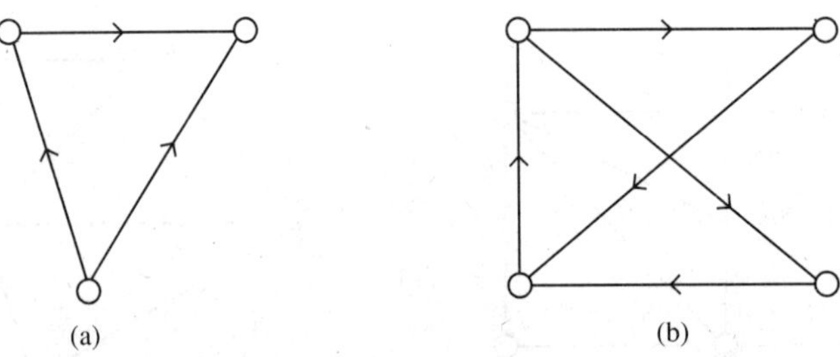

Figure 15.79

15.5 Find any three connected sub graphs of the following graph with six vertices. In addition, find cut-points and bridges in them.

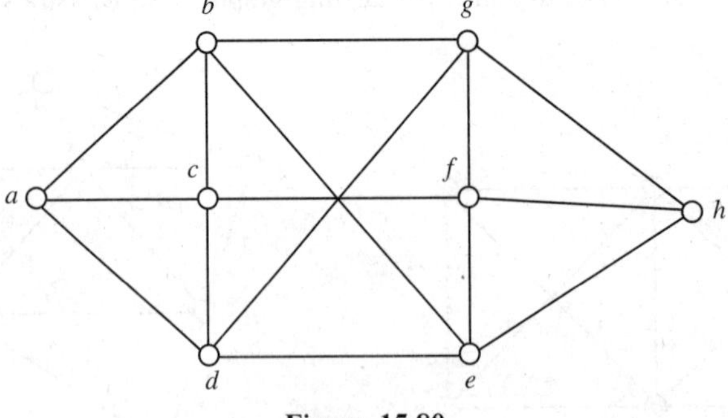

Figure 15.80

15.6 State and prove edge version of Menger's theorem for an undirected graph.

15.7 Find the maximum flow in the following capacited network.

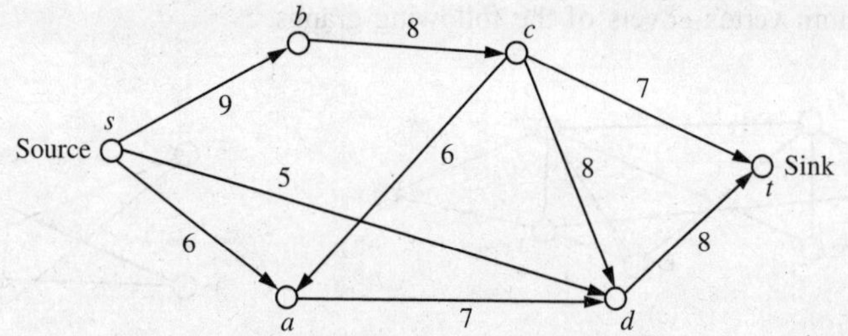

Figure 15.81

15.8 Find the maximum flow in the following capacited network.

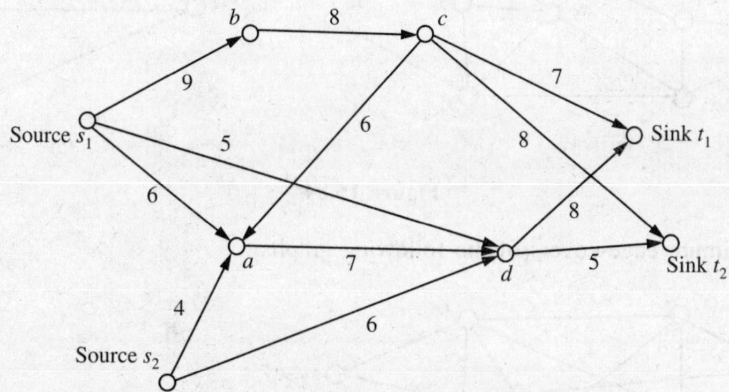

Figure 15.82

15.9 Find all maximum matchings in the following graphs.

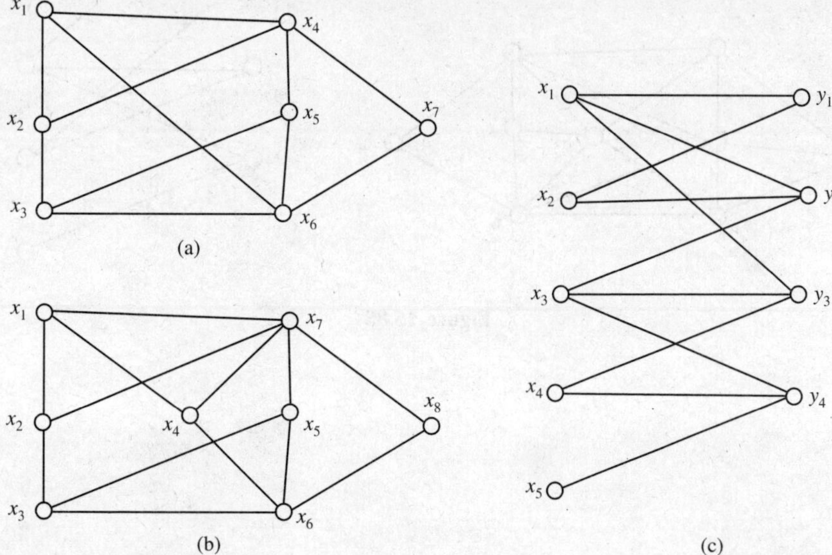

Figure 15.83

15.10 Find all minimum vertex-covers of the following graphs.

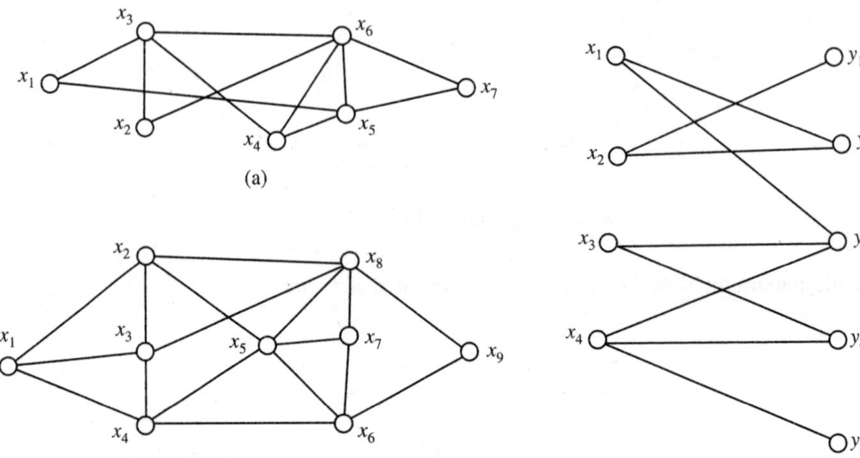

Figure 15.84

15.11 Find all minimum edge-covers of the following graphs.

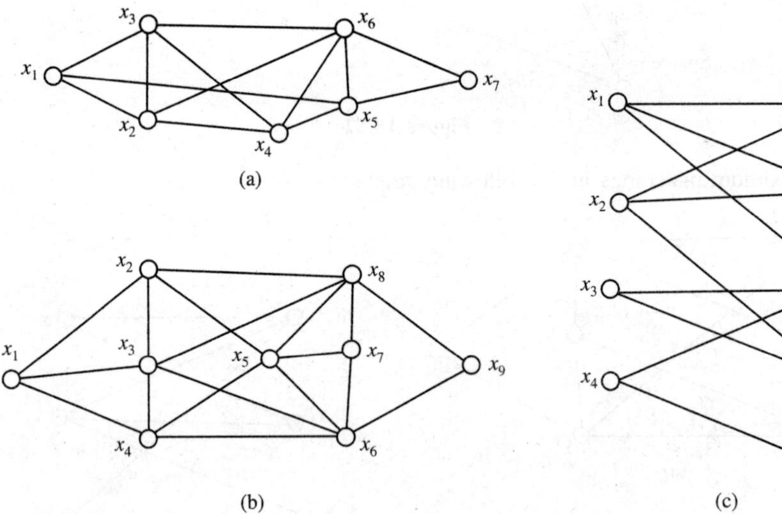

Figure 15.85

Note: *Solution of dodecahedron puzzle discussed in Section 15.5.

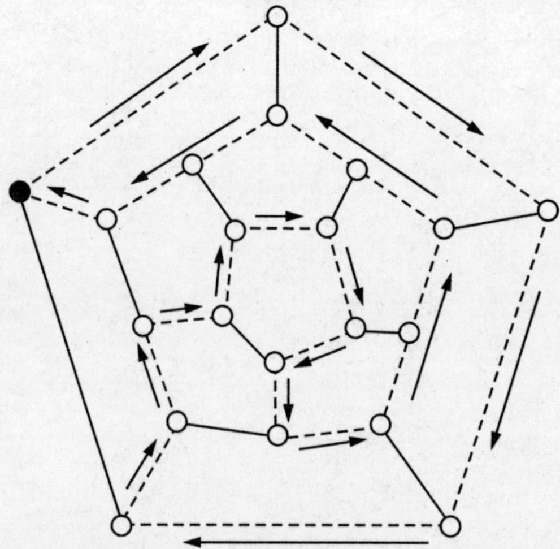

Figure 15.86 Dodecahedron puzzle.

Part E

AUTOMATA THEORY AND GRAMMAR

16

Finite State Automaton

16.1 Introduction

Automata theory is the study of abstract mathematical models called automatons or automata. These models are used to solve computational problems with discrete inputs and outputs. Figure 16.1 illustrates a finite state automaton with finite number of states (represented by small circles) and transitions (represented by arrows). When the automaton receives an input symbol, it makes a transition (or shift) from one state to another according to its transition function. Automata theory can also simulate formal language theory. Automata are often classified by the class of formal languages they recognize. An automaton can be used to recognize a formal language.

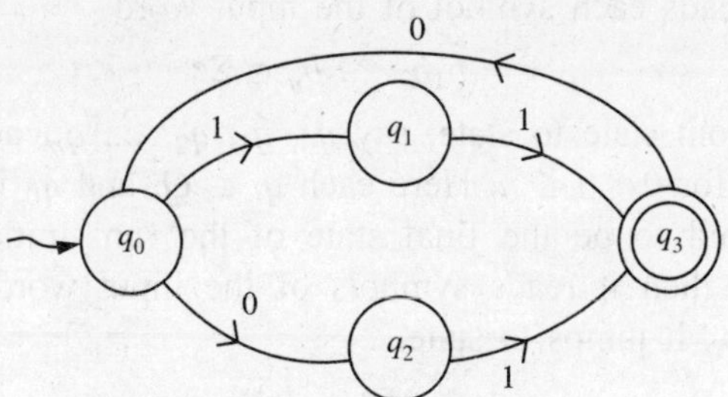

Figure 16.1 Finite state automaton.

16.2 Finite State Automaton

An automaton contains a finite set of states with one of these states as the **starting state**. Some of the states are designated as **accepting states** (using double circles in the diagram). The automaton runs on some given sequence or string of input symbols called words over some finite alphabet. At each time, it receives one input symbol from the input word and jumps or transits to the next state depending on some rule or function. This function, in terms of the current state and input symbol, is called transition function. The automaton reads the input word one symbol after another in the sequence and moves from state to state, until the word is read completely. Once the input word has been read, the automaton is said to have stopped and reached the final state. Depending on the final state, the automaton either accepts or rejects the input word. If the final state is an accepting state, then the automaton accepts the word. Otherwise, the word is rejected. The set of all words accepted by an automaton is called the language of the automaton.

Formally, the automaton is represented by the 5 tuples $< q_0, Q, \Sigma, \delta, F >$, where:

- $q_0 \in Q$ is the starting state, i.e., the state of the automaton when no input has been processed yet. It is marked with an incoming arrow.
- Q is a finite set of states.
- Σ is a finite set of symbols, called the alphabet of the automaton.
- δ is the transition function for input symbols, i.e., $\delta: Q \times \Sigma \rightarrow Q$.
- F is a subset of Q (i.e. $F \subseteq Q$), having all the accepting states of the automaton.

Some important terms related to the finite state automaton are defined below:

Input word

An automaton reads a finite string of symbols

$$w = b_1 b_2 \ldots b_n$$

where $b_i \in \Sigma$, is an input symbol. This string is called input word. The set of all words is denoted by Σ^*.

Run

In a run, the automaton reads each symbol of the input word

$$w = b_1 b_2 \ldots. b_n \in \Sigma^*$$

in sequence and jumps from state to state, say $q_0, q_1, q_2, \ldots, q_n$, according to the transition function $q_i = \delta(q_{i-1}, b_i)$ for $0 < i \leq n$. Here each $q_i \in Q$, and q_0 is the starting state of the automaton. Also, q_n is said to be the final state of the run. Initially, the automaton is at the starting state q_0, and then it reads symbols of the input word in sequence. When the automaton reads symbol b_i, it jumps to state

$$q_i = \delta(q_{i-1}, b_i) .$$

Accept

A word $w \in \Sigma^*$ is said to be accepted by the automaton if

$$q_n \in F.$$

Reject

A word $w \in \Sigma^*$ is said to be rejected by the automaton if

$$q_n \notin F.$$

Language recognized

A formal language is the set of words over some alphabet.

The language $L \subset \Sigma^*$ recognized by an automaton is the set of all words which are accepted by the automaton.

Transition function

A rule that decides the next state of the automaton on receiving an input symbol is termed transition function. Table (or more specifically transition table) is an appropriate model to represent transition function. Table 16.1 gives a transition function. Further, its graphical model is given in Figure 16.2.

Table 16.1 Transition Table

State	*Input*	
	a	*b*
q_0 (starting state)	q_1	q_2
q_1	q_2	—
$*q_2$ (accepting state)	q_1	q_0

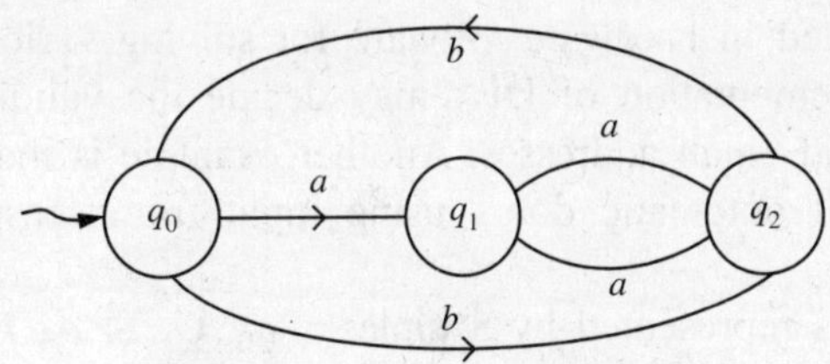

Figure 16.2 Transition function.

Remark: An accepting state is shown with an asterisk(*) in the the transition table.

16.3 Types of Automaton

There are various models of automaton depending on different types of applications. Some of these models, along with the language recognized by each are listed in Table 16.2. In this chapter, we will discuss only important models, which are useful in solving some fundamental problems of computer science.

Table 16.2 Models of Automaton

Automata	*Language recognized*
Deterministic finite automata (DFA)	Regular languages
Nondeterministic finite automata (NFA)	Regular languages
Nondeterministic finite automata with ε transitions (NFA-ε or ε-NFA)	Regular languages
Pushdown automata (PDA)	Context-free languages
Linear bounded automata (LBA)	Context-sensitive language
Turing machines	Recursively enumerable languages
Nondeterministic Büchi automata	Omega regular languages
Nondeterministic/Deterministic Rabin automata	Omega regular languages
Nondeterministic/Deterministic Streett automata	Omega regular languages
Nondeterministic/Deterministic parity automata	Omega regular languages
Nondeterministic/Deterministic Muller automata	Omega regular languages

16.4 Deterministic Finite State Automaton

A deterministic finite state automaton, also known as deterministic finite automaton (DFA), is an automaton that accepts strings of finite input symbols. Deterministic means that there is only one outgoing directed edge leading to the next state for an input symbol. Further, there is no transition on ε (empty string). It has a starting state, which is marked with an incoming arrow where computations start, and a set of accepting states (denoted by a double circles) which help in defining successful computation. Upon reading a symbol, a DFA jumps deterministically from one state to another by using the transition function. DFAs recognize the set of languages, known as regular languages, which are useful for doing lexical analysis and pattern matching. A DFA can be implemented in hardware/software for solving various specific problems. For instance, a software implementation of DFA may decide the validity of user-inputs such as names, phone numbers and email addresses. Another example is the digital logic circuits that control the functioning of automatic doors using input from sensors to decide whether to perform a state transition.

Formally, the DFA is represented by 5 tuples $< q_0, Q, \Sigma, \delta, F >$, where:

- $q_0 \in Q$ is the starting state.
- Q is a finite set of states.
- Σ is a finite set of symbols, called the alphabet DFA.
- δ is the transition function for input symbols, i.e., $\delta: Q \times \Sigma \to Q$.
- F is a subset of Q (i.e. $F \subseteq Q$), the accepting states of the DFA.

Remark: Another function $\hat{\delta}$, which is used in derivations is defined as:

$$\hat{\delta}: Q \times \Sigma^* \to Q$$

Let M be an automaton and

$$w = b_1 b_2 \ldots b_n$$

be an input string over the alphabet Σ. The automaton M accepts the string w if a sequence of states,

$$q_0, p_1, \ldots, p_n$$

exists in Q with the following conditions:

1. $\delta(q_0, b_1) = p_1$
2. $p_{i+1} = \delta(p_i, b_{i+1})$, for $i = 0, ..., n-1$
3. $p_n \in F$.

By the first condition, it is clear that the machine starts in the starting state q_0. The second condition states that given each input symbol of string w, the machine will move from one state to other according to the transition function δ. The last condition says that the machine accepts w if last input symbol of w causes the machine to reach one of its accepting states. Otherwise, the input string is rejected by the DFA. The set of strings accepted by DFA 'M' is called the language recognized by M, and is denoted by $L(M)$.

EXAMPLE 16.1 Consider a deterministic finite state automaton M given in Figure 16.3. Its transition function is given in Table 16.3. This DFA accepts words of an even number of 1s.

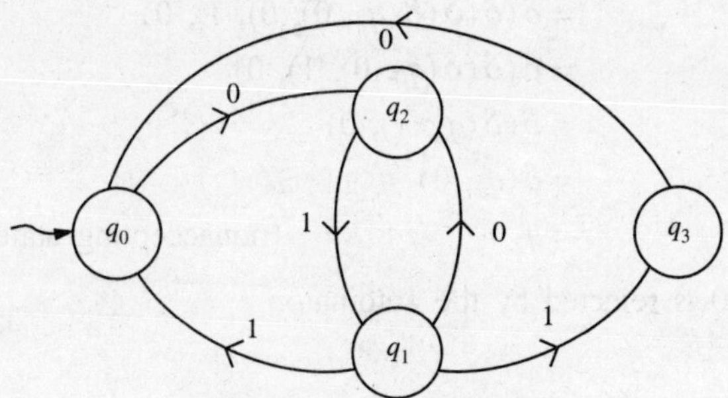

Figure 16.3 State diagram.

Here $M = (Q, \Sigma, \delta, q_0, F)$,
where

- $Q = \{q_0, q_1, q_2, q_3\}$,
- $\Sigma = \{0, 1\}$,
- Starting state is q_0,
- $F = \{q_3\}$, and
- δ is defined by the following state transition table:

Table 16.3 Transition Function of M

State	*Input* 0	1
q_0 (starting state)	q_2	q_1
q_1	q_2	q_3
q_2	–	q_1
*q_3 (accepting state)	q_0	–

Here q_0 is the starting state, and q_3 is the final (accepting) state of DFA. Strings accepted by the automaton are 11, 011, 110011, and 0110011 etc. Let us trace the derivation steps of the automaton running on the string w_1 = 110011.

$$\begin{aligned}\hat{\delta}(q_0, 110011) &= \delta(\delta(\delta(\delta(\delta(\delta(q_0, 1), 1), 0), 0), 1), 1)\\ &= \delta(\delta(\delta(\delta(\delta(q_1, 1), 0), 0), 1), 1)\\ &= \delta(\delta(\delta(\delta(q_3, 0), 0), 1), 1)\\ &= \delta(\delta(\delta(q_0, 0), 1), 1)\\ &= \delta(\delta(q_2, 1), 1)\\ &= \delta(q_1, 1)\\ &= {}^*q_3 \qquad \{\text{accepting state}\}\end{aligned}$$

Therefore, $w_1 = 110011$ is accepted by the automaton.

Again take another string $w_2 = 110010$. The derivation steps are as under:

$$\begin{aligned}\hat{\delta}(q_0, 110010) &= \delta(\delta(\delta(\delta(\delta(\delta(q_0, 1), 1), 0), 0), 1), 0)\\ &= \delta(\delta(\delta(\delta(\delta(q_1, 1), 0), 0), 1), 0)\\ &= \delta(\delta(\delta(\delta(q_3, 0), 0), 1), 0)\\ &= \delta(\delta(\delta(q_0, 0), 1), 0)\\ &= \delta(\delta(q_2, 1), 0)\\ &= \delta(q_1, 0)\\ &= q_2 \qquad \{\text{nonaccepting state}\}\end{aligned}$$

Therefore, $w_2 = 110010$ is rejected by the automaton.

16.5 Nondeterministic Finite State Automaton

A nondeterministic finite state automaton, also known as nondeterministic finite automaton (NFA) is an automaton that accepts strings of finite input symbols. Unlike DFA, it may have more than one outgoing edge from a state for one input symbol. Thus, the next state may be a subset of states. This subset is an element of the power set of the set of states.

Formally, the NFA is represented by 5 tuples $< q_0, Q, \Sigma, \delta, F >$, where:

- $q_0 \in Q$ is the starting state.
- Q is a finite set of states.
- Σ is a finite set of symbols, called the alphabet of NFA.
- δ is the transition function for input symbols, i.e., $\delta : Q \times \Sigma \to \rho(Q)$.
- F is a subset of Q (i.e. $F \subseteq Q$), the accepting states of the NFA.

Here, $\rho(Q)$ denotes the power set of Q.

Remark: As in DFA, the definition of δ is extended to $\hat{\delta}$ for NFA to take input from Σ^*. That is,

$$\hat{\delta} : Q \times \Sigma^* \to P(Q).$$

Let M be a nondeterministic automaton and

$$w = b_1 b_2 \ldots b_n$$

be an input string over the alphabet Σ. Then M accepts the string w if a sequence of states, $q_0, p_1, \ldots, p_n$, exists in Q, such that:

1. $\delta(q_0, b_1) = p_1$,
2. $p_{i+1} = \delta(p_i, b_{i+1})$, for $i = 0, ..., n-1$
3. $p_n \in F$.

Like DFA, NFA starts in the starting state q_0. It moves from one state to other according to the transition function δ. It accepts w if the last input symbol of w causes the automaton to reach one of the accepting states. Otherwise, the input string is rejected. The set of strings accepted by NFA 'M' is called the language recognized by M, and is denoted by $L(M)$.

EXAMPLE 16.2 Consider a nondeterministic finite state automaton M given in Figure 16.4. Its transition function is given in Table 16.4.

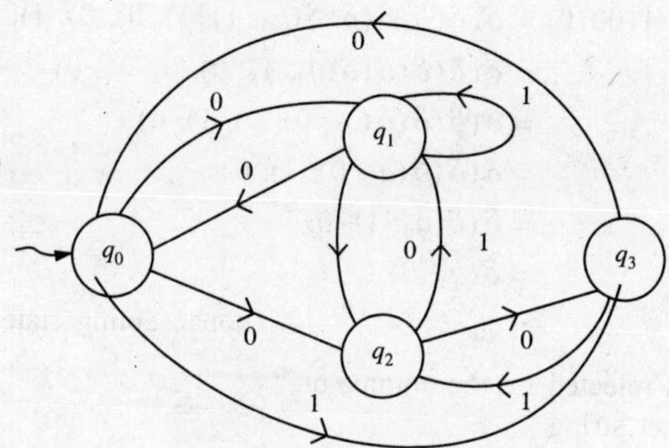

Figure 16.4 State diagram of M.

Here, $M = (Q, \Sigma, \delta, q_0, F)$, where

- $Q = \{q_0, q_1, q_2, q_3\}$,
- $\Sigma = \{0, 1\}$,
- Starting state is q_0,
- $F = \{q_3\}$, and
- δ is defined by the following state transition table:

Table 16.4 Transition Table of M

State	*Input*	
	0	1
q_0 (starting state)	$\{q_1, q_2\}$	$\{q_3\}$
q_1	$\{q_0, q_2\}$	$\{q_1\}$
q_2	$\{q_3\}$	$\{q_1\}$
$*q_3$ (accepting state)	$\{q_0\}$	$\{q_2\}$

Here q_0 is the starting state and q_3 is the final (accepting) state of NFA. Let us find the derivation steps of the automaton running on the string $w_1 = 110011$.

$$\hat{\delta}(q_0, 110011) = \delta(\delta(\delta(\delta(\delta(\delta(q_0, 1), 1), 0), 0), 1), 1)$$
$$= \delta(\delta(\delta(\delta(\delta(q_3, 1), 0), 0), 1), 1)$$

$$= \delta(\delta(\delta(\delta(q_2, 0), 0), 1), 1)$$
$$= \delta(\delta(\delta(q_3, 0), 1), 1)$$
$$= \delta(\delta(q_0, 1), 1)$$
$$= \delta(q_3, 1)$$
$$= q_2 \qquad \{\text{nonaccepting state}\}$$

Therefore, $w_1 = 110011$ is rejected by the automaton.

Consider another string

$$w_2 = 110010$$

The derivation steps are as follows:

$$\hat{\delta}(q_0, 110010) = \delta(\delta(\delta(\delta(\delta(\delta(q_0, 1), 1), 0), 0), 1), 0)$$
$$= \delta(\delta(\delta(\delta(\delta(q_3, 1), 0), 0), 1), 0)$$
$$= \delta(\delta(\delta(\delta(q_2, 0), 0), 1), 0)$$
$$= \delta(\delta(\delta(q_3, 0), 1), 0)$$
$$= \delta(\delta(q_0, 1), 0)$$
$$= \delta(q_3, 0)$$
$$= q_0 \qquad \{\text{nonaccepting state}\}$$

Hence, $w_2 = 110010$ is rejected by the automaton.

Again, take another string

$$w_3 = 11001$$

The derivation steps are as under:

$$\hat{\delta}(q_0, 11001) = \delta(\delta(\delta(\delta(\delta(q_0, 1), 1), 0), 0), 1)$$
$$= \delta(\delta(\delta(\delta(q_3, 1), 0), 0), 1)$$
$$= \delta(\delta(\delta(q_2, 0), 0), 1)$$
$$= \delta(\delta(q_3, 0), 1)$$
$$= \delta(q_0, 1)$$
$$= {}^*q_3 \qquad \{\text{accepting state}\}$$

Therefore, $w_3 = 11001$ is accepted by the automaton.

16.6 Nondeterministic Finite State Automaton with ε

A nondeterministic finite state automaton with ε (denoted as NFA-ε) is a finite state automaton that accepts strings of finite input symbols. It is an extension of the NFA model. Like NFA, a nondeterministic finite state automaton with ε may have more than one possible outgoing edges from a state on one input symbol. In addition, it may have transitions on ε (or empty) input. If there is an outgoing edge labelled ε from a state p to a state q, then the automaton may move from state p to q without consuming any input symbol.

Formally, the NFA-ε is represented by 5 tuples $< q_0, Q, \Sigma, \delta, F >$, where:

- $q_0 \in Q$ is the starting state.
- Q is a finite set of states.
- Σ is a finite set of symbols, called the alphabet NFA-ε.
- δ is the transition function for input symbols, i.e.,

$$\delta : Q \times \Sigma \cup \{\varepsilon\} \to P(Q)$$

- F is a subset of Q (i.e. $F \subseteq Q$), the accepting states of the NFA-ε.

Note: Definition of $\hat{\delta}$ is same as in NFA, discussed in Section 16.5.
Let M be a non-deterministic finite state automaton with ε,

and
$$w = b_1 b_2 \ldots b_n$$

be an input string over the alphabet Σ. Then M accepts the string w if a sequence of states, $q_0, p_1, \ldots, p_n$, exists in Q such that:

1. $\delta(q_0, b_1) = p_1$,
2. $p_{i+1} = \delta(p_i, b_{i+1})$, for $i = 0, \ldots, n-1$
3. $p_n \in F$.

Like NFA, NFA-ε starts in the starting state q_0. It moves from one state to another according to the transition function δ. It accepts w if the last input symbol of w causes the automaton to reach one of the accepting states. Otherwise, the input string is rejected by the NFA-ε. The set of strings accepted by NFA-ε 'M' is called the language recognized by M, and is denoted by $L(M)$.

EXAMPLE 16.3 Consider a nondeterministic finite state automaton M with ε given in Figure 16.5. Its transition function is given in Table 16.5.

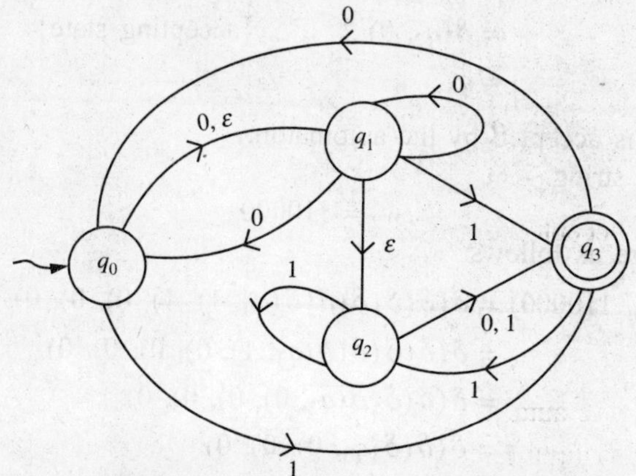

Figure 16.5 State diagram of M with ε.

Here, $M = (Q, \Sigma, \delta, q_0, F)$, where

- $Q = \{q_0, q_1, q_2, q_3\}$,
- $\Sigma = \{0, 1\}$,
- Starting state is q_0,
- $F = \{q_3\}$, and
- δ is defined by the following state transition table:

Table 16.5 Transition Function of M with ε

State	*Input*		
	0	1	ε
q_0 (starting state)	$\{q_1\}$	$\{q_3\}$	$\{q_1\}$
q_1	$\{q_1\}$	$\{q_3\}$	$\{q_2\}$
q_2	$\{q_3\}$	$\{q_2, q_3\}$	–
$*q_3$ (accepting state)	$\{q_0\}$	$\{q_2\}$	–

Here q_0 is the starting and q_3 is final state of NFA-ε. Let us find the derivation steps of the automaton running on string

$$w_1 = 1100110$$

$$\begin{aligned}
\hat{\delta}(q_0, 1100110) &= \delta(\delta(\delta(\delta(\delta(\delta(\delta(q_0, 1), 1), 0), 0), 1), 1), 0) \\
&= \delta(\delta(\delta(\delta(\delta(\delta(q_3, 1), 0), 0), 1), 1), 0) \\
&= \delta(\delta(\delta(\delta(\delta(q_2, 0), 0), 1), 1), 0) \\
&= \delta(\delta(\delta(\delta(q_3, 0), 1), 1), 0) \\
&= \delta(\delta(\delta(q_0, 1), 1), 0) \\
&= \delta(\delta(q_3, 1), 0) \\
&= \delta(q_2, 0) \qquad \{\text{accepting state}\} \\
&= q_3
\end{aligned}$$

Hence, $w_1 = 1100110$ is accepted by the automaton.

Consider another string

$$w_2 = 110000$$

The derivation steps are as follows:

$$\begin{aligned}
\hat{\delta}(q_0, 110000) &= \delta(\delta(\delta(\delta(\delta(\delta(q_0, 1), 1), 0), 0), 0), 0) \\
&= \delta(\delta(\delta(\delta(\delta(q_3, 1), 0), 0), 0), 0) \\
&= \delta(\delta(\delta(\delta(q_2, 0), 0), 0), 0) \\
&= \delta(\delta(\delta(q_3, 0), 0), 0) \\
&= \delta(\delta(q_0, 0), 0) \\
&= \delta(q_1, 0) \\
&= q_1 \qquad \{\text{nonaccepting state}\}
\end{aligned}$$

Therefore, $w_2 = 110000$ is rejected by the automaton.

Again take another string

$$w_3 = 11001.$$

The derivation steps are as under:

$$\begin{aligned}\hat{\delta}(q_0, 11001) &= \delta(\delta(\delta(\delta(\delta(q_0, 1), 1), 0), 0), 1)\\ &= \delta(\delta(\delta(\delta(q_3, 1), 0), 0), 1)\\ &= \delta(\delta(\delta(q_2, 0), 0), 1)\\ &= \delta(\delta(q_3, 0), 1)\\ &= \delta(q_0, 1)\\ &= {}^*q_3 \qquad \{\text{accepting state}\}\end{aligned}$$

Therefore, $w_3 = 11001$ is accepted by the automaton.

16.7 Equivalence of NFA and DFA

It is easy to prove the equivalence of nondeterministic and deterministic finite state automaton. Both the models have the same computational capability. It means that the set of strings accepted by both NFA and DFA are same. The formal proof showing the equivalence can be seen by the following theorem.

THEOREM 16.1: If L is a set of strings accepted by a nondeterministic finite state automaton, then there exists an equivalent deterministic finite state automaton, which accepts L.

Proof: Consider an NFA $M = (Q, \Sigma, \delta, q_0, F)$ that accepts L.

Let $M' = (Q', \Sigma, \delta', q_0', F')$ be defined as:

- Q' = power set of M, which is the set of states of Q'.
- q_0' is the starting state of Q'.
- F' is the set of states of Q' containing an accepting state of M.
- Define δ' as

$$\delta'([q_1, q_2, \ldots, q_i], b) = [p_1, p_2, \ldots, p_j]$$

if and only if

$$\delta(\{q_1, q_2, \ldots, q_i\}, b) = \{p_1, p_2, \ldots, p_j\}$$

where $[q_1, q_2, \ldots, q_i]$

and $[p_1, p_2, \ldots, p_j]$

are two states.

Observe that $[q_1, q_2, \ldots, q_i]$ is a single state of the DFA, where $q_1, q_2, \ldots, q_i$ are the states of NFA.

Also $$q_0' = [q_0].$$

Here, $$\delta'([q_1, q_2, \ldots, q_i], b) = [p_1, p_2, \ldots, p_j]$$

is computed by applying δ to each of $q_1, q_2, \ldots, q_i$ and taking the union, i.e.

$$\{\delta(q_1, b) \cup \delta(q_2, b) \cup \cdots \cup \delta(q_i, b)\} = \{p_1, p_2, \ldots, p_j\}.$$

Then $\{p_1, p_2, \ldots, p_j\}$ is represented by one state $[p_1, p_2, \ldots, p_j]$ in DFA.

Now, we can show by induction that

$$\delta'(q_0', x) = [q_1, q_2, \ldots, q_i],$$

if and only if

$$\delta(q_0, x) = \{q_1, q_2, \ldots, q_i\} \qquad \{\text{Hypothesis}\}$$

(i) **Base case:** Let x be an empty string, i.e., $|x| = 0$. Then $q_0' = [q_0]$.

(ii) **General case:** Let the hypothesis be true for strings of length n or less, and also let xa be a string of length $n + 1$ with $a \in \Sigma$.

Then $$\delta'(q_0', xa) = \delta'(\delta'(q_0', x), a).$$

Now, by the hypothesis,

$$\delta'(q_0', x) = [p_1, p_2, \ldots, p_j],$$

if and only if

$$\delta(q_0, x) = \{p_1, p_2, \ldots, p_j\}.$$

But, by the definition of δ',

$$\delta'([p_1, p_2, \ldots, p_j], a) = [r_1, r_2, \ldots, r_k]$$

if and only if

$$\delta(\{p_1, p_2, \ldots, p_j\}, a) = \{r_1, r_2, \ldots, r_k\}.$$

Therefore, $$\delta'(q_0', xa) = [r_1, r_2, \ldots, r_k]$$

if and only if

$$\delta(q_0, xa) = \{r_1, r_2, \ldots, r_k\}$$

That proves the hypothesis. Also, $\delta'(q_0', x)$ is in F' if $\delta(q_0, x)$ contains a state of Q, that is, in F.

Thus, NFA and DFA are equivalent.

EXAMPLE 16.4 Let $M = (Q, \Sigma, \delta, q_0, F)$ be an NFA, where

$$Q = \{q_0, q_1\}$$
$$\Sigma = \{0, 1\}$$
$$F = \{q_1\}$$

and δ is given by the transition table, in Table 16.6.

Table 16.6 Transition Table of Example 16.4

State	*Input*	
	0	1
q_0 (starting state)	$\{q_0, q_1\}$	$\{q_1\}$
$*q_1$	φ	$\{q_0, q_1\}$

We can construct an equivalent DFA $M' = (Q', \Sigma, \delta', q_0', F')$ for M, where

$$Q' = \{[q_0], [q_1], [q_0, q_1], \varphi\}, \qquad \{Q' \text{ is the power set of } Q\},$$
$$q_0' = [q_0],$$
and
$$F' = \{[q_1], [q_0, q_1]\}.$$

Here φ is the null or reject state. It is optional and may be deleted finally. Transition function δ' may be computed by the following way:

$$\delta'([q_0], 0) = [q_0, q_1] \text{ since } \delta(q_0, 0) = \{q_0, q_1\},$$
$$\delta'([q_0], 1) = [q_1] \text{ since } \delta(q_0, 1) = \{q_1\},$$
$$\delta'([q_1], 0) = \varphi \text{ since } \delta(q_1, 0) = \varphi,$$
$$\delta'([q_1], 1) = [q_0, q_1] \text{ since } \delta(q_1, 1) = \{q_0, q_1\},$$
$$\delta'([q_0, q_1], 0) = [q_0, q_1] \text{ since } \delta(\{q_0, q_1\}, 0) = \{\delta(q_0, 0) \cup \delta(q_1, 0)\}$$
$$= \{q_0, q_1\},$$
$$\delta'([q_0, q_1], 1) = [q_0, q_1] \text{ since } \delta(\{q_0, q_1\}, 1) = \{\delta(q_0, 1) \cup \delta(q_1, 1)\}$$
$$= \{q_0, q_1\}$$

It is summarized in Table 16.7.

Table 16.7 Transition Function δ'

State	*Input*	
	0	1
$[q_0]$ (starting state)	$[q_0, q_1]$	$[q_1]$
$*[q_1]$	φ	$[q_0, q_1]$
$*[q_0, q_1]$	$[q_0, q_1]$	$[q_0, q_1]$

16.8 Equivalence of NFA and NFA-ε

Like nondeterministic and deterministic finite state automata, the computational capabilities of NFA and NFA-ε are same. It means both NFA and NFA-ε accept the same sets of strings. The formal proof showing the equivalence can be viewed by the following construction.

Construction of NFA from NFA-ε

Suppose,

$$M = (Q, \Sigma, \delta, q_0, F)$$

is an NFA-ε that accepts L. Let

$$M' = (Q, \Sigma, \delta', q_0, F')$$

be the equivalent NFA. Then δ' and F' can be defined as:

- $F' = F \cup \{p \mid p \text{ belongs to } \varepsilon\text{-closure}(q) \text{ for every } q \text{ in } F\}$
- Define δ' as

$$\delta'(q, b) = \{p_1, p_2, \ldots, p_j\}$$

if and only if

$$\delta(\{\varepsilon\text{-closure}(q)\}, b) = \{p_1, p_2, \ldots, p_j\}$$

where ε-closure(r) is the set of all states, which are reachable from r on input ε. Also for each state q, q belongs to ε-closure(q).

The names of the states will remain same.

EXAMPLE 16.5 Consider an NFA with ε given by the transition Table 16.8 with its transition diagram in Figure 6.6. Let us construct an equivalent NFA for this automaton.

Table 16.8 Transition Table of NFA-ε

State	*Input*			
	0	1	2	ε
q_0 (starting state)	q_0	φ	φ	q_1
q_1	φ	q_1	φ	q_2
$*q_2$	φ	φ	q_2	φ

Note: Here ϕ is null or reject state.

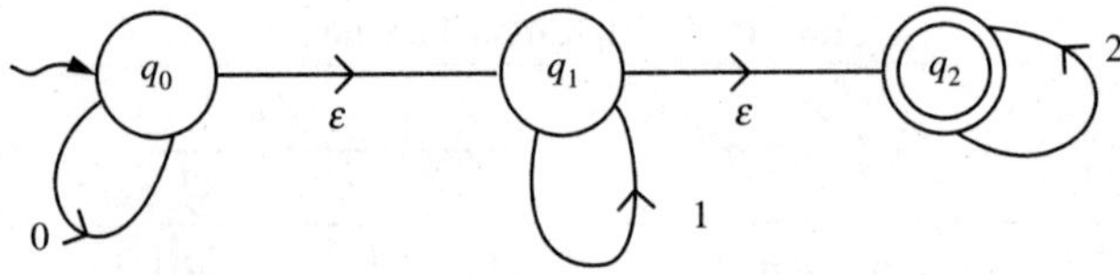

Figure 16.6 Transition diagram of NFA-ε.

Computation of the states of NFA:

1. ε-closure(q_0) = $\{q_0, q_1, q_2\}$
   ε-closure(q_1) = $\{q_1, q_2\}$
   ε-closure(q_2) = $\{q_2\}$

2. $\delta'(q_0, 0)$ = ε-closure(δ(ε-closure(q_0), 0))
   = ε-closure ($\delta(\{q_0, q_1, q_2\}, 0)$)
   = ε-closure($\{\delta(q_0, 0)\} \cup \{\delta(q_1, 0)\} \cup \{\delta(q_2, 0)\}$)
   = ε-closure($\{q_0\} \cup \varphi \cup \varphi$)
   = ε-closure($\{q_0\}$)
   = ε-closure(q_0)
   = $\{q_0, q_1, q_2\}$

3. $\delta'(q_0, 1)$ = ε-closure(δ(ε-closure(q_0), 1))
   = ε-closure($\delta(\{q_0, q_1, q_2\}, 1)$)
   = ε-closure($\{\delta(q_0, 1)\} \cup \delta\{(q_1, 1)\} \cup \{\delta(q_2, 1)\}$)
   = ε-closure($\varphi \cup \{q_1\} \cup \varphi$)

$= \varepsilon\text{-closure}(\{q_1\})$
$= \varepsilon\text{-closure}(q_1)$
$= \{q_1, q_2\}$

4. $\delta'(q_0, 2) = \varepsilon\text{-closure}(\delta(\varepsilon\text{-closure}(q_0), 2))$
$= \varepsilon\text{-closure}(\delta(\{q_0, q_1, q_2\}, 2)$
$= \varepsilon\text{-closure}(\{\delta(q_0, 2)\} \cup \{\delta(q_1, 2)\} \cup \{\delta(q_2, 2)\})$
$= \varepsilon\text{-closure}(\varphi \cup \varphi \cup \{q_2\})$
$= \varepsilon\text{-closure}(\{q_2\})$
$= \varepsilon\text{-closure}(q_2)$
$= \{q_2\}$

5. $\delta'(q_1, 0) = \varepsilon\text{-closure}(\delta(\varepsilon\text{-closure}(q_1), 0))$
$= \varepsilon\text{-closure}(\delta(\{q_1, q_2\}, 0))$
$= \varepsilon\text{-closure}(\{\delta(q_1, 0)\} \cup \{\delta(q_2, 0)\})$
$= \varepsilon\text{-closure}(\varphi \cup \varphi)$
$= \varphi$

6. $\delta'(q_1, 1) = \varepsilon\text{-closure}(\delta(\varepsilon\text{-closure}(q_1), 1))$
$= \varepsilon\text{-closure}(\delta(\{q_1, q_2\}, 1))$
$= \varepsilon\text{-closure}(\{\delta(q_1, 1)\} \cup \{\delta(q_2, 1)\})$
$= \varepsilon\text{-closure}(\{q_1\} \cup \varphi)$
$= \varepsilon\text{-closure}(\{q_1\})$
$= \varepsilon\text{-closure}(q_1)$
$= \{q_1, q_2\}$

7. $\delta'(q_1, 2) = \varepsilon\text{-closure}(\delta(\varepsilon\text{-closure}(q_1), 2))$
$= \varepsilon\text{-closure}(\delta(\{q_1, q_2\}, 2))$
$= \varepsilon\text{-closure}(\{\delta(q_1, 2)\} \cup \{\delta(q_2, 2)\})$
$= \varepsilon\text{-closure}(\varphi \cup \{q_2\})$
$= \varepsilon\text{-closure}(\{q_2\})$
$= \varepsilon\text{-closure}(q_2)$
$= \{q_2\}$

8. $\delta'(q_2, 0) = \varepsilon\text{-closure}(\delta(\varepsilon\text{-closure}(q_2), 0))$
$= \varepsilon\text{-closure}(\delta(\{q_2\}, 0))$
$= \varepsilon\text{-closure}(\varphi)$
$= \varphi$

9. $\delta'(q_2, 1) = \varepsilon\text{-closure}(\delta(\varepsilon\text{-closure}(q_2), 1))$
$= \varepsilon\text{-closure}(\delta(\{q_2\}, 1))$
$= \varphi$

10. $\delta'(q_2, 2) = \varepsilon\text{-closure}(\delta(\varepsilon\text{-closure}(q_2), 2))$
$= \varepsilon\text{-closure}(\delta(\{q_2\}, 2))$
$= \varepsilon\text{-closure}(\{q_2\})$
$= \varepsilon\text{-closure}(q_2)$
$= \{q_2\}$

All states of NFA will be final (accepting) states. Finally, the transition diagram and transition function of the NFA are given in Figure 16.7 and Table 16.9, respectively.

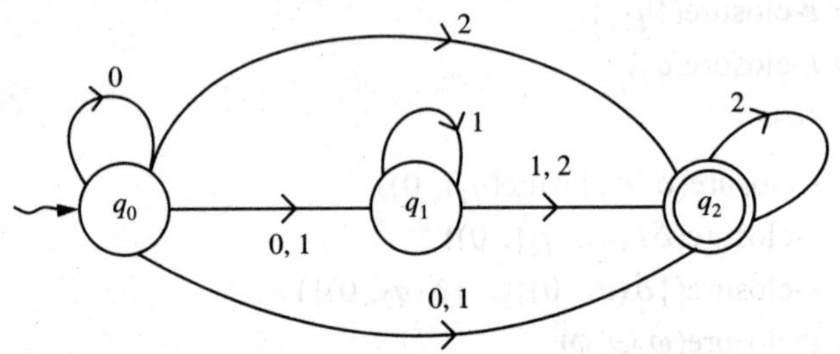

Figure 16.7 Transition diagram of NFA.

Table 16.9 Transition Table of NFA

State	*Input*		
	0	1	2
$*q_0$ (starting state)	$\{q_0, q_1, q_2\}$	$\{q_1, q_2\}$	$\{q_2\}$
$*q_1$	φ	$\{q_1, q_2\}$	$\{q_2\}$
$*q_2$	φ	φ	$\{q_2\}$

16.9 Equivalence of NFA-ε and DFA

From the previous discussion, it is obvious that NFA-ε and DFA are equivalent. Therefore, we can construct DFA from NFA-ε. We will discuss the procedures for computing ε-closure, DFA states and transition function. Minimization of states will also be discussed.

Let N be an NFA-ε and D be the DFA constructed for N. Each state of D will be a set of states of N. The initial state of D is ε-closure(q_0). The accepting states of D are those states which contain at least one accepting state of N.

16.9.1 Computation of ε-closure

Let q, t and u be some states of N. The ε-closure of a state is computed by the following rules.

1. State q is added to ε-closure(q).
2. If t is in ε-closure(q) and there is an edge with input ε from t to u, then u is added to ε-closure(q), if it is not already there.
3. Repeat from (2) until no more states can be added to ε-closure(q).

4. If S is a set of states, then ε-closure(S) will be the union containing ε-closure(s) of each states s in S, i.e., $\varepsilon\text{-closure}(S) = \bigcup_{s} \varepsilon\text{-closure}(s)$, such that s is in S.

16.9.2 Computation of DFA States and Transition Function

Suppose there are n states in N. Then the number of possible states in D may be 2^n. Also, if q_0 is the starting state of N then the starting state of DFA will be ε-closure(q_0).

The following procedure, which is written in pseudo-Pascal code, may be used for computing the DFA states and transition function.

```
/*************Procedure for computing the DFA states and transition function *************/
  Let x be an unmarked state of D.
    begin
      x := ε-closure(q0);
        while there is an unmarked state x = {x1, x2, ..., xn} in D do
          begin
            mark x;
              for each input symbol b do
                begin
                      let S be the set of states to which there is a
                      transition on b from some state xi in x;
                      y := ε-closure(S);
                      add y as an unmarked state of D if it is not already there in D;
                      also add an edge from x to y with input b if not already present;
                end;
          end;
    end;
/*******************************************************************************************/
```

16.9.3 Minimization of DFA States

The DFA, D constructed by the procedure given above may not be optimum (i.e. having minimum number of states). There may be DFAs with lesser number of states than D, and yet accepting the same set of strings. Therefore, we will discuss a procedure given below to minimize the number of states of D.

Let Q be the set of states of D. Also, let F be the set containing all its accepting states. Partition Q into two groups, one having accepting states and the other nonaccepting states. Let us denote the partition by symbol Π.

1. Initial partition is

$$\Pi : Q_1 = F,\ Q_2 = Q - F.$$

2. For each group Q_i of Π do

    begin

    partition Q_i into subgroups such that the two states x_1 and x_2 of Q_i are in the same subgroup if and only if for all inputs, x_1 and x_2 have transitions to states in the same group of Π. Call the new refinement containing all subgroups as Π_{new}.

end;

3. If $\Pi_{new} \neq \Pi$ then replace Π by Π_{new} and repeat from step 2, else exit. Π_{new} is the final partition.
4. In each group of Π_{new}, choose the names of the states representing the groups of new DFA. Alternatively, assign new names to DFA states. Delete states which are not reachable from initial state, if any. Also, delete all the states, which are nonaccepting and have transitions to themselves on all input symbols.

EXAMPLE 16.6 Consider the following NFA-ε given in Figure 16.8.

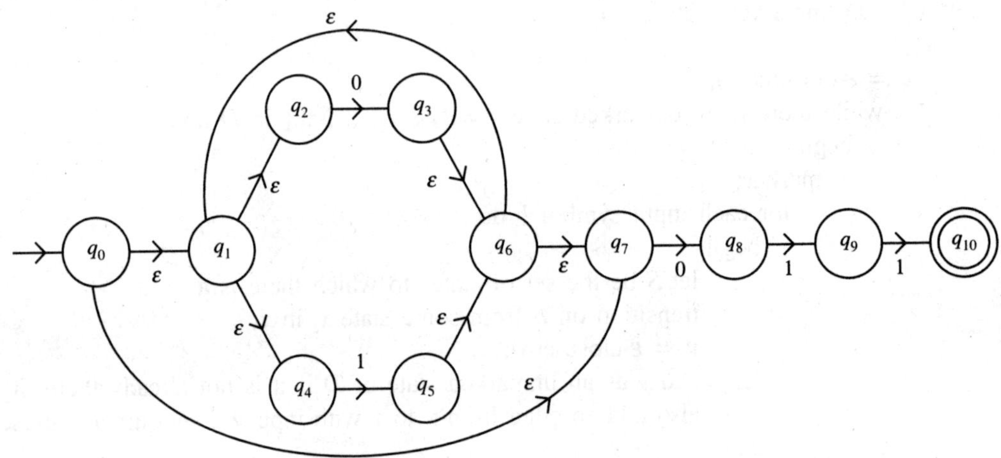

Figure 16.8 State diagram of NFA-ε.

Let us construct an equivalent minimum state DFA for it.

Here, $M = (Q, \Sigma, \delta, q_0, F)$,

$Q = \{q_0, q_1, q_2, q_3, q_4, q_5, q_6, q_7, q_8, q_9, q_{10}\}$,

$F = \{q_{10}\}$

The transition table of the NFA-ε is given in Table 16.10.

Table 16.10 Transition Table of NFA-ε

State	*Input*		
	0	1	ε
q_0(starting state)	–	–	$\{q_1, q_7\}$
q_1	–	–	$\{q_2, q_4\}$
q_2	$\{q_3\}$	–	–
q_3	–	–	$\{q_6\}$
q_4	–	$\{q_5\}$	–
q_5	–	–	$\{q_6\}$

(*Contd.*)

Table 16.10 Transition Table of NFA-ε (*Contd.*)

State	*Input*		
	0	1	ε
q_6	–	–	$\{q_1, q_7\}$
q_7	$\{q_8\}$	–	–
q_8	–	$\{q_9\}$	–
q_9	–	$\{q_{10}\}$	–
$*q_{10}$	–	–	–

Let the equivalent DFA be

$$M' = (Q', \Sigma, \delta', q_0', F')$$

We will name the DFA states by capital letters A, B, C etc.

1. **Computation of DFA states and transition function**

Use algorithm given in section 16.9.2.

Starting state of DFA $= \varepsilon\text{-closure}(q_0)$
$= \{q_0, q_1, q_2, q_4, q_7\}$
$= A$ (say)

Find transitions from A on each input symbol.

Set $x = A$
$= \{q_0, q_1, q_2, q_4, q_7\}$,

and mark A.

Input 0: States q_2 and q_7 in x have transitions to states q_3 and q_8.

Therefore, $y = \varepsilon\text{-closure}(q_3, q_8)$
$= \{q_1, q_2, q_3, q_4, q_6, q_7, q_8\}$
$= B$ (say).

Input 1: State q_4 in x has transition to state q_5

Therefore, $y = \varepsilon\text{-closure}(q_5)$
$= \{q_1, q_2, q_4, q_5, q_6, q_7\}$
$= C$ (say).

Now, find transition from B on each input symbol.

Set $x = B$
$= \{q_1, q_2, q_3, q_4, q_6, q_7, q_8\}$

Mark B.

Input 0: States q_2 and q_7 in x have transitions to states q_3 and q_8.

Therefore, $y = \varepsilon\text{-closure}(q_3, q_8)$
$= \{q_1, q_2, q_3, q_4, q_6, q_7, q_8\}$
$= B$.

Input 1: State q_4 and q_8 in x have transitions to states q_5 and q_9.

Therefore,
$$y = \varepsilon\text{-closure } (\{q_5, q_9\})$$
$$= \{q_1, q_2, q_4, q_5, q_6, q_7, q_9\}$$
$$= D \text{ (say)}.$$

Continue computing the states to get the following DFA:

$Q' = \{A, B, C, D, E\}$, where A, B, C and D are as named above, and state E, which is ε-closure $(\{q_5, q_{10}\})$, is $\{q_1, q_2, q_4, q_5, q_6, q_7, q_{10})$

In this DFA, starting state $q_0' = A$,

and accepting state $F' = \{E\}$.

Transition function, δ' is given by the transition diagram (Figure 16.9) and transition Table (Table 16.11).

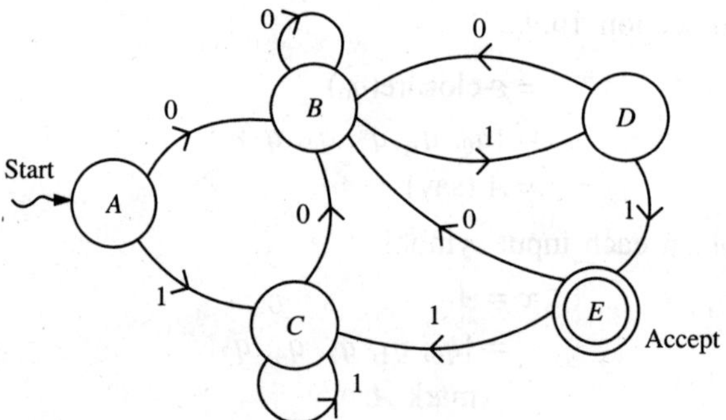

Figure 16.9 Transition diagram.

Table 16.11 Transition Table δ'

State	Input	
	0	1
A (starting state)	*B*	*C*
B	*B*	*D*
C	*B*	*C*
D	*B*	*E*
**E*	*B*	*C*

2. **Minimization of DFA States**

Initially, $\Pi : Q_1 = (E), Q_2 = (A, B, C, D)$

The first group Q_1 has only one state, so we can partition only Q_2.

On input 0, each state in Q_2 has transition to B which is in the same group, therefore no refinement is possible. However, on input 1, states A, B and C have transitions to states C and

D, which are in the same group; but *D* has transition to state *E*, which is in different group. Thus, *D* can be separated from the group.

Therefore, $\Pi_{new} = (E), (D), (A, B, C)$.

Since, $\Pi_{new} \neq \Pi$,

therefore, further refinement is needed. Only the group (*A*, *B*, *C*) can be partitioned.

On input 0 states A, B and C have transitions to state B, which is in the same group. However, on input 1, states A and C have transitions to C, whereas B has transition to D. Therefore, D can be separated.

Thus, $\Pi_{new} = (E), (D), (B), (A, C)$.

Again $\Pi_{new} \neq \Pi$,

therefore, try partitioning (*A*, *C*). We can see that no further refinement is possible. Thus, this is the final partition. Choose the name *A* for the group (*A*, *C*). Replace *C* by *A* in the DFA. In the transition table, two rows, row 1 and row 3 will become identical, so, delete one row. Finally, we get minimized DFA given in Table 16.12. Its corresponding transition diagram is given in Figure 16.10.

Table 16.12 Transition Table of Minimized DFA

State	*Input* 0	1
A (starting state)	*B*	*A*
B	*B*	*D*
D	*B*	*E*
**E*	*B*	*A*

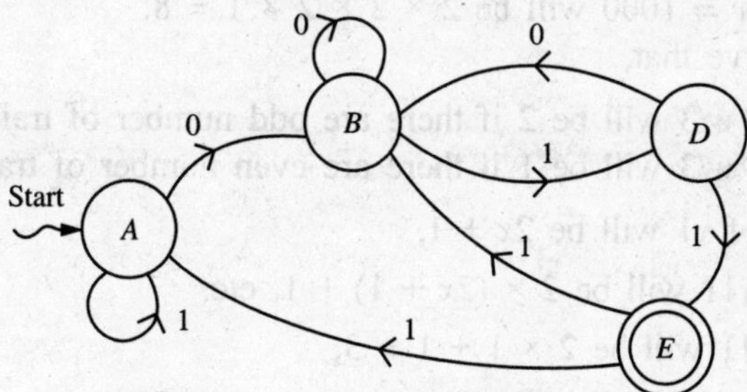

Figure 16.10 Transition diagram of DFA.

16.10 Finite State Machine (FSM)

Finite state machine is the extension of finite state automaton model. Finite state machines are those automatons that also generate some outputs. We will discuss two well-known models, Moore machine and Mealy machine.

16.10.1 Moore Machine

In Moore[$] machine, output is associated with the states. Formally, Moore machine is defined as a six tuples,

$$M = (Q, \Sigma, \Delta, \delta, \lambda, q_0)$$

where Q, Σ, δ, and q_0 are same as defined earlier in finite state automaton. Symbol Δ is the output alphabet. In Moore machine, each state is considered as an accepting state.
The output function λ is given by

$$\lambda : Q \to \Delta.$$

For instance, say input string is

$$w = b_1 b_2 b_3 \ldots b_n.$$

Then output will be

$$w' = \lambda(q_0)\lambda(q_1)\lambda(q_2) \ldots \lambda(q_n),\ n \geq 0,$$

where $q_0, q_1, q_2, \ldots, q_n$ is the sequence of states such that

$$\delta(q_{i-1}, b_i) = q_i, \text{ for } 1 \leq i \leq n.$$

In this model, the output string will contain one symbol more than the number of symbols in the input string. It is $\lambda(q_0)$ which is generated without consuming any input.

EXAMPLE 16.7 Let us design a Moore machine that computes residue mod 3 for input, a binary string.

Let x be some binary number.

Then the value of $x0$ will be $2 \times x$,
value of $x00$ will be $2 \times (2 \times x)$, etc.

For example, value of w = 10 will be $2 \times 1 = 2$,
value of w = 100 will be $2 \times (2 \times 1) = 4$,
value of w = 1000 will be $2 \times 2 \times 2 \times 1 = 8$.

From here, we can observe that,

(i) the remainder in $w/3$ will be 2 if there are odd number of trailing zeroes in w.
(ii) the remainder in $w/3$ will be 1 if there are even number of trailing zeroes in w.

In a similar way, value of $x1$ will be $2x + 1$,
value of $x11$ will be $2 \times (2x + 1) + 1$, etc.

For example, value of 11 will be $2 \times 1 + 1 = 3$,
value of 111 will be $2 \times (2 \times 1 + 1) + 1 = 7$,
value of 1111 will be $2 \times (2 \times (2 \times 1 + 1) + 1) + 1 = 15$, etc.

[$] **Edward Forrest Moore** (1925–2003) was an American professor of mathematics and computer science. He is known as the inventor of the Moore finite state machine. Moore received a B.S. in chemistry from Virginia Polytechnic Institute in Blacksburg, VA in 1947 and a Ph.D. in Mathematics from Brown University in Providence, RI in 1950. He worked at UIUC from 1950 to 1952 and was a visiting lecturer at MIT and Harvard university, simultaneously in the years 1952 and 1953. Later, he worked at Bell Labs for about 10 years. After that, he was a professor at the University of Wisconsin–Madison from 1966 until he retired in 1985. Moore died in 2003.

Hence, we can deduce that

(i) the remainder in $w/3$ will be 0 if there are even number of trailing 1s in w.
(ii) the remainder in $w/3$ will be 1 if there are odd number of trailing 1s in w.

Further, if x is divided by three, then there could be three remainders 0, 1 or 2. Thus, there will be three states in the Moore machine.

Considering the above facts, the Moore machine designed is given below in Figure 16.11 along with its transition table (Table 16.13).

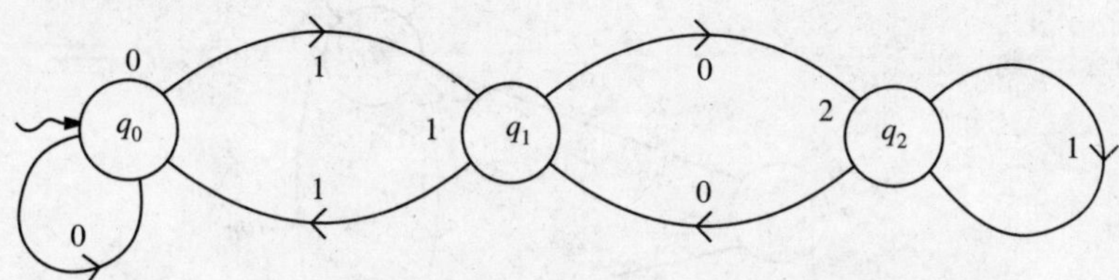

Figure 16.11 Moore machine for residue mod 3.

Table 16.13 Transition Table of Moore Machine for residue mod 3

Current state	*Output*	*Input* 0	1
q_0	0	q_0	q_1
q_1	1	q_2	q_0
q_2	2	q_1	q_2

16.10.2 Mealy Machine

In Mealy machine, output symbols are attached with the edges of the machine.

Formally, Mealy machine is defined as a six tuple

$$M = (Q, \Sigma, \Delta, \delta, \lambda, q_0)$$

where symbols have usual meaning. Like Moore machine, each state in Mealy machine is considered as accepting state

The output function λ is given by

$$\lambda : Q \times \Sigma \rightarrow \Delta.$$

For input string

$$w = b_1 b_2 b_3 \cdots b_n,$$

output will be

$$w' = \lambda(q_0, b_1)\ \lambda(q_1, b_2)\lambda(q_2, b_3)\ \ldots\ \lambda(q_{n-1}, b_n),\ n \geq 0,$$

where $q_0, q_1, q_2, \ldots, q_n$ is the sequence of states,

such that $\delta(q_{i-1}, b_i) = q_i,\ 1 \leq i \leq n.$

Note that the input and output strings will contain equal number of symbols.

EXAMPLE 16.8 A Mealy machine

$$M = (Q, \Sigma, \Delta, \delta, \lambda, q_0)$$

with $Q = \{q_0, q_1, q_2\}, \Sigma = \{0, 1\}$

and $\Delta = \{y, n\}$

is given in Figure 16.12 with transition Table 16.14.

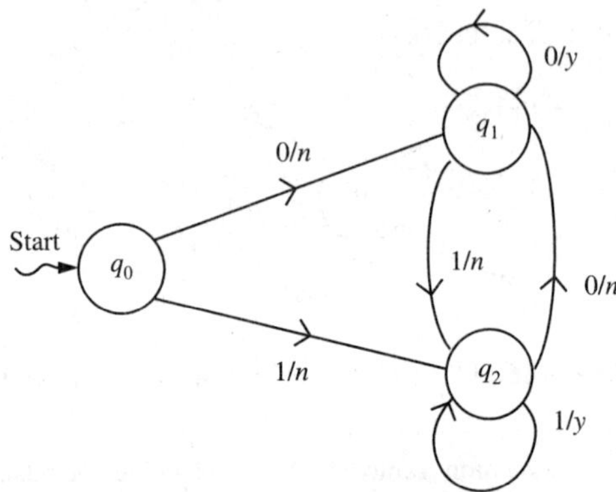

Figure 16.12 Mealy machine.

Table 16.14 Transition Table of Mealy Machine

	Input			
	0		1	
Current state	*Next state*	*Output*	*Next state*	*Output*
q_0	q_1	n	q_2	n
q_1	q_1	y	q_2	n
q_2	q_1	n	q_2	y

16.11 Construction of Moore Machine from Mealy Machine

We have seen that the output string of a Moore machine contains one symbol more than the Mealy machine. This is due to the output symbol attached with the starting state of the Moore machine, which produces the first output symbol on blank input, i.e., without consuming any input. If this output is ignored, then the two models are equivalent. Therefore, a Moore machine can be constructed from a Mealy machine and vice versa. The procedure is quite simple. The number of states in the Moore machine will depend on the number of output symbols in the Mealy machine. If there are n symbols in the output alphabet, then create n Moore states for each Mealy state. Attach output symbols with the states, and join arcs using transition function of Mealy machine. Note that there will be n starting states, but retain only one, depending on the output desired at this state, and delete others.

EXAMPLE 16.9 Moore machine, constructed for Mealy machine of Figure 16.12 is given in Figure 16.13.

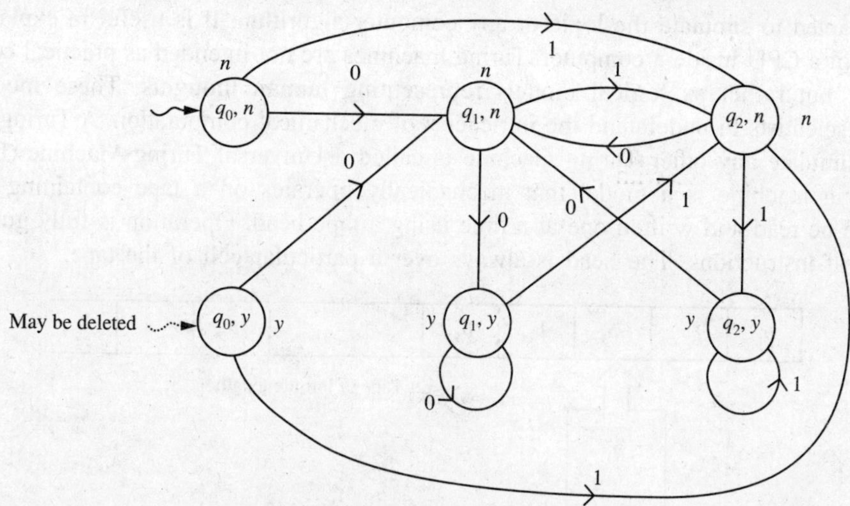

Figure 16.13 Construction of Moore machine.

16.12 Construction of Mealy Machine from Moore Machine

Construction from Moore machine to Mealy machine is very straightforward and needs very less effort. Simply shift the output symbols from the states to the incoming edges. The output symbol at starting state is neglected for outgoing edges.

EXAMPLE 16.10 Mealy machine constructed for Moore machine of Figure 16.11 is given in Figure 16.14.

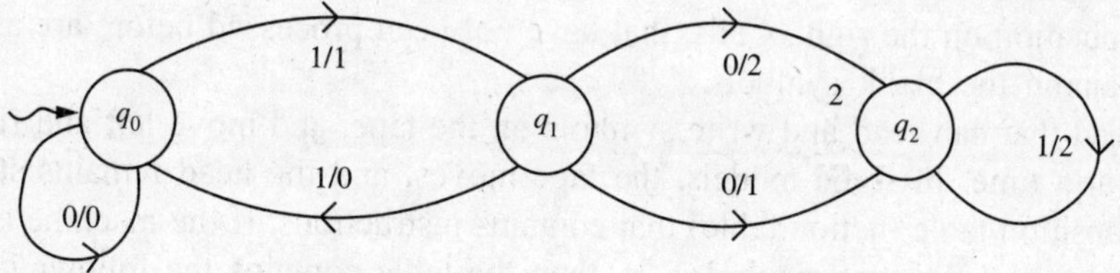

Figure 16.14 Construction of Mealy machine.

16.13 Turing Machine (TM)

Turing machine model was introduced by Alan Turing# in 1936. It is a theoretical model that manipulates symbols on a tape according to some prescribed rules. In spite of its simplicity, it

---

#**Alan Mathison Turing** (1912–1954), was an English mathematician, logician, and computer scientist. He was highly influential in the development of computer science, providing a formalization of the concepts of "algorithm" and "computation" with the Turing machine, which played a significant role in the creation of the modern computer.

(Contd.)

can be adapted to simulate the logic of any computer algorithm. It is useful in explaining the functions of a CPU inside a computer. Turing machines are not intended as practical computing machines, but rather as logical models representing human thoughts. These models help computer scientists to understand the intricacies of mechanical computation. A Turing machine that can simulate any other Turing machine is called a Universal Turing Machine (UTM).

Turing machine is a model that mechanically operates on a tape containing symbols, which can be read and written one at a time using a tape head. Operation is fully guided by a finite set of instructions. The head is always over a particular cell of the tape.

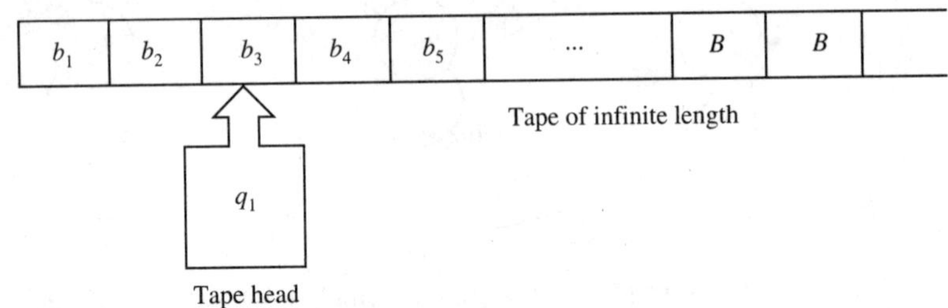

Figure 16.15 Turing machine model.

A Turing machine model is given in Figure 16.15. Here, machine is in state q_1 and the tape head is over the cell containing symbol b_3. Tape is considered of infinite length on the right. More precisely, a Turing machine consists of:

1. A tape which is divided into cells. Each cell contains a symbol from some finite alphabet. The alphabet contains a special symbol blank, 'B' along with other symbols. The tape is assumed to be infinite on the right, but has a left most cell containing a special symbol, although there is a variant TM that contains tape, which is infinite on both sides. A Turing machine is always supplied with as much tape as it needs for its computation on the right. Cells, that have not been processed before are assumed to be containing the blank symbols.
2. A head that can read and write symbols on the tape, and move left and right over one cell at a time. In some models, the tape moves, and the head remains stationary.
3. A transition table (action table) that contains instructions. If the machine is currently in state(q_i), and the tape symbol is b_j, then the table contains the following sequence of actions:
   - Write, erase or modify the tape symbol.

(*Contd.*)

During the Second World War, Turing worked for the British Government at code breaking centre. After the war, he worked at the National Physical Laboratory, where he created one of the first designs for a stored-program computer.

In 1926, at the age of 14, Turing was admitted to a school. His first day of the school term coincided with the General Strike in Britain, but he was so determined to attend his first day that he rode his bicycle unaccompanied more than 97 km from Southampton to school, stopping overnight at an inn. Turing travelled to the United States in November 1942 and worked with U.S. Navy cryptanalysts. Turing died in 1954.

- Move the head left (L) or right (R).
- Assume the same or a new state.

In some models of the Turing machine, if the table contains no data for the current combination of symbol and state, then the machine will halt. However, there are models that require all entries to be filled.

4. A state register that stores the state of the transition table. The state register is initialized with the start state.

Formally, Turing machine is a 7-tuple

$$M = (Q, \Gamma, \Sigma, \delta, q_0, F, B),$$

where

- Q is a finite nonempty set of states.
- Γ is a finite nonempty set of tape symbols.
- $\Sigma \subseteq \Gamma$ is the set of input symbols.
- $\delta : Q \times \Gamma \to Q \times \Gamma \times \{L, R\}$ is a transition function, where L means left shift, and R means right shift.
- $q_0 \in Q$ is the initial state.
- $F \subseteq Q$ is the set of final or accepting states.
- $B \in \Gamma$ is a special symbol called blank.

EXAMPLE 16.11 A 3-state TM is given in transition Table 16.15.

Here,

$Q = \{q_0, q_1, q_2, \text{Halt}\}$
$\Gamma = \{B, 1\}$
$\Sigma = \{1\}$
q_0 = initial state
F = set of accepting states {Halt}

Initially, all tape cells are marked with B.

Table 16.15 Transition Table Format 1

State Table for three state, two symbols TM

Input symbol	*Current state q_0*			*Current state q_1*			*Current state q_2*		
	Write symbol	*Move tape*	*Next state*	*Write symbol*	*Move tape*	*Next state*	*Write symbol*	*Move tape*	*Next state*
B	1	R	q_1	1	L	q_0	1	L	q_1
1	1	L	q_2	1	R	q_1	1	R	Halt

Two more formats for representing transition function are given in Tables 16.16 and 16.17. Format given in Table 16.15 is more compact, but feasible only if number of states are small.

Table 16.16 Transition Table Format 2

State	Action on input					
	B			1		
	Next state	*Next symbol on tape*	*Head movement*	*Next state*	*Next symbol on tape*	*Head movement*
q_0	q_1	1	*R*	q_2	1	*L*
q_1	q_0	1	*L*	q_1	1	*R*
q_2	q_1	1	*L*	Halt	1	*R*

Table 16.17 Transition Table Format 3

State	Action on input				
	0	1	*X*	*Y*	*B*
q_0	(q_1, X, R)	–	–	(q_1, X, R)	–
q_1	(q_2, X, L)	(q_0, Y, R)	–	(q_2, Y, R)	–
q_2	(q_1, X, R)	–	(q_0, Y, L)	(q_1, X, L)	–

Description of Table 16.17: Let the machine be in state q_0 and input be 0. Then the state is changed to q_1, 0 is replaced by X and the head moves towards right. Other entries can be interpreted in a similar way.

16.14 Instantaneous Description of Turing Machine (ID)

The ID of a TM is given by $\beta_1 q \beta_2$, where q is the current state, and $\beta_1\beta_2 \in \Gamma^*$ is a string on the tape up to the rightmost non blank symbol.

Let the ID of the TM is

$$X_1\ X_2\ X_3\ \ldots\ X_{i-1}\ qX_i\ \ldots\ X_n$$

then the transition function

$$\delta(q, X_i) = (p, Y, L),\ i > 1$$

will result the TM in

$$\text{ID} = X_1\ X_2\ X_3\ \ldots\ qX_{i-1}\ YX_{i+1}\ \ldots\ X_n.$$

More appropriately,

$$X_1\ X_2\ X_3\ \ldots\ X_{i-1}\ qX_i\ \ldots X_n \underset{M}{\Rightarrow} X_1\ X_2\ X_3\ \ldots\ pX_{i-1}\ YX_{i+1}\ \ldots\ X_n.$$

If $\delta(q, X_i) = (p, Y, R)$, $i > 1$, then we get

$$X_1\ X_2\ X_3\ \ldots\ X_{i-1}\ qX_i\ \ldots\ X_n \underset{M}{\Rightarrow} X_1\ X_2\ X_3\ \ldots X_{i-1}\ YpX_{i+1}\ \ldots\ X_n.$$

If one ID results from the other by applying some finite number of transitions, then this is represented by the following symbol:

$$\text{ID}_1 \overset{*}{\underset{M}{\Rightarrow}} \text{ID}_2$$

The language accepted by Turing machine M is denoted by $L(M)$, and it is the set of those words in Σ^* which cause M to enter the accepting state.

That is, $L(M) = \{w \mid w \in \Sigma^* \text{ and } q_0w \overset{*}{\underset{M}{\Rightarrow}} \beta_1 p \beta_2 \text{ for some } p \in F \text{ and } \beta_1\beta_2 \in \Gamma^*\}$.

Whenever an input is accepted by a TM, it halts, i.e., has no next move.

EXAMPLE 16.12 Let us design a TM that accepts $\{0^n1^n \mid n \geq 1\}$.

Let $M = (Q, \Gamma, \Sigma, \delta, q_0, F, B)$ be a TM. Consider a tape that initially contains 0^n1^n followed by infinite number of blanks. The aim is to design a transition function for M that accepts strings containing a sequence of 0s followed by an equal number of 1s. Think about the logic that implements this design. Imagine that M repeatedly replaces the leftmost 0 by X, moves right up to the leftmost 1 and replaces it by Y. Next, it moves towards left to find the rightmost X, and then moves one cell right to find the leftmost 0. This cycle is repeated. If after changing a 1 to Y, M finds no more 0s, then M checks that no more 1s remain on the tape. If so, it accepts the string. However, while searching for a 1, if M finds a blank then M halts without accepting the string.

Formally, we must have

$$Q = \{q_0, q_1, q_2, q_3, q_4\},$$
$$\Gamma = \{0, 1, X, Y, B\},$$
$$\Sigma = \{0, 1\},$$
$$q_0 = \text{starting state},$$
$$F = \{q_4\}.$$

Transition function δ is given in Table 16.18.

Table 16.18 Transition Table δ

	Action on input				
State	0	1	X	Y	B
q_0	(q_1, X, R)	–	–	(q_3, Y, R)	–
q_1	$(q_1, 0, R)$	(q_2, Y, L)	–	(q_1, Y, R)	–
q_2	$(q_2, 0, L)$	–	(q_0, X, R)	(q_2, Y, L)	–
q_3	–	–	–	(q_3, Y, R)	(q_4, B, R)
$^*q_4$	–	–	–	–	–

Now, let us test the machine for string $w = 0^21^2$, i.e., $w = 0011$.

$$q_0 0011 \underset{M}{\Rightarrow} X\, q_1\, 011 \underset{M}{\Rightarrow} X0\, q_1 11 \underset{M}{\Rightarrow} X\, q_2\, 0Y1$$
$$\underset{M}{\Rightarrow} q_2\, X\, 0Y1 \underset{M}{\Rightarrow} X\, q_0\, 0Y1 \underset{M}{\Rightarrow} X X q_1\, Y1$$
$$\underset{M}{\Rightarrow} XXY\, q_1\, 1 \underset{M}{\Rightarrow} XX\, q_2\, YY \underset{M}{\Rightarrow} X\, q_2\, XYY$$
$$\underset{M}{\Rightarrow} XX\, q_0\, YY \underset{M}{\Rightarrow} XXY\, q_3\, Y \underset{M}{\Rightarrow} X\, XYYq_3B$$
$$\underset{M}{\Rightarrow} XXYYBq_4$$

Machine is now in the state q_4, which is the accepting state and no input symbol other than blank is left on the tape, therefore, string is accepted. Further, readers may test it on the strings 001100, 010011, 000111 etc. The first two strings will be rejected, while the third will be accepted.

16.15 Partial Recursive Functions

Partial recursive functions are those functions, which are computed by Turing machines that may or may not halt on a given input. It means, a machine may reach its accepting state even without halting. On the other hand, a function is total recursive, if it is computed by a Turing machine that always halts. An arithmetic function, like $n!$ is total recursive.

16.16 Solved Problems

PROBLEM 16.1 Construct deterministic finite state automaton M that accepts words of an even number of 0s.

Solution

The DFA M given in Figure 16.16 accepts words of even number of 0s. Its transition function is given in Table 16.19.

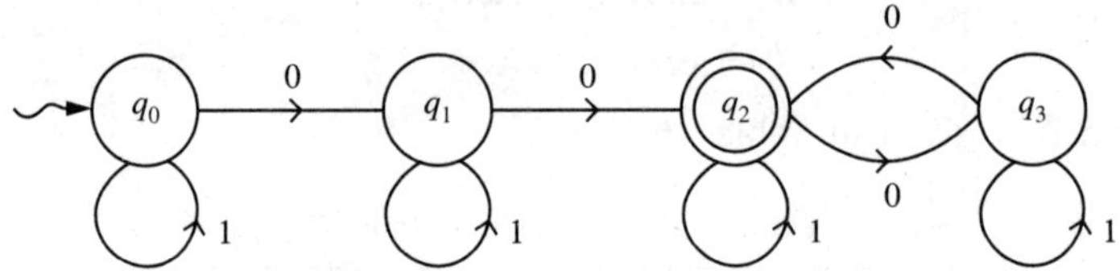

Figure 16.16 State diagram of DFA M.

Here, $M = (Q, \Sigma, \delta, q_0, F)$
where

- $Q = \{q_0, q_1, q_2, q_3\}$
- $\Sigma = \{0, 1\}$
- Starting state is q_0
- $F = \{q_2\}$, and
- δ is defined by the following state transition table:

Table 16.19 Transition Table of DFA M

State	*Input*	
	0	1
q_0 (starting state)	q_1	q_0
q_1	q_2	q_1
$*q_2$(accepting state)	q_3	q_2
q_3	q_2	q_3

Here q_0 is the starting state and q_2 is the final (accepting) state of DFA. The strings accepted by the automaton are 00, 100, 0011, and 001100 etc.

PROBLEM 16.2 Consider an NFA-ε given by transition Table 16.20 with its transition diagram in Figure 16.17. Construct an equivalent NFA for this automaton.

Table 16.20 Transition Table of NFA-ε

State	*Input*			
	0	1	2	ε
q_0 (starting state)	q_0	φ	φ	$\{q_1, q_2\}$
q_1	φ	q_1	φ	q_2
$*q_2$	φ	q_2	q_2	φ

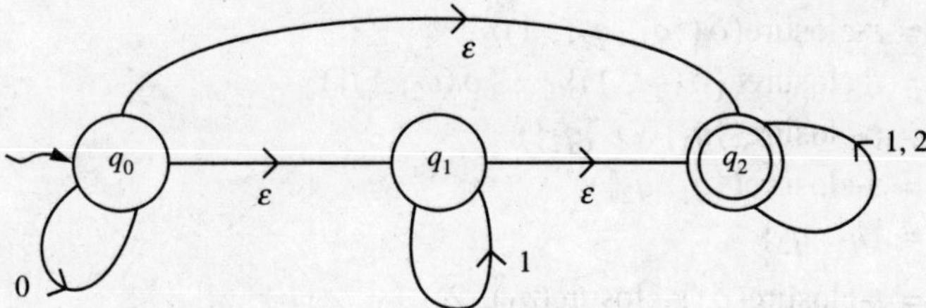

Figure 16.17 Transition diagram of NFA-ε.

Solution

Computation of the states of NFA:

(i) $\varepsilon\text{-closure}(q_0) = \{q_0, q_1, q_2\}$
$\varepsilon\text{-closure}(q_1) = \{q_1, q_2\}$
$\varepsilon\text{-closure}(q_2) = \{q_2\}$

(ii) $\delta'(q_0, 0) = \varepsilon\text{-closure}(\delta(\varepsilon\text{-closure}(q_0), 0))$
$= \varepsilon\text{-closure}(\delta(\{q_0, q_1, q_2\}, 0)$
$= \varepsilon\text{-closure}(\{\delta(q_0, 0)\} \cup \{\delta(q_1, 0)\} \cup \{\delta(q_2, 0)\})$
$= \varepsilon\text{-closure}(\{q_0\} \cup \varphi \cup \varphi)$
$= \varepsilon\text{-closure}(\{q_0\})$
$= \varepsilon\text{-closure}(q_0)$
$= \{q_0, q_1, q_2\}$

(iii) $\delta'(q_0, 1) = \varepsilon\text{-closure}(\delta(\varepsilon\text{-closure}(q_0), 1))$
$= \varepsilon\text{-closure}(\delta(\{q_0, q_1, q_2\}, 1))$
$= \varepsilon\text{-closure}(\{\delta(q_0, 1)\} \cup \delta\{(q_1, 1)\} \cup \{\delta(q_2, 1)\})$
$= \varepsilon\text{-closure}(\varphi \cup \{q_1\} \cup \{q_2\})$
$= \varepsilon\text{-closure}(\{q_1, q_2\})$
$= \{q_1, q_2\}$

(iv) $\delta'(q_0, 2) = \varepsilon\text{-closure}(\delta(\varepsilon\text{-closure}(q_0), 2))$
$= \varepsilon\text{-closure}(\delta(\{q_0, q_1, q_2\}, 2)$
$= \varepsilon\text{-closure}(\{\delta(q_0, 2)\} \cup \{\delta(q_1, 2)\} \cup \{\delta(q_2, 2)\})$
$= \varepsilon\text{-closure}(\varphi \cup \varphi \cup \{q_2\})$
$= \varepsilon\text{-closure}(\{q_2\})$
$= \{q_2\}$

(v) $\delta'(q_1, 0) = \varepsilon\text{-closure}(\delta(\varepsilon\text{-closure}(q_1), 0))$
$= \varepsilon\text{-closure}(\delta(\{q_1, q_2\}, 0))$
$= \varepsilon\text{-closure}(\{\delta(q_1, 0)\} \cup \{\delta(q_2, 0)\})$
$= \varepsilon\text{-closure}(\varphi \cup \varphi)$
$= \varphi$

(vi) $\delta'(q_1, 1) = \varepsilon\text{-closure}(\delta(\varepsilon\text{-closure}(q_1), 1))$
$= \varepsilon\text{-closure}(\delta(\{q_1, q_2\}, 1))$
$= \varepsilon\text{-closure}(\{\delta(q_1, 1)\} \cup \{\delta(q_2, 1)\})$
$= \varepsilon\text{-closure}(\{q_1\} \cup \{q_2\})$
$= \varepsilon\text{-closure}(\{q_1, q_2\})$
$= \{q_1, q_2\}$

(vii) $\delta'(q_1, 2) = \varepsilon\text{-closure}(\delta(\varepsilon\text{-closure}(q_1), 2))$
$= \varepsilon\text{-closure}(\delta(\{q_1, q_2\}, 2))$
$= \varepsilon\text{-closure}(\{\delta(q_1, 2)\} \cup \{\delta(q_2, 2)\})$
$= \varepsilon\text{-closure}(\varphi \cup \{q_2\})$
$= \varepsilon\text{-closure}(\{q_2\})$
$= \varepsilon\text{-closure}(q_2)$
$= \{q_2\}$

(viii) $\delta'(q_2, 0) = \varepsilon\text{-closure}(\delta(\varepsilon\text{-closure}(q_2), 0))$
$= \varepsilon\text{-closure}(\delta\ (\{q_2\}, 0))$
$= \varepsilon\text{-closure}(\varphi)$
$= \varphi$

(ix) $\delta'(q_2, 1) = \varepsilon\text{-closure}(\delta(\varepsilon\text{-closure}(q_2), 1))$
$= \varepsilon\text{-closure}(\delta(\{q_2\}, 1))$
$= \{q_2\}$

(x) $\delta'(q_2, 2) = \varepsilon\text{-closure}(\delta(\varepsilon\text{-closure}(q_2), 2))$
$= \varepsilon\text{-closure}(\delta(\{q_2\}, 2))$
$= \varepsilon\text{-closure}(\{q_2\})$
$= \varepsilon\text{-closure}(q_2)$
$= \{q_2\}$

All states are final (accepting) states. Finally, the transition diagram and transition function of the NFA are given in Figure 16.18 and Table 16.21, respectively.

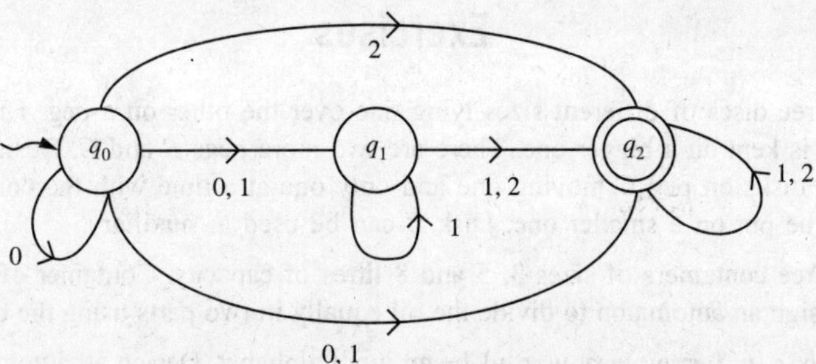

Figure 16.18 Transition diagram of NFA.

Table 16.21 Transition Table of NFA

State	Input		
	0	1	2
$*q_0$ (starting state)	$\{q_0, q_1, q_2\}$	$\{q_1, q_2\}$	$\{q_2\}$
$*q_1$	φ	$\{q_1, q_2\}$	$\{q_2\}$
$*q_2$	φ	$\{q_2\}$	$\{q_2\}$

PROBLEM 16.3 Design a Mealy machine that performs binary addition.

Solution

Let $M = (Q, \Sigma, \Delta, \delta, \lambda, q_0)$ be a Mealy machine,

where $Q = \{q_0, q_1, q_2\}$,

$\Sigma = \{bb, 00, 01, 10, 11\}$ and

$\Delta = \{b, 0, 1\}$.

Three states of the machine will be no carry(q_0), carry(q_1) and stop(q_2). Symbol b stands for blank or empty input. If machine is in carry state, then 1 is automatically added to the input string. The machine designed is given in Figure 16.19.

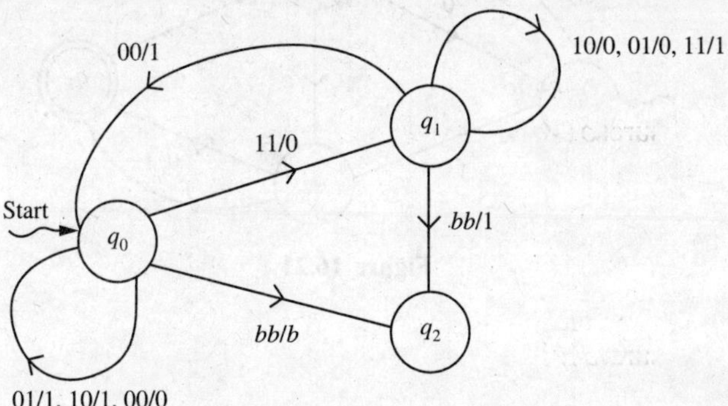

Figure 16.19 Mealy machine.

Exercises

16.1 There are three discs of different sizes lying one over the other on a peg *A* in such a way that a smaller disc is kept on a bigger one. There are two more pegs *B* and *C*. Design a DFA to transfer all the three disks on peg *C* moving one and only one at a time with the condition that a bigger disc cannot be put on a smaller one. Disk *B* can be used as auxiliary.

16.2 There are three containers of sizes 3, 5 and 8 litres of capacity. Container of size 8 litre is filled with oil. Design an automaton to divide the oil equally in two parts using the other two containers.

16.3 Let $\Sigma = \{a, b, e, h, i, s, m, n, o, w, t, y\}$ be an input alphabet. Design an automaton that recognizes meaningful sentences of the English language, e.g., he is a boy, etc.

16.4 Construct DFA for the following NFA.

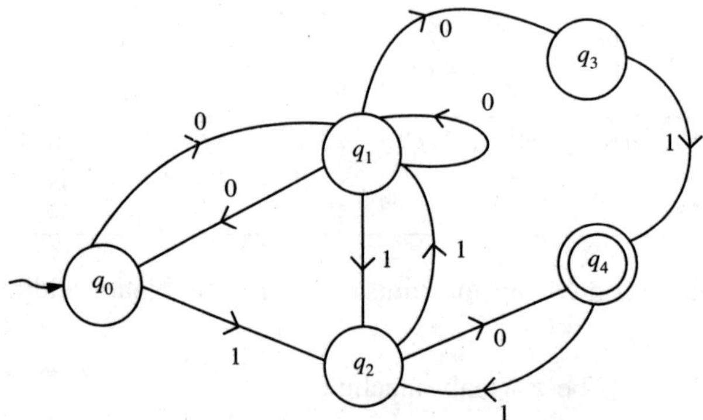

Figure 16.20

16.5 Construct minimum state DFA for the following NFA-ε.

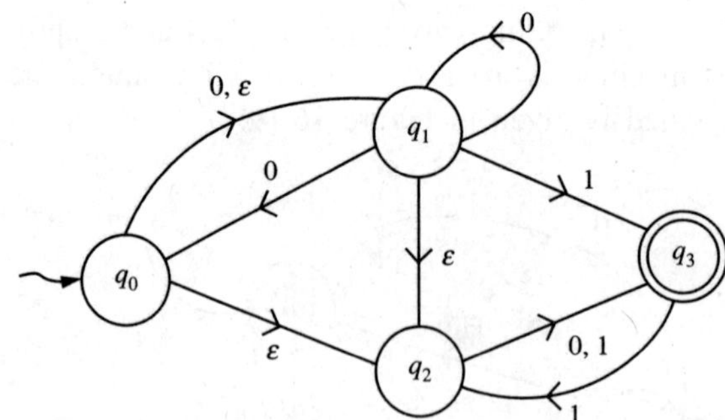

Figure 16.21

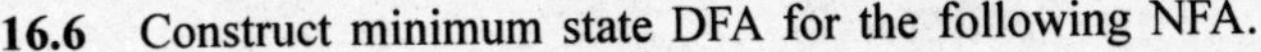

16.6 Construct minimum state DFA for the following NFA.

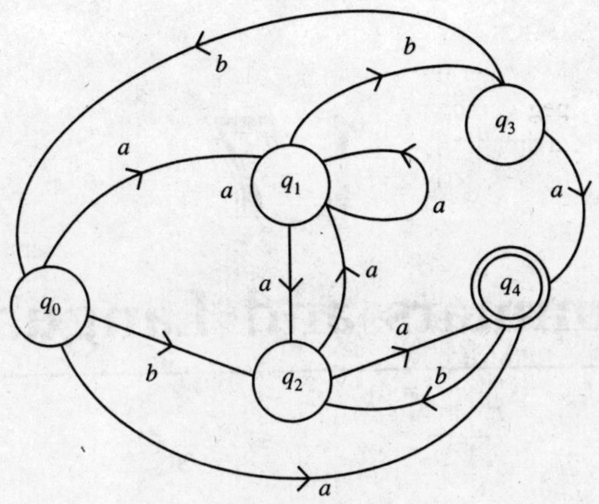

Figure 16.22

16.7 Design a Moore machine for computing residue-mod 4. Convert it to a Mealy machine.

16.8 Design a Turing machine for proper subtraction, i.e., *m-n*, where *m* and *n* are integers and $m > n$.

17
Grammars and Languages

17.1 Introduction

In this chapter, we will study some essential features of grammars and languages. Language theory is an important part of computer science. Context free grammar is a fundamental model for designing a programming language. Syntax analysis is done with the help of a parser, therefore, some parsers will also be discussed here.

A **formal language** is a set of words defined over a finite set of symbols called alphabet. It is often defined by means of a formal grammar (called formation rules). Words that belong to a language are called well-formed words. Formal languages are studied by linguists and computer scientists. Formal languages are often used as foundations for other constructs involving semantics. In computer science, these languages are used for defining data formats and the syntax of programming languages. The study of languages plays a crucial role in the development of translators like compilers, interpreters, assemblers, etc. Formal languages can also be used in logic to represent the syntax of formal theorems or arguments.

17.2 Regular Language

In computer science, a regular language is a formal language that is accepted by a finite state automaton. It is described by a formal regular expression. Let Σ be the finite alphabet. Then the regular expressions denoting regular languages are defined recursively as follows:

- ϕ is a regular expression denoting regular language empty.
- ε is a regular expression denoting regular language $\{\varepsilon\}$.

- Each $a \in \Sigma$ (a belongs to Σ) is a regular expression denoting the language $\{a\}$.
- If r and s are regular expressions denoting languages L_r and L_s, then
  1. $(r) \mid (s)$ is a regular expression denoting language $L_r \cup L_s$ (union).
  2. $(r) \cdot (s)$ is a regular expression denoting language $L_r \cdot L_s$ (concatenation).
  3. r^* (Kleene star) is a regular expression denoting language L_r^*.
- Nothing else is a regular expression.

EXAMPLE 17.1

1. $r = a^*$ is a regular expression denoting language

$$L_r = \{\varepsilon, a, aa, aaa, \ldots\}.$$

2. $r = a^+$
   $= aa^*$ is a regular expression representing regular language

$$L_r = \{a, aa, aaa, \ldots\}.$$

3. $r = (a \mid b)$ is a regular expression denoting language containing all strings over Σ.
4. All finite languages are regular.

Note: A simple example of a language that is not regular is the set of strings $\{a^n b^n \mid n \geq 0\}$. However, $L = \{a^m b^n \mid m \geq 0, n \geq 0\}$ is regular.

17.3 Regular Language and Finite Automaton

Finite automaton models are used to represent regular languages. In fact, a language is said to be regular, if it is accepted by an automaton. Thus, regular language and finite state automaton are equivalent. Construction of a finite state automaton for a regular expression is described by the following procedure.

1. If $r = \varepsilon$ is a regular expression, then it has the automaton

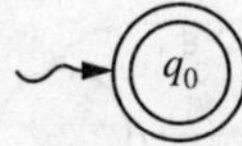

2. If $r = \phi$ is a regular expression, then the automaton will be

3. For regular expression $r = a$, the automaton will be

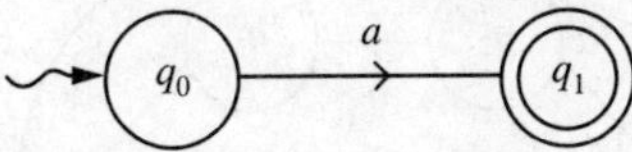

4. Let M_1 and M_2 be two automatons representing regular expressions r_1 and r_2, respectively, then the automaton for the following combinations will be

(i) $r_1 \mid r_2$

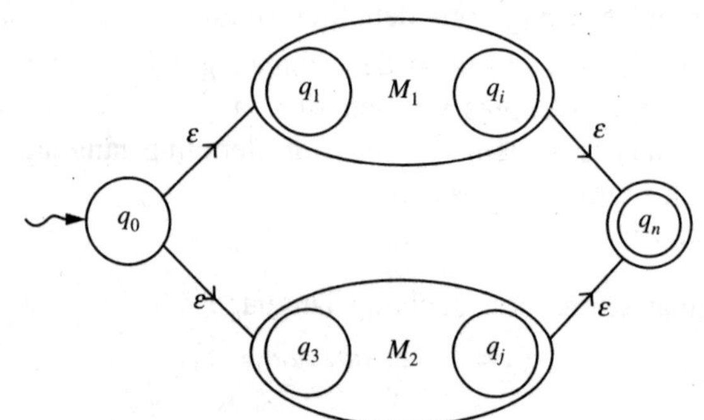

(ii) $r_1 \cdot r_2$

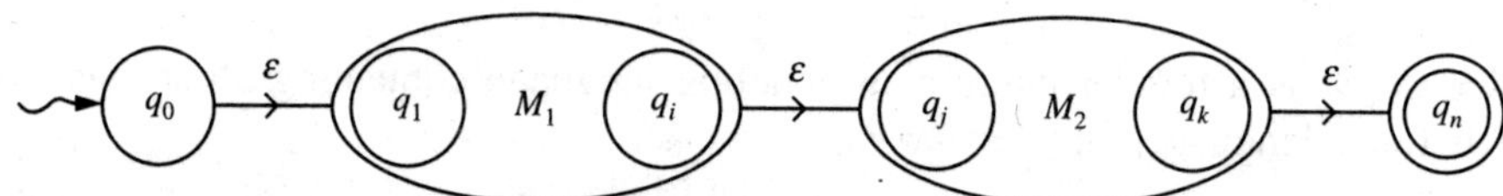

(iii) r_1^*

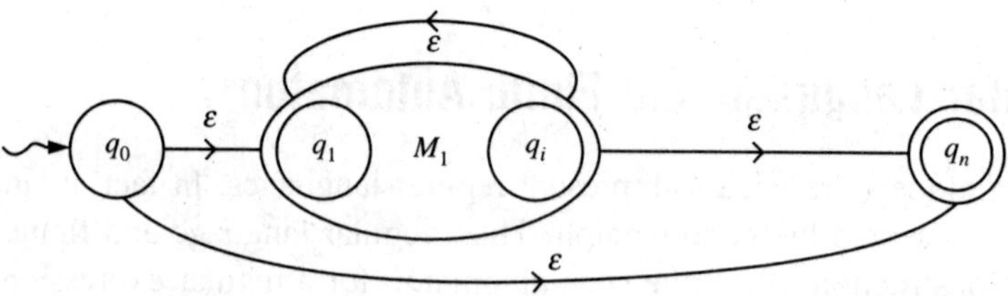

EXAMPLE 17.2

1. Automaton for regular expression $r = 0^*1$:

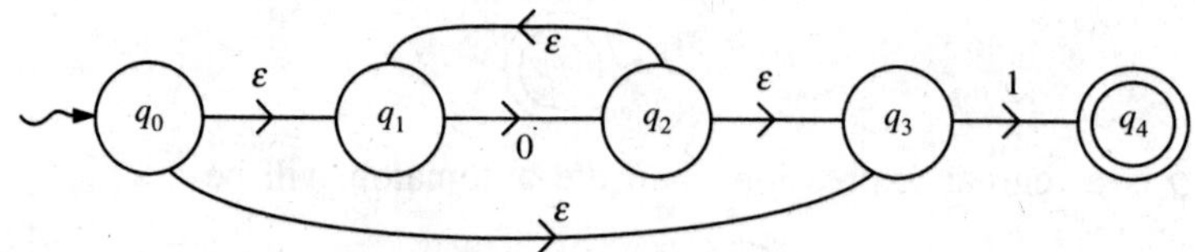

2. Automaton for regular expression $r = (0 \mid 1)^*a$:

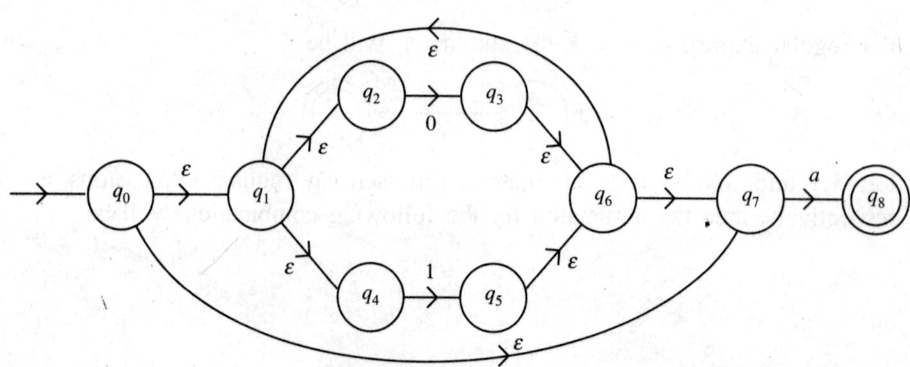

17.4 Construction of Regular Expression from a DFA

We have seen how to construct a nondeterministic finite state automaton (NFA-ε) for a regular expression. In Chapter 16, we have discussed the conversion of NFA-ε to DFA. Now, here we will discuss the construction of regular expression for a DFA. Let L be the set of words accepted by a deterministic finite state automaton

$$M = (\{q_1, q_2, q_3, \ldots, q_n\}, \Sigma, \delta, q_1, F).$$

Let us take a notation r_{ij}^k to denote the set of all strings which take the DFA from state q_i to state q_j without going through any state numbered higher than k, i.e., intermediate states through which automaton moves may have numbers less than or equal to k. Let the set of accepting states be

$$F = \{q_{f1}, q_{f2}, q_{f3}, \ldots, q_{fn}\}.$$

It's obvious that the collection of all the strings accepted by the automaton will be given by the following expression.

$$R = r_{1f1}^n + r_{1f2}^n + \cdots + r_{1fp}^n. \tag{17.1}$$

Now, the aim is to find each regular expression involved in Eq. (17.1), which can be computed recursively by the following relation. Proof of this relation is omitted here, though it is easy to prove it by induction taking r_{ij}^0 as the basis which may give ε or a single symbol as the regular expression.

Consider
$$r_{ij}^k = r_{ik}^{k-1}\,(r_{kk}^{k-1})^*\,r_{kj}^{k-1} + r_{ij}^{k-1} \tag{17.2}$$

where
$$r_{ij}^0 = \{a \mid d(q_i, a) = q_j, \text{if } i \neq j\}$$
$$= \{a + \{\varepsilon\} \mid \delta(q_i, a) = q_j, \text{if } i = j\}$$

Further, during the computation process, following identities can be used to simplify the regular expressions.

1. $r + s = s + r$
2. $(r + s) + t = r + (s + t)$
3. $(rs)t = r(st)$
4. $r(s + t) = rs + rt$
5. $(r^*)^* = r^*$
6. $\phi^* = \varepsilon$
7. $\varepsilon + r^* = r^*$
8. $(r^*s^*)^* = (r + s)^*$

Proof of these identities is trivial, and therefore, left for the readers to construct.

EXAMPLE 17.3 Let us construct the regular expression for the following automaton:

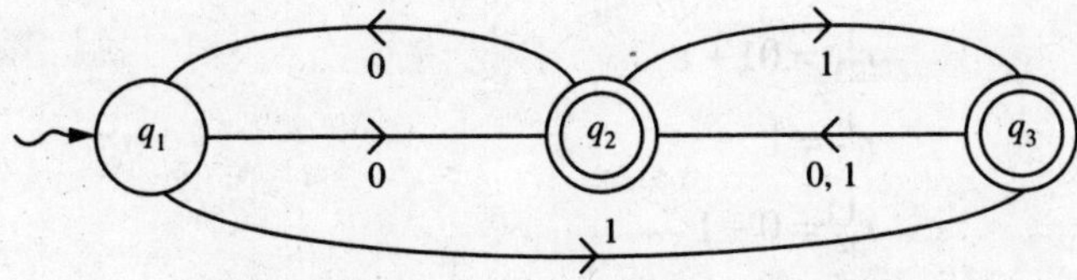

Figure 17.1 State diagram of automaton.

Given $$Q = \{q_1, q_2, q_3\}$$
and $$F = \{q_2, q_3\}.$$
The regular expression will be given by the following expression:

$$r = r_{12}^3 + r_{13}^3 \tag{17.3}$$

Now, compute r_{12}^3 and r_{13}^3 using recursive relations.

$$r_{12}^3 = r_{13}^2 (r_{33}^2)^* \; r_{32}^2 + r_{12}^2 \tag{17.4}$$

Further,

$$r_{13}^2 = r_{12}^1 (r_{22}^1)^* \; r_{23}^1 + r_{13}^1$$

$$r_{33}^2 = r_{32}^1 (r_{22}^1)^* \; r_{23}^1 + r_{33}^1$$

$$r_{32}^2 = r_{32}^1 (r_{22}^1)^* \; r_{22}^1 + r_{32}^1$$

$$r_{12}^2 = r_{12}^1 (r_{22}^1)^* \; r_{22}^1 + r_{12}^1$$

which will need the following:

$$r_{12}^1 = r_{11}^0 \,(r_{11}^0)^* \; r_{12}^0 + r_{12}^0$$

$$r_{22}^1 = r_{21}^0 \,(r_{11}^0)^* \; r_{12}^0 + r_{22}^0$$

$$r_{23}^1 = r_{21}^0 \,(r_{11}^0)^* \; r_{13}^0 + r_{23}^0$$

$$r_{13}^1 = r_{11}^0 \,(r_{11}^0)^* \; r_{13}^0 + r_{13}^0$$

$$r_{32}^1 = r_{31}^0 \,(r_{11}^0)^* \; r_{12}^0 + r_{32}^0$$

$$r_{33}^1 = r_{31}^0 \,(r_{11}^0)^* \; r_{13}^0 + r_{33}^0 .$$

Also,

$$r_{11}^0 = \varepsilon; \quad r_{12}^0 = 0; \quad r_{32}^0 = 0 + 1$$

$$r_{22}^0 = \varepsilon; \quad r_{13}^0 = 1; \quad r_{32}^0 = 0 + 0$$

$$r_{33}^0 = \varepsilon; \quad r_{23}^0 = 1; \quad r_{32}^0 = 0 + \phi .$$

Using these values, we get

$$r_{12}^1 = \varepsilon(\varepsilon)^* \, 0 + 0$$

$$= 0.$$

Similarly,

$$r_{22}^1 = 00 + \varepsilon$$

$$r_{23}^1 = 01 + \varepsilon$$

$$r_{13}^1 = 1$$

$$r_{32}^1 = 0 + 1$$

$$r_{33}^1 = \phi(\varepsilon)^* \, 1 + \varepsilon = \phi + \varepsilon.$$

Again using these values, we get

$$r_{13}^2 = 0*1$$

$$r_{33}^2 = (0+1)0^*1+\varepsilon$$

$$r_{32}^2 = (0+1)(00)^*$$

$$r_{12}^2 = 0(00)^*$$

Finally,

$$\begin{aligned} r_{12}^3 &= 0^*1((0+1)0^*1+\varepsilon)^*(0+1)(00)^* + 0(00)^* \\ &= 0^*1((0+1)0^*1)^*(0+1)(00)^* + 0(00)^* \end{aligned}$$

$$\begin{aligned} r_{13}^3 &= 0^*1((0+1)0^*1+\varepsilon)^*((0+1)0^*1+\varepsilon) + 0^*1 \\ &= 0^*1((0+1)0^*1)^* \end{aligned}$$

Hence,

$$\begin{aligned} r &= r_{12}^3 + r_{13}^3 \\ &= 0^*1((0+1)0^*1)^*(0+1)(00)^* + 0(00)^* + 0^*1((0+1)0^*1)^* \\ &= 0^*1((0+1)0^*1)^*((0+1)(00)^* + \varepsilon) + 0(00)^* \end{aligned}$$

17.5 Pumping Lemma

The pumping lemma is a tool for proving a language non-regular. It describes an essential property which is the repetition of certain part of a word of a regular language. Informally, it says that a sufficiently long word in a regular language will have a section of the word, which may be repeated an arbitrary number of times to produce a new word belonging to the same language. Formally, the lemma is stated as:

Let L be a regular language. Then there exists an integer $n \geq 1$ depending only on L, such that every string z ($|z| \geq n$) in L can be written as $z = uvw$ (i.e., z can be divided into three substrings), satisfying the following conditions:

1. $|uv| \leq n$,
2. $|v| \geq$ and
3. for all $i \geq 0$, $uv^iw \in L$.

Furthermore, n is not greater than the number of states of the smallest automaton FA accepting L.

Here, v is the substring that can be pumped (repeated or removed any number of times, and the resulting string is always in L). From condition (2), it is obvious that the substring v to be pumped must be of length at least one. There is no restriction on u and w. It is trivial to say, that the finite languages satisfy the pumping lemma by having n equal to the maximum string length in L plus one.

The pumping lemma is often used to prove that a particular language is non-regular. It is proved by contradiction. For instance, the language $L = \{0^n1^n \mid n \geq 0\}$ over the alphabet $\Sigma = \{0, 1\}$ can be shown to be non-regular. Let z, u, v, w, n, and i be as used in the statement for the pumping lemma. Let $z \in L$ be given by $z = 0^n1^n$. By the pumping lemma, there must be some decomposition $z = uvw$ with $|uv| \leq n$, $|v| \geq 1$ such that uv^iw is in L for every $i \geq 0$. Using $|uv| \leq n$, v only consists of instances of 0. Moreover, because $|v| \geq 1$, it contains at least one instance of the symbol 0. Now, pump v, i.e. uv^2w has more instances of the symbol 0 than the symbol 1, since we have added some instances of 0 without adding instances of 1. Therefore, uv^2w is not in L. It is a contradiction. Therefore, the assumption that L is regular is incorrect. Therefore, L is not regular.

Note: A typical method for proving a language regular is to construct either a finite state automaton or a regular expression. If it is possible for a language to construct a finite state automaton or a regular expression, then it is called regular.

17.6 Unrestricted Grammar

A formal grammar in which there are no restrictions on the left and right sides of the grammar's production rules is called an unrestricted grammar. This is the most general class of grammars and can generate arbitrary recursively enumerable languages. An **unrestricted grammar** is a 4-tuple,

$$G = (N, T, P, S)$$

where N is a set of non-terminal symbols, T a set of terminal symbols, S a start symbol, and P a set of production rules of the form $\alpha \to \beta$. Here α (a non-empty string) and β are strings of symbols in $N \cup T$. There are no real restrictions on the types of production rules for unrestricted grammar. An unrestricted grammar characterizes recursively enumerable language. This is the same as saying that for every unrestricted grammar G, there exists some Turing machine capable of recognizing $L(G)$ and vice-versa.

17.7 Context-sensitive Grammar

Noam Chomsky introduced the concept of context-sensitive grammar in the 1950s for describing the syntax of natural language. A **context-sensitive grammar (CSG)** is a formal grammar in which the left hand side and the right hand side of a production rule (or rewriting rule) may be surrounded by a context of terminal and non-terminal symbols. The context-sensitive grammars are more general than context-free grammars. A formal language that can be described by a context-sensitive grammar is called a context-sensitive language.

Formally, a grammar $G = (N, T, P, S)$ is context-sensitive if the production rules in P are of the form

$$\alpha A \beta \to \alpha \gamma \beta$$

where

1. A is a single non-terminal,
2. $\alpha, \beta \in (N \cup T)^*$, i.e., α and β are strings of non-terminals and terminals and
3. $\gamma \in (N \cup T)^+$, i.e., γ is a non-empty string of non-terminals and terminals.

In addition, for any production rule of the form $\alpha \to \beta$ of a context-sensitive grammar, $|\alpha| \leq |\beta|$. Here $|\alpha|$ and $|\beta|$ denote the lengths of the strings respectively.

The following grammar generates the context-sensitive language $\{a^n b^n c^n \mid n \geq 1\}$.

$$S \rightarrow aSBC \mid \text{aBC}$$
$$CB \rightarrow HB$$
$$HB \rightarrow HC$$
$$HC \rightarrow BC$$
$$aB \rightarrow ab$$
$$bB \rightarrow bb$$
$$bC \rightarrow bc$$
$$cC \rightarrow cc$$

The derivation for $a^3b^3c^3$ is:

$$\begin{array}{llll} S \Rightarrow aSBC & \Rightarrow aaSBCBC & \Rightarrow aaaBCBCBC \\ \Rightarrow aaaBHBCBC & \Rightarrow aaaBHCCBC & \Rightarrow aaaBBCCBC \\ \Rightarrow aaaBBCHBC & \Rightarrow aaaBBCHCC & \Rightarrow aaaBBCBCC \\ \Rightarrow aaaBBHBCC & \Rightarrow aaaBBHCCC & \Rightarrow aaaBBBCCC \\ \Rightarrow aaabBBCCC & \Rightarrow aaabbBCCC & \Rightarrow aaabbbCCC \\ \Rightarrow aaabbbcCC & \Rightarrow aaabbbccC & \Rightarrow aaabbbccc \end{array}$$

17.8 Context-free Grammar (CFG)

Context-free grammars play an important role in the description and design of programming languages and compilers. They are also used for analyzing the syntax of natural languages. Every production rule of a context-free grammar is of the form:

$$A \rightarrow \alpha$$

where, A is a single non-terminal symbol, and α is a string of terminals and/or non-terminals (α may also be empty).

These production rules, when applied successively produce a parse tree in which the interior nodes are non-terminal symbols. The leaves are the terminal symbols, and each node expands further using appropriate production into the next level of the tree. The tree describes the nesting structure of the expression.

In a context-free grammar, the left hand side of a production rule is always a single non-terminal symbol. In a context-sensitive grammar, it may be a string of terminal and/or non-terminal symbols. The grammars are called context-free, because the replacement during the derivation can be made for the non-terminal on the left hand side of a production rule without considering the surrounding symbols in the string. Therefore, the context in which the symbol occurs is not important. The context-free grammars are simple and suitable for the construction of efficient parsing algorithms. Formally, a context-free grammar G is defined by 4-tuple:

$$G = (N, T, P, S)$$

where,

1. N is a finite set of non-terminals or variables. Each variable represents a different type of phrase in the sentence.

2. T is a finite set of terminals, disjoint from N. The terminals make the actual content of the sentence. The set of terminals is the alphabet of the language defined by the context-free grammar G.
3. P is a set of production or rewriting rules. A production rule is of type:

$$A \to \alpha$$

4. S is the start variable (or start symbol) of the grammar.

A language L is said to be a context-free language (CFL), if there exists a CFG, G such that

$$L(G) = \{w \in T^* \mid S \overset{*}{\Rightarrow} w\}$$

EXAMPLE 17.4 Consider a context free grammar

$$G = < N, T, P, S >$$

where $N = \{ S \}$

and $T = \{(\ ,\ )\}$.

Obviously, there are two terminal symbols '(' and ')', and one non-terminal symbol 'S'. Also, S is the start symbol. This grammar is used for parenthesis matching.

The production rules are:

$$S \to SS$$
$$S \to (S)$$
$$S \to (\ )$$

These three production rules can jointly be written as:

$$S \to SS \mid (S) \mid (\ )$$

Starting with S and applying the rules, one can construct any string of matching parenthesis. For instance,

$$S \Rightarrow SS \Rightarrow SSS \Rightarrow (S)SS \Rightarrow ((S))SS \Rightarrow ((SS))SS$$
$$\Rightarrow (((\ )S))SS \Rightarrow (((\ )(\ )))SS \Rightarrow (((\ )(\ )))(\ )S \Rightarrow (((\ )(\ )))(\ )(\ )$$

Here string $w = (((\ )(\ )))(\ )(\ )$ has been obtained or derived from the grammar. This process is called **derivation.**

EXAMPLE 17.5 Consider a context-free grammar for the language consisting of all strings over $T = \{0, 1\}$, containing an unequal number of 0's and 1's:

$$S \to X \mid Y$$
$$X \to Z0X \mid Z0Z$$
$$Y \to Z1Y \mid Z1Z$$
$$Z \to 0Z1Z \mid 1Z0Z \mid \varepsilon$$

Here, the non-terminal Z can generate all strings with the same number of 0's as 1's, the non-terminal X generates all strings with more 0's than 1's and the non-terminal Y generates all strings with fewer 0's than 1's.

17.8.1 Derivations

There are two common ways to derive a given string from the given grammar: Leftmost derivation and rightmost derivation. If in a derivation, the leftmost non-terminal is always replaced first, then it is called the leftmost derivation.

For instance, if we take the following grammar:

1. $A \rightarrow A + A$
2. $A \rightarrow 1$
3. $A \rightarrow a$

Then the leftmost derivation of the string $w = 1 + 1 + a$ will be:

$$\begin{aligned} A &\Rightarrow A + A \\ &\Rightarrow \boldsymbol{A + A} + A \\ &\Rightarrow \mathbf{1} + A + A \\ &\Rightarrow 1 + \mathbf{1} + A \\ &\Rightarrow 1 + 1 + \boldsymbol{a} \end{aligned}$$

Similarly, in the rightmost derivation, the rightmost non-terminal is always replaced first. The rightmost derivation of the string $w = 1 + 1 + a$ will be:

$$\begin{aligned} A &\Rightarrow A + A \\ &\Rightarrow A + \boldsymbol{A + A} \\ &\Rightarrow A + A + \boldsymbol{a} \\ &\Rightarrow A + \mathbf{1} + a \\ &\Rightarrow \mathbf{1} + 1 + a \end{aligned}$$

17.8.2 Parse Tree

The parse tree is a tree representation of the derivation of a string. It is also called derivation tree.

A parse tree of a string $w = 010$, derived from the grammar of Example 17.5 is given in Figure 17.2.

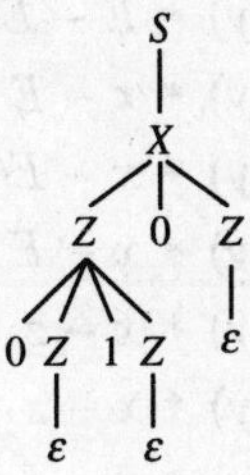

Figure 17.2 Parse tree of string $w = 010$.

EXAMPLE 17.6 A context-free grammar, that pairs up characters is as follows.

$$S \rightarrow 0S1 \mid 01$$

This grammar generates the language

$$L = \{0^n1^n \mid n \geq 1\}.$$

which is not regular (already proved by using pumping lemma for regular languages).

A parse tree for $w = 0011$ is as follows.

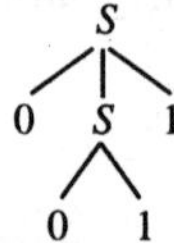

Figure 17.3 Parse tree for $w = 0011$.

EXAMPLE 17.7 Given below is a context-free grammar for infix algebraic expressions in variables x, y and z.

$$E \to E + E$$
$$E \to E - E$$
$$E \to E * E$$
$$E \to E/E$$
$$E \to (E)$$
$$E \to x$$
$$E \to y$$
$$E \to z$$

In short, $E \to E + E \mid E - E \mid E * E \mid E/E \mid (E) \mid x \mid y \mid z$.

This grammar can generate the string $w = (x + y) * x - z * y/(x + x)$. The left most derivation is as follows:

$$\begin{aligned}
E &\Rightarrow E - E \\
&\Rightarrow E * E - E \\
&\Rightarrow (E) * E - E \\
&\Rightarrow (E + E) * E - E \\
&\Rightarrow (x + E) * E - E \\
&\Rightarrow (x + y) * E - E \\
&\Rightarrow (x + y) * x - E \\
&\Rightarrow (x + y) * x - E/E \\
&\Rightarrow (x + y) * x - E * E/E \\
&\Rightarrow (x + y) * x - z * E/E \\
&\Rightarrow (x + y) * x - z * y/E \\
&\Rightarrow (x + y) * x - z * y/(E) \\
&\Rightarrow (x + y) * x - z * y/(E + E) \\
&\Rightarrow (x + y) * x - z * y/(x + E) \\
&\Rightarrow (x + y) * x - z * y/(x + x)
\end{aligned}$$

Parse tree for this derivation is given in Figure 17.4.

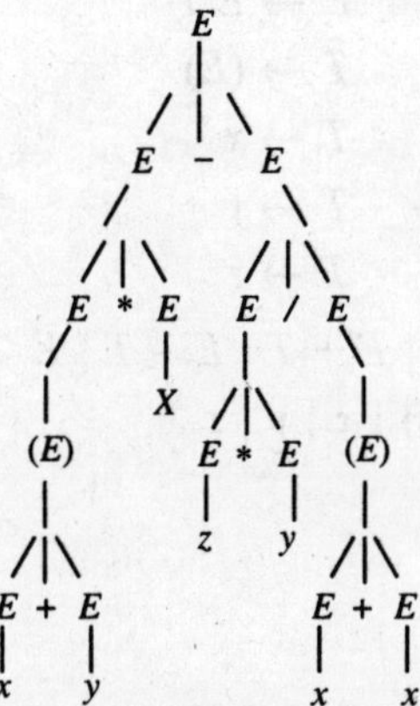

Figure 17.4 Parse tree of the string $w = (x + y) * x - z * y/(x + x)$.

Starting at the root E, this tree is expanded gradually to obtain the given expression. Changing the order of expansion will produce a different derivation, but the same parse tree. The parse tree will only change, if we choose a different production rule to apply at some position in the tree.

17.8.3 Ambiguous Grammar

A grammar is said to be ambiguous if, for a string, there exists more than one parse trees. An ambiguous grammar may derive ambiguous strings, which may be undesirable words of a language. Thus, if a grammar is ambiguous, then it has to be modified by making necessary changes in the production rules.

For instance, consider a string

$$w = x + y * z$$

for the grammar given in Example 17.7. This grammar has two parse trees given in Figure 17.5. Thus, the grammar is ambiguous.

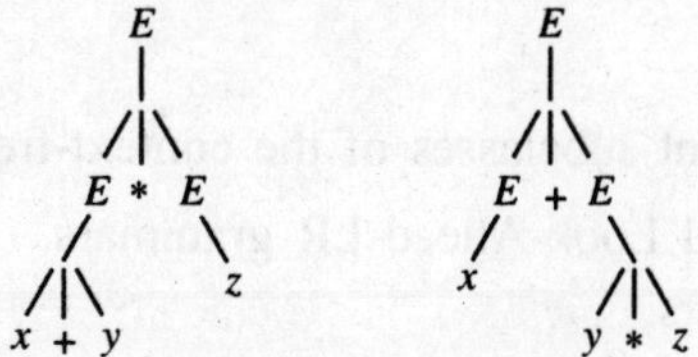

Figure 17.5 Two parse trees for w.

However, a slight modification will make this grammar unambiguous which is as follows:

$$E \rightarrow T$$

$$E \rightarrow E + T$$

$$E \rightarrow E - T$$

$$E \rightarrow E^* T$$
$$E \rightarrow E/T$$
$$T \rightarrow (E)$$
$$T \rightarrow x$$
$$T \rightarrow y$$
$$T \rightarrow z$$

In short,
$$E \rightarrow T \mid E + T \mid E - T \mid E^* T \mid E/T$$
$$T \rightarrow (E) \mid x \mid y \mid z$$

Here, E is the start symbol.

17.9 Regular Grammar

A grammar is said to be regular if the productions are of the form:

$$S \rightarrow a$$
$$S \rightarrow aB$$
$$S \rightarrow \varepsilon$$

EXAMPLE 17.8 Consider the following grammar:

$$S \rightarrow a$$
$$S \rightarrow aS$$
$$S \rightarrow bS$$

or in short,
$$S \rightarrow a \mid aS \mid bS$$

Here, S is the non-terminal, while a and b are terminals. It is a regular grammar. The language generated by this grammar contains all non-empty strings of symbols a and b, which end in a. Every regular grammar corresponds to a regular expression, and hence, there exists a nondeterministic finite automaton for it. The language generated by regular grammar is called regular language.

17.10 LR Grammar

There are a number of important subclasses of the context-free grammars. For instance,

- LR(k), Simple LR, and Look-Ahead LR grammars,
- LL and LL(k) grammars.

These classes are important in parsing the strings and are deterministic in nature. Here, we will discuss LR(0) and LR(1) grammars because of their importance in compiler design. A grammar for which it is possible to construct an **LR parsing table** with each entry in the table uniquely defined, is said to be an LR grammar. A **parser** for a grammar G is a software program that takes as input a string w, and produces a parse tree as the output for w, if w is a sentence of G. On the other hand, if w is not a sentence of G, then it generates an error message. An LR parser scans the input from **left to right** to construct a rightmost derivation. The main parts of an LR parser are a driver routine and a parsing table. The driver routine is almost the same for all LR parsers but the parsing Table is different for each. There are automatic LR parser generators, which take LR grammar as input and generate the LR parsing table. The procedure for creating a parsing table and parsing an input is depicted in the block diagram given in Figure 17.6.

An LR parser can be constructed for a context-free grammar by a parser generator. It can also be written by a programmer, manually. However, it is difficult to write it manually, therefore, it is usually constructed by a parser generator, which is also called a compiler-compiler. A context-free grammar is named LR(k) if there exists an LR(k) parser for it. An LR parser performs bottom-up parsing. A deterministic context-free language is a language for which there exists some LR(k) grammar. Every LR(k) grammar for $k > 1$ can be transformed mechanically into an LR(1) grammar. However, an LR(0) grammar for the same language may not exist. An LR parser refers to a class of parsers that can process almost every programming

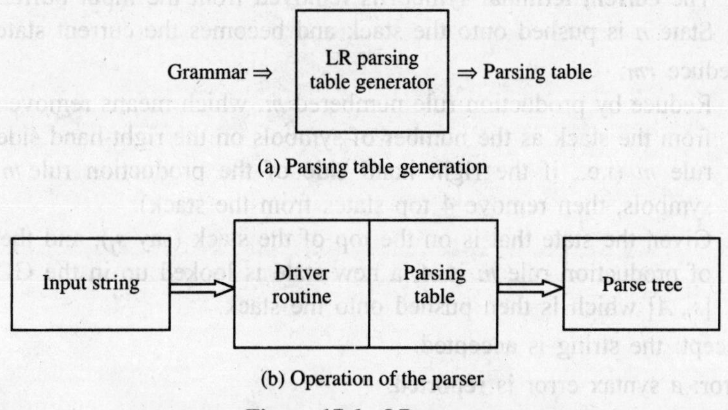

Figure 17.6 LR parser.

language, as long as the parsing table for a programming language is given.

Some well-known parsers are SLRs (simple LR parsers), LALR (look-ahead LR parsers), and canonical LR parsers. Canonical LR and LALR parsers are more powerful than SLR parsers. A block diagram of LR parser is given in Figure 17.7.

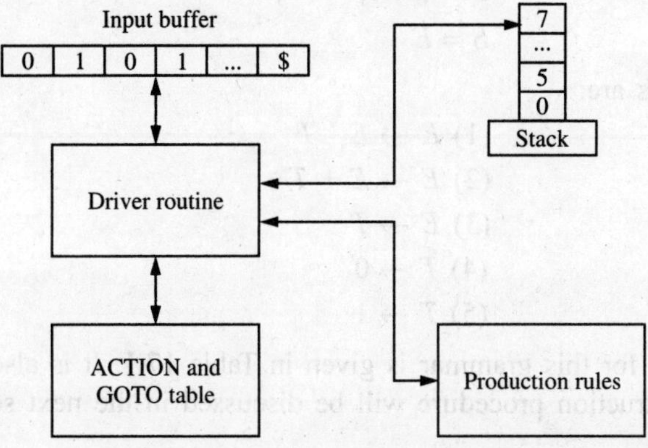

Figure 17.7 Block diagram of LR parser.

A parser consists of:

- an input buffer
- a stack that contains states
- an ACTION and GOTO Table
- a set of production rules

The LR parsing algorithm works as follows:

1. Stack is initialized to [**0**]. The current state will always be at the top of the stack.
2. Parser reads the current state on the stack and the current terminal symbol on the input buffer. Let it be $[s_i, b]$, where s_i is a state and b is the leftmost input symbol on input buffer. Then an action is looked up in the ACTION and GOTO table. There are four cases:
   (i) Shift *sn*:
   - The current terminal symbol is removed from the input buffer.
   - State n is pushed onto the stack and becomes the current state.

   (ii) Reduce *rm*:
   - Reduce by production rule numbered m, which means remove as many states from the stack as the number of symbols on the right-hand side of production rule m (i.e., if the right hand side of the production rule m consists of 4 symbols, then remove 4 top states from the stack).
   - Given the state that is on the top of the stack (say s_j), and the left hand side of production rule m is A, a new state is looked up in the GOTO table, i.e., $[s_j, A]$ which is then pushed onto the stack.

   (iii) Accept: the string is accepted.

   (iv) Error: a syntax error is reported.
3. Step 2 is repeated until either the string is accepted or a syntax error is reported.

EXAMPLE 17.9 To understand the above algorithm, consider the following grammar:

$$G = (N, T, P, S)$$

$$N = \{E, B\}$$

$$T = \{0, 1, ^*, + \}$$

and

$$S = E$$

where production rules are:

$$(1)\ E \rightarrow E\ ^*\ T$$

$$(2)\ E \rightarrow E + T$$

$$(3)\ E \rightarrow T$$

$$(4)\ T \rightarrow 0$$

$$(5)\ T \rightarrow 1$$

The LR parsing Table for this grammar is given in Table 17.1. It is also called ACTION and GOTO table. Its construction procedure will be discussed in the next section.

Table 17.1 ACTION and GOTO Table

State	ACTION					GOTO	
	*	+	0	1	#	E	T
0	–	–	s1	s2	–	3	4
1	r4	r4	r4	r4	r4	–	–
2	r5	r5	r5	r5	r5	–	–
3	s5	s6	–	–	Accept	–	–
4	r3	r3	r3	r3	r3	–	–
5	–	–	s1	s2	–	–	7
6	–	–	s1	s2	–	–	8
7	r1	r1	r1	r1	r1	–	–
8	r2	r2	r2	r2	r2	–	–

The ACTION and GOTO table (Table 17.1) contains three types of actions:

- **Shift**, written as '*sn*', indicates that the current input symbol has to be removed from the input buffer and state *n* has to be pushed onto the stack.
- **Reduce**, written as '*rm*', indicates that a reduction with production rule *m* should be performed.
- **Accept**, indicates that the parser accepts the string on the input buffer.

Note: Blank cells represent an Error in the input string.

Parsing process: Now, consider the input string $w = 1 + 1\#$, where # (hash) is a special symbol showing the end of the input string. The complete parsing of string *w* is shown in Table 17.2. Here, the rightmost element in column named **stack,** is the element currently at the top of the stack. The column under **Input Stream** contains the given string to be parsed. The required action, determined from the ACTION and GOTO Table 17.1, is explained in the last column under **Action**.

Table 17.2 Derivation Steps of Parsing $w = 1 + 1\#$

S. No.	Stack	Input stream	Action
1	0	1 + 1#	s2: shift state 2 on stack
2	0, 2	+ 1#	r5: reduce by production rule 5
3	0, 4	+ 1#	r3: reduce by production rule 3
4	0, 3	+ 1#	s6: shift state 6 on stack
5	0, 3, 6	1#	s2: shift state 2 on stack
6	0, 3, 6, 2	#	r5: reduce by production rule 5
7	0, 3, 6, 8	#	r2: reduce by production rule 2
8	0, 3	#	accept

The parser starts with the stack containing the initial state ('0'). Therefore, the top of the stack is 0, i.e., **[0]**.

The leftmost symbol of the input string that the parser reads is '1'. In order to determine the next action (shift, reduce, accept or error), consult entry [0, 1] in the ACTION and GOTO table. It is s_2, i.e., shift 2, and so, state 2 is pushed onto the stack. Also, delete 1 from the input buffer. The resulting stack is [0, 2] with 2 on top of the stack and input buffer contains + 1#.

Now, entry [2, +] in the ACTION and GOTO table contains r_5, i.e., reduce with production rule 5, which is $T \rightarrow 1$. Pop one symbol from the stack because the right hand side of the production rule 5 has one symbol. Now, state 0 is on top of the stack and left hand side of production rule 5 contains T, therefore, consult entry [0, T] in ACTION and GOTO table. It is 4, hence push 4 onto the stack. The resulting stack is [0, 4]. Rest of the entries can be interpreted in a similar way. It is worth mentioning here, that if any grammar contains a production rule with ε on the right hand side, for instance $A \rightarrow \varepsilon$, and the action is reduce, then during parsing, add the value stored in the cell [x, A] of ACTION and GOTO table to the stack instead of deleting from the stack. Here x is currently on top of the stack.

17.11 Construction of LR(0) Parsing Table

In this section, we will define some terms used in the construction of parsing table. The procedure to construct a parsing table will also be discussed.

17.11.1 Item

An LR (0) item is a production rule with a dot anywhere on the right hand side.
For instance, the production rule $E \rightarrow E + T$ has the following four **items**:

$$E \rightarrow .\, E + T$$
$$E \rightarrow E\, .\, + T$$
$$E \rightarrow E +\, .\, T$$
$$E \rightarrow E + T.$$

Production rule of the form $E \rightarrow \varepsilon$ has only a single item $E \rightarrow .$. If dot is the rightmost symbol of an item then it is called **complete item**. For example, $E \rightarrow E + T.$ is a complete item.

17.11.2 Handle

A handle is a substring of terminals and non-terminals on the right hand side of a production rule, such that its replacement, by the left hand side of the production rule, leads the input to the start symbol. For example, consider the derivation:
$E \Rightarrow E + T \Rightarrow E * T + \text{T}$. Here $E * T$ is a handle.

17.11.3 Viable Prefix

A viable prefix is a substring of the right hand side of a derivation that does not cross the right end of the handle. Thus, E, $E\,*$, and $E * T$ are all viable prefixes.

17.11.4 Valid Item

An item

$$A \Rightarrow \alpha \cdot \beta$$

is said to be valid for a viable prefix ρ, if there is a rightmost derivation of the form

$$E \overset{*}{\Rightarrow} uAv \Rightarrow u\alpha\beta v$$

Here, u and v may be strings of terminals and non-terminals, and

$$\rho = u\alpha.$$

7.11.5 Conditions for LR (0) Grammar

1. Start symbol of the grammar should not appear on the right hand side of a production rule. Whenever this rule is violated, augment the grammar with an extra production rule. For instance, in the grammar given in Example 17.9, start symbol E appears on the right hand side. Therefore, augment the grammar with an additional production rule, $S \rightarrow E\#$.

   Then the grammar becomes:

$$S \rightarrow E\#$$
$$E \rightarrow E * T$$
$$E \rightarrow E + T$$
$$E \rightarrow T$$
$$T \rightarrow 0$$
$$T \rightarrow 1$$

   Here S is the new start symbol. This production rule is used for reduction, only if the parser accepts the input string.

2. For every viable prefix ρ, if $E \rightarrow \alpha \cdot$ is a complete valid item, then neither any other complete item nor any item with terminal on the right of a dot is valid for 5.

17.11.6 CLOSURE ( ) Set

Let H be the set of items for a grammar G. Then CLOSURE (H) is a set of items constructed by the following rules:

1. Every item in H will be in CLOSURE (H).
2. If $E \rightarrow u \cdot Fv$ is in CLOSURE (H) and $F \rightarrow \alpha$ is a production rule, then add $F \rightarrow \cdot\, \alpha$ to CLOSURE (H), if it is not already there.

17.11.7 Construction of DFA (deterministic finite state automaton) for LR (0)

A finite state automaton that has transitions on terminals as well as non-terminals of a grammar can be constructed using sets of items. Each state of the automaton is a collection of items, which are computed by using the procedure given for CLOSURE ( ) set. The start or first state of the automaton is always the CLOSURE ( ) of the first item of the production rule containing start symbol.

Consider a grammar G:

$$S \to E\#$$
$$E \to E * T$$
$$E \to E + T$$
$$E \to T$$
$$T \to 0$$
$$T \to 1$$

The first item is $S \to . E\#$. Then CLOSURE ($S \to . E\#$) will be the starting state of DFA and it will contain the following items,

State q_0:

$$S \to . E\#$$
$$E \to . E * T$$
$$E \to . E + T$$
$$E \to . T$$
$$T \to . 0$$
$$T \to . 1$$

Beginning at the start state (q_0), determine all states that can be reached from this state. The possible transitions from a set of items can be found by looking at the symbols, both terminals and non-terminals that we find right after the dots. In this case, terminals '0' and '1', and the non- terminals E and T are immediately to the right of the dot. These symbols will be the input symbols on state q_0 for making transitions to other states. In moving from one state to the other, the symbol immediately right to the dot in an item is taken as the input, and then this item becomes the first item of the next state with position of dot shifted one place to the right. Formally, we define a transition function:

$$\delta([A \to X . BC], B) = ([A \to XB . C])$$

where A is an arbitrary non-terminal, and X, B and C may be terminals or non-terminals.

Next, compute the CLOSURE($[A \to XB . C]$) to find the next state.

For terminal '0', the first item of the next state will be $T \to 0$ ..

$$\text{CLOSURE } (T \to 0 .) = T \to 0 .$$

Therefore, next state of DFA (say q_1) will be $T \to 0$ ..

State q_1: {transition from q_0 on input 0}

$$T \to 0 .$$

Similarly,

State q_2: {transition from q_0 on input 1}

$$T \to 1 .$$

State q_3: {transition from q_0 on input E}

$$S \rightarrow E \,.\, \#$$
$$E \rightarrow E \,.\, {}^{*}\, T$$
$$E \rightarrow E \,.\, + T$$

State q_4: {transition from q_0 on input T}

$$E \rightarrow T \,.$$

Note that the CLOSURE( ) may not always add new items in a set of items. Also, from a set with a complete item, like q_1, no further transitions are made. We continue this process until no more states are created. For the states q_1, q_2, and q_4, there will be no transitions since the dot is the last symbol, i.e., these states contain complete items. From state q_3, there will be transitions on input '#', '*' and '+'.

State q_5: {transition from q_3 on input *}

$$E \rightarrow E^{*} \,.\, T$$
$$T \rightarrow \,.\, 0$$
$$T \rightarrow \,.\, 1$$

State q_6: {transition from q_3 on input + }

$$E \rightarrow E + \,.\, T$$
$$T \rightarrow \,.\, 0$$
$$T \rightarrow \,.\, 1$$

State q_7: {transition from q_3 on input #}

$$S \rightarrow E \# \,.$$

Other states are:

State q_8: {transition from q_5 on input T}

$$E \rightarrow E^{*}\, T \,.$$

State q_9: {transition from q_6 on input T}

$$E \rightarrow E + T \,.$$

No other new state is possible. Also, assign the numbers to the production rules.

The numbered production rules are:

1. $S \rightarrow E\#$
2. $E \rightarrow E^{*}\, T$
3. $E \rightarrow E + T$
4. $E \rightarrow T$
5. $T \rightarrow 0$
6. $T \rightarrow 1$

The transition diagram of DFA is shown in Figure 17.8, and its corresponding transition table is given in Table 17.3.

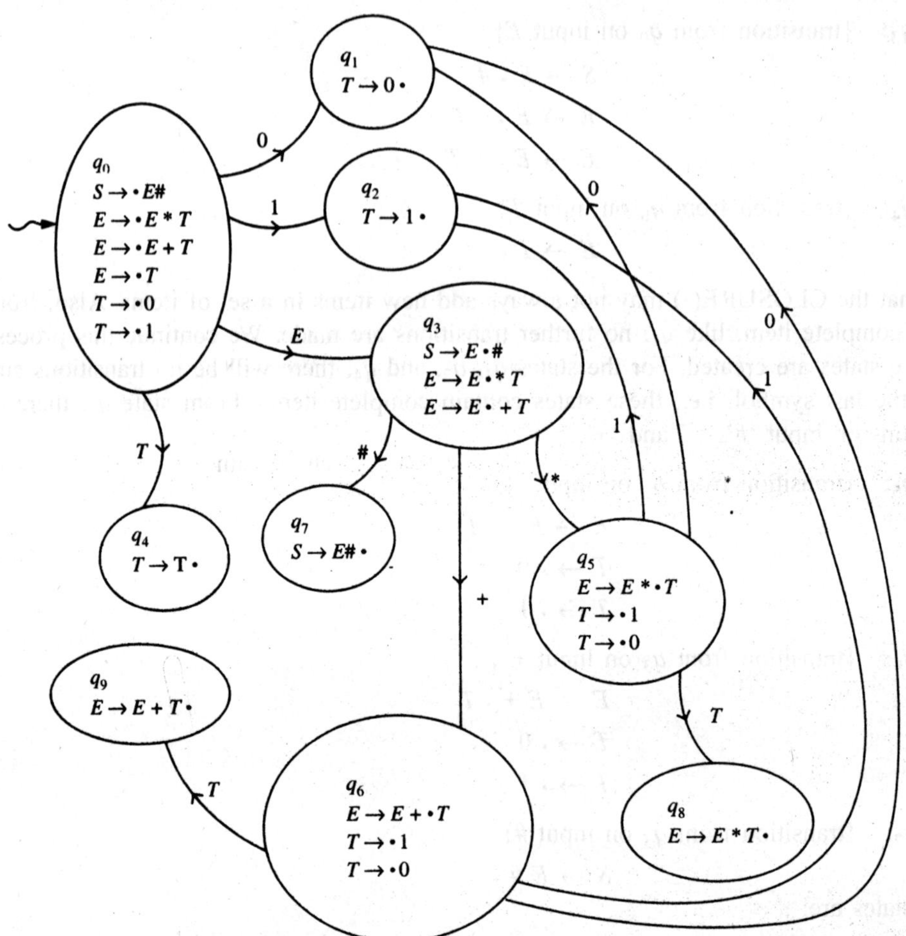

Figure 17.8 Transition diagram of DFA.

Table 17.3 Transition Table of DFA

State		+	0	1	#	E	T
	Input symbol						
q_0			q_1	q_2		q_3	q_4
q_1							
q_2							
q_3	q_5	q_6			q_7		
q_4							
q_5			q_1	q_2			q_8
q_6			q_1	q_2			q_9
q_7							
q_8							
q_9							

17.11.8 Constructing the ACTION and GOTO Table

For the grammar given in Table 17.3, the ACTION and GOTO table is constructed by the following procedure:

1. Columns containing non-terminals in the transition table become GOTO part of the ACTION and GOTO table. For instance, there is q_3 in the cell [q_0, E] in Table 17.3, therefore, the corresponding entry in GOTO part of the ACTION and GOTO table will be q_3.
2. Columns containing terminals become ACTION part of the ACTION and GOTO table by replacing letter q by s in each cell. For instance, there is q_1 in the cell [q_0, 0] in Table 17.3, therefore, the corresponding entry in ACTION part of the ACTION and GOTO table will be $s1$. It is obtained by replacing letter q by s. Here, s stands for shift, and 1 is the state number. In general, sn means shift n.
3. Column for '#' contains Accept for the state which contains $S \to E\,\#\bullet$.
4. If a state q contains an item of the form $A \to \alpha\bullet$, and $A \to \alpha$ is a production rule numbered n with $n > 0$, then all the cells of the row for state q in the ACTION table are filled with the reduce action (rn).

The ACTION and GOTO table constructed finally, is given in Table 17.4.

Table 17.4 ACTION and GOTO Table

	ACTION					*GOTO*	
State	*	+	0	1	#	*E*	*T*
q_0			*s*1	*s*2		q_3	q_4
q_1	*r*5	*r*5	*r*5	*r*5	*r*5		
q_2	*r*6	*r*6	*r*6	*r*6	*r*6		
q_3	*s*5	*s*6			Accept		
q_4	*r*4	*r*4	*r*4	*r*4	*r*4		
q_5			*s*1	*s*2			q_8
q_6			*s*1	*s*2			q_9
q_7							
q_8	*r*2	*r*2	*r*2	*r*2	*r*2		
q_9	*r*3	*r*3	*r*3	*r*3	*r*3		

It is easy to observe that Table 17.1 and Table 17.4 accept the same language. The minor difference is due to the changes in the names of the states, and the addition of production rule 1 in later grammar.

17.11.9 Conflicts in the Parsing Table

Sometimes, during construction, one may encounter more than one entry in one cell of the parsing table of a given grammar. For example, when reduce action is added to some cell of the ACTION and GOTO table, it may happen that the same cell is already filled with a shift action and vice versa. This is undesirable, and is called a shift-reduce conflict. Similarly, there may be a reduce-reduce conflict. A grammar with such conflicts is said to be non-LR (0). An example of a non-LR (0) grammar with a shift-reduce conflict is:

$$E \to 1\,E$$
$$E \to 1$$

Its augmented grammar will be,

$$A \to E\#$$
$$E \to 1\,E$$
$$E \to 1$$

The DFA states will be:

State q_0

$$A \to .\,E\#$$
$$E \to .\,1\,E$$
$$E \to .\,1$$

State q_1

$$E \to 1\,.\,E$$
$$E \to 1\,.$$
$$E \to .\,1\,E$$
$$E \to .\,1$$

There is a shift-reduce conflict in state q_1, because there is a complete item $E \to 1$., which forces parser for a reduce action, and item $E \to .\ 1\ E$ forces for a shift.
Another example of a non-LR(0) grammar with a reduce-reduce conflict is:

$$S \to A\ 1$$
$$S \to B\ 1$$
$$A \to 1$$
$$B \to 1$$

In this case, we get a state:

State q_1

$$A \to 1\,.$$
$$B \to 1\,.$$

There is a reduce-reduce conflict due to the presence of two complete items.

17.12 Construction of LR(1) Parsing Table

Observe that in LR(0) parser, reduce action fills the entire row in ACTION and GOTO Table, causing the reduction to occur regardless of the next symbol in the input stream, i.e., look-ahead zero symbols. This is the reason why it is called LR(0) parser. In LR(1) parser, a look-ahead set containing terminals, is attached with each item. This look-ahead set guides the parser generator in taking decisions for selecting cells to fill with reduce action.

The DFA of an LR(1) parser is constructed in a similar way as that of LR(0), except that an item in a state also contains a look-ahead set. Thus, an LR(1) item consists of two parts: an LR(0) item, and a look-ahead set. For example,

LR(1) item: $[S \to A\ .\ B,\ \{d\}]$. Here $\{d\}$ is a look-ahead set

It would mean that the parser has read a string corresponding to A, and now expects a string corresponding to B followed by the terminal 'd'.

LR(1) parsers can deal with a very large class of grammars. The look-ahead set of an LR(1) item is used only when considering reduce actions (i.e., when the dot is at the rightmost end). We will discuss various terms related with the construction of LR(1) parser.

17.12.1 Valid LR(1) Item

An LR(1) item $[B \rightarrow \alpha \cdot \beta, \{b\}]$ is said to be valid for a viable prefix $\rho = u\alpha$, if there is a rightmost derivation of the form:

$$E \overset{*}{\Rightarrow} uB\tau \Rightarrow u\alpha\beta\tau$$

Here u may be string of terminals and non-terminals,

and $$\rho = u\alpha.$$

Also, either

1. b is the first symbol of τ, or
2. if $\tau = \varepsilon$, then b = #.

17.12.2 Conditions for LR (1) Grammar

The first condition of LR(1) is similar to the first condition of LR(0). However, the second is different due to the look-ahead set of LR(1) item. The conditions are:

1. Start symbol of the grammar should not appear on the right hand side of a production rule. If this rule is violated then augment the grammar with an extra production rule of the form
$$S \rightarrow E\#.$$
2. Whenever a set of LR(1) items(say H), valid for some viable prefix ρ, has a complete item
$$[E \rightarrow \alpha \cdot, \{a_1\ a_2\ a_3\ \ldots\ a_n\}], \text{ then}$$
   (i) any other complete item $[E \rightarrow \alpha \cdot, \{b_1\ b_2\ b_3\ \ldots b_n\}]$ must have a disjoint look-ahead set.

   (ii) none of the a_i's must appear to the immediate right of a dot in any item of H.

17.12.3 FIRST( )

The function FIRST( ) is defined as:

1. If B is a terminal then FIRST(B) = $\{B\}$.
2. If B is a non-terminal, and $B \rightarrow b\alpha$ is a production rule, then add b to FIRST(B).
3. If $B \rightarrow \varepsilon$ is a production rule then add # to FIRST(B).
4. If $B \rightarrow X_1X_2X_3 \ldots X_n$ is a production rule then for all i, such that $X_1X_2X_3 \ldots X_{i-1} \overset{*}{\Rightarrow} \varepsilon$, add every non ε symbol in FIRST(X_i) to FIRST(B). If $X_1X_2X_3 \ldots X_n \overset{*}{\Rightarrow} \varepsilon$ then add # to FIRST(B).

17.12.4 CLOSURE ( )

1. If H is an LR(1) item, then H belongs to CLOSURE (H).
2. If $[E \to \alpha \cdot B\beta, \{a\}]$ is an LR(1) item, and $B \to \gamma$ is a production rule of the grammar, then add the item $[B \to \cdot\, \gamma, \{b\}]$ for all b in FIRST(β) to the CLOSURE ($[E \to \alpha \cdot B\beta, \{a\}]$). However, if $\beta \Rightarrow \varepsilon$, then b is a_i for $1 \le i \le n$.

17.12.5 Shift Action

If $[A \to \alpha \cdot b\beta, \{a\}]$ is in state q_k and there is a transition from state q_k to state q_m with label b, then add the action "shift m" in the cell $[k, b]$.

17.12.6 Reduce Actions

If $[A \to \alpha \cdot, \{a\}]$ is in state q_k, then add the action 'Reduce by rule $A \to \alpha$' to the cell ACTION$[q_k, a]$. The GOTO part of the table is same as before in LR(0).

17.12.7 DFA Construction for LR(1) Parser

The procedure for the construction of a DFA for LR(1) parser is almost similar to the procedure for DFA of LR(0). Start with the first state q_0 having its first LR(1) item with start symbol on left and look-ahead set = {#}, and then use CLOSURE ( ) function to compute the remaining items of the state q_0. This state will be the start state of the DFA. For transitions from one state to other, we define a transition function

$$\delta([A \to X \cdot BC, \{x_1, x_2, x_3, .., x_n\}], B) = ([A \to XB \cdot A, \{x_1, x_2, x_3, .., x_n\}]$$

where A is arbitrary non-terminal and $\{x_1, x_2, x_3, .., x_n\}$ is an arbitrary look-ahead set. Symbols X, B and C may be terminals and/or non-terminals.
Clearly, when there is a transition from one state to other, look-ahead set does not change.
Let us understand the procedure of DFA construction with the help of an example.
Consider the following grammar G. We will construct DFA for this grammar.

$$S \to A\#$$
$$A \to BA \mid \varepsilon$$
$$B \to aB \mid b$$

The DFA states will be named as q_0, q_1 and q_2, etc. Each DFA state will be a set of LR(1) items. The transition diagram and transition table constructed finally, are given in Figure 17.9 and Table 17.5, respectively.

The first state q_0 will contain $[S \to \cdot\, A\#, \{\#\}]$ as an LR(1) item including others which can be obtained by computing CLOSURE ($[S \to \cdot\, A\#, \{\#\}]$). We see that there are production rules $A \to BA$ and $A \to \varepsilon$, therefore, add $[A \to \cdot\, BA, \{\#\}]$ and $[A \to \cdot, \{\#\}]$ to the CLOSURE ($[S \to \cdot\, A\#, \{\#\}]$). Note that the look-ahead set remains same because there is symbol # after A in the item $[S \to \cdot\, A\#, \{\#\}]$. Further, items $[B \to \cdot\, aB, \{a, b, \#\}]$ and $[B \to \cdot\, b, \{a, b, \#\}]$ will be added to the CLOSURE ($[A \to \cdot\, BA, \{\#\}]$) because of the production rules $B \to \cdot\, aB$ and $B \to \cdot\, b$. The look-ahead set, in both items is $\{a, b, \#\}$ because

$$\text{FIRST}(A) = \text{FIRST}(B) = \{a, b, \#\}$$

Thus, the state q_0 will be:

State q_0:

$$S \to \cdot\, A\#,\ \{\#\}$$
$$A \to \cdot\, B\, A,\ \{\#\}$$
$$A \to \cdot,\ \{\#\}$$
$$B \to \cdot\, aB,\ \{a, b, \#\}$$
$$B \to \cdot\, b,\ \{a, b, \#\}$$

Now, we see that in the item $[S \to \cdot\, A\#, \{\#\}]$, symbol A appears before dot (.), therefore, we get next state q_1 with LR(1) item $[S \to A \cdot \#, \{\#\}]$ by using transition function δ, already discussed. Input on state q_0 to reach state q_1 is taken as A. Thus, the state q_1 will be:

State q_1: {transition from q_0 on input A}

$$S \to A \cdot \#,\ \{\#\}$$

Similarly, we can continue to get other states.

State q_2: {transition from q_0 on input B}

$$A \to B \cdot A,\ \{\#\}$$
$$A \to \cdot\, B\, A,\ \{\#\}$$
$$A \to \cdot,\ \{\#\}$$
$$B \to \cdot\, aB,\ \{a, b, \#\}$$
$$B \to \cdot\, b,\ \{a, b, \#\}$$

State q_3: {transition from q_0 on input a}

$$B \to a \cdot B,\ \{a, b, \#\}$$
$$B \to \cdot\, aB,\ \{a, b, \#\}$$
$$B \to \cdot\, b,\ \{a, b, \#\}$$

State q_4: {transition from q_0 on input b}

$$B \to b\, \cdot,\ \{a, b, \#\}$$

State q_5: {transition from q_2 on input A}

$$A \to BA\, \cdot,\ \{\#\}$$

State q_6: {transition from q_3 on input B}

$$B \to aB\, \cdot,\ \{a, b, \#\}$$

Finally, we get the DFA states q_0, q_1, q_2, q_3, q_4, q_5, q_6 and q_7. The resulting DFA is given in Figure 17.9.

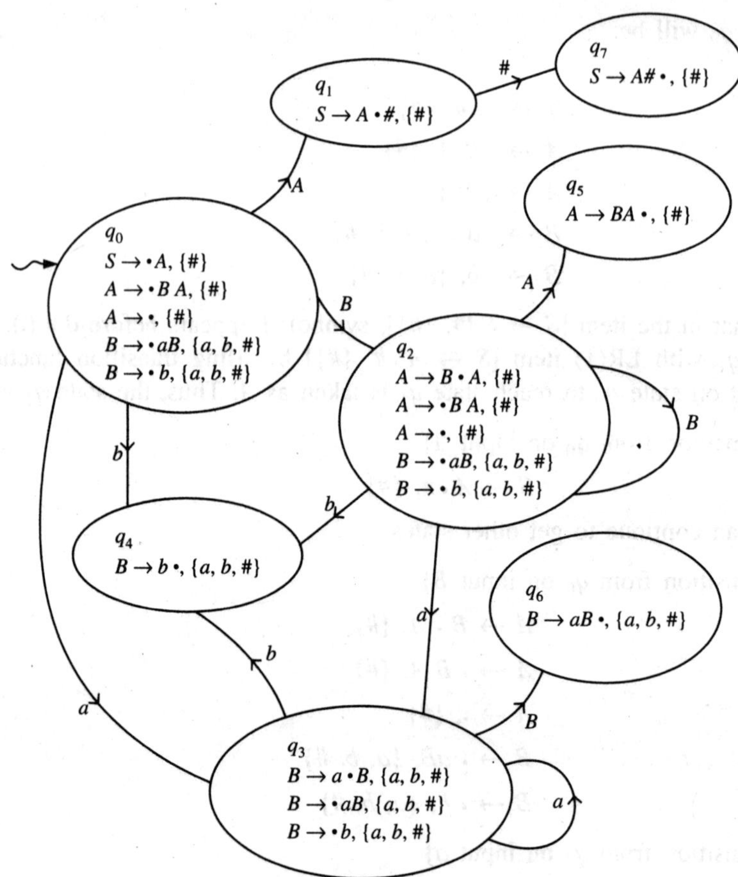

Figure 17.9 Transition diagram of DFA.

Table 17.5 Transition Table of DFA

States	Input				
	a	*b*	#	*A*	*B*
q_0	q_3	q_4	–	q_1	q_2
q_1	–	–	q_7	–	–
q_2	q_3	q_4	–	q_5	q_2
q_3	q_3	q_4	–	–	q_6
q_4	–	–	–	–	–
q_5	–	–	–	–	–
q_6	–	–	–	–	–
q_7	–	–	–	–	–

17.12.8 ACTION and GOTO Table

Let us use Table 17.5 for the construction of ACTION and GOTO table. The procedure used is described below.

1. Columns for non-terminals become GOTO part of the table.
2. Columns for terminals become ACTION part of the table with entries as shift actions (sn), where s stands for shift, and n, the state number.
3. Column for '#' contains Accept for state that contains $S \to A \cdot \#, \{\#\}$.
4. If a state q in DFA contains an item of the form $A \to \alpha \cdot, T$ and $A \to \alpha$ is a production rule numbered n with $n > 0$, then the cells, corresponding to the input symbols in look-ahead set T of the row of state q (in the ACTION part of the Table), are filled with the reduce action (rn).

The production rules are numbered as:

1. $S \to A$
2. $A \to BA$
3. $A \to \varepsilon$
4. $B \to aB$
5. $B \to b$

The ACTION and GOTO table constructed using this procedure is given in Table 17.6.

The entry in cell $[q_2, \#]$ of ACTION and GOTO table is r_3. Since, state q_2 contains a complete item $A \to \cdot, \{\#\}$, and there is a production rule $A \to \varepsilon$ numbered 3 in the list, therefore, action is r_3 (reduce by production rule 3). This action is entered in the column under the symbol '#', because '#' is the only symbol in the look-ahead set of this item. During parsing, whenever this action comes, instead of popping symbols off the stack, entry in cell $[x, A]$ is added to the stack, where x is currently at top of the stack.

Table 17.6 Construction of ACTION and GOTO Table Using Production Rules

	ACTION			*GOTO*		
State	*a*	*b*	*#*	*S*	*A*	*B*
q_0	s3	s4	r3	–	q_1	q_2
q_1	–	–	Accept	–	–	–
q_2	s3	s4	r3	–	q_5	q_2
q_3	s3	s4	–	–	–	q_6
q_4	r5	r5	r5	–	–	–
q_5	–	–	r2	–	–	–
q_6	r4	r4	r4	–	–	–
q_7	–	–	–	–	–	–

Let us parse a string

$$w = aabbaab\#$$

The derivation steps are shown in Table 17.7. The format of this Table is the same as given in Table 17.2.

Table 17.7 Derivation steps of W = *aabbaab#*

S. No.	Stack	Input stream	Action
1	q_0	*aabbaab#*	s3: shift state q_3 on the stack
2	q_0, q_3	*abbaab#*	s3: shift state q_3 on the stack
3	q_0, q_3, q_3	*bbaab#*	s4: shift state q_4 on the stack
4	q_0, q_3, q_3, q_4	*baab#*	r5: reduce by production rule 5
5	q_0, q_3, q_3, q_6	*baab#*	r4: reduce by production rule 4
6	q_0, q_3, q_6	*baab#*	r4: reduce by production rule 4
7	q_0, q_2	*baab#*	s4: shift state q_4 on the stack
8	q_0, q_2, q_4	*aab#*	r5: reduce by production rule 5
9	q_0, q_2, q_2	*aab#*	s3: shift state q_3 on the stack
10	q_0, q_2, q_2, q_3	*ab#*	s3: shift state q_3 on the stack
11	q_0, q_2, q_2, q_3, q_3	*b#*	s4: shift state q_4 on the stack
12	$q_0, q_2, q_2, q_3, q_3, q_4$	#	r5: reduce by production rule 5
13	$q_0, q_2, q_2, q_3, q_3, q_6$	#	r4: reduce by production rule 4
14	q_0, q_2, q_2, q_3, q_6	#	r4: reduce by production rule 4
15	q_0, q_2, q_2, q_2	#	r3: reduce by production rule 3
16	q_0, q_2, q_2, q_2, q_5	#	r2: reduce by production rule 2
17	q_0, q_2, q_2, q_5	#	r2: reduce by production rule 2
18	q_0, q_2, q_5	#	r2: reduce by production rule 2
19	q_0, q_1	#	accept

We interpret some of the entries of the derivation table given above. The parser starts with the stack containing the initial state ('q_0').

The leftmost symbol of the input string that the parser reads is '*a*'. In order to determine the next action (shift, reduce, accept or error), consult entry [q_0, *a*] in the ACTION and GOTO table. It is s_3, i.e., shift 3, and so, state q_3 is pushed onto the stack. Also, delete *a* from the input stream. The resulting stack is q_0, q_3 with q_3 on top of the stack and input buffer containing *abbaab#*.

Again, entry [q_3, *a*] in the ACTION and GOTO Table contains s_3, so, shift q_3 on the stack and remove *a* from the input stream.

Let us interpret the action at S. No. 4. The leftmost symbol on the input stream is *b*, and top of the stack contains q_4, therefore, consult the entry [q_4, *b*] in the ACTION and GOTO table. it is r_5, i.e., reduce with production rule 5, which is $B \to b$. Pop one symbol from the stack, since the right hand side of the production rule 5 has one symbol. Now, state q_3 is on top of the stack and left hand side of production rule 5 contains *B*, therefore, consult entry [q_3, *B*] in ACTION and GOTO table. It is q_6, and hence, push q_6 on to the stack. The resulting stack is (q_0, q_3, q_3, q_6).

Let us interpret the action at S. No. 15. Here, top of the stack contains q_2, and symbol on input is #, therefore, consult entry [q_2, #]. It is r_3, i.e., reduce by production rule 3, which is

$$A \to \varepsilon$$

In this production rule, the right hand side is ε, therefore, instead of deleting from the stack, we add q_5, which is stored at [q_2, *A*] to the stack. In general, whenever the right hand side of a production rule is ε, instead of deleting symbols from the stack, add the appropriate symbol to the stack. Rest of the entries can be interpreted in a similar way.

17.13 Decision Algorithms for CFL

There are some basic questions about CFL (context-free language), such as: whether a given CFL is empty, finite or infinite. There are many more such questions, some of which may be answerable and some may not. However, we discuss the algorithms for deciding whether a CFL is empty, finite or infinite. These are discussed below.

17.13.1 CFL is Empty

A CFL is said to be empty if there is no string in the language. On the other hand, if start symbol of a CFG generates some strings then the corresponding CFL is said to be non-empty.

For instance, $A \rightarrow \varepsilon$ generates an empty CFL.

Obviously, it is easy to decide the emptiness of a CFL.

17.13.2 CFL is Finite

A CFL is said to be finite if it contains finite number of strings.

For instance, the following grammar generates finite CFL.

CFG $G1$:

$$S \rightarrow AB$$
$$A \rightarrow BC \mid a$$
$$B \rightarrow CC \mid b$$
$$C \rightarrow a$$

We can draw a directed graph to determine the finiteness of a CFL. If the directed graph of a CFG has no loops, or closed paths then the corresponding CFL is finite. The directed graph of CFG $G1$ is given in Figure 17.10.

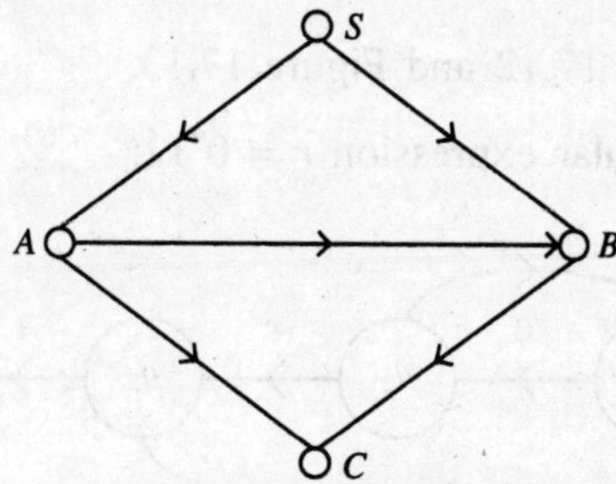

Figure 17.10 A finite CFL.

Note that there is no loop in the graph, therefore, the corresponding CFL is finite.

17.13.3 CFL is Infinite

A CFL is said to be infinite if it contains infinite number of strings.

For instance, the following grammar generates an infinite CFL.

CFG $G2$:

$$S \to AB$$
$$A \to BC \mid a$$
$$B \to CC \mid b$$
$$C \to AB \mid a$$

The directed graph of CFG $G2$ is given below in Figure 17.11.

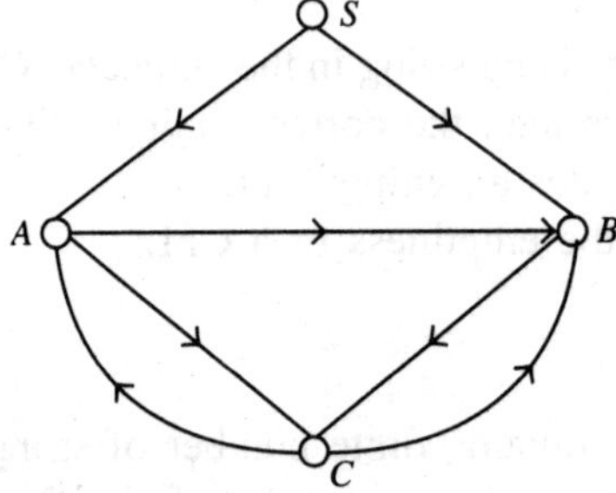

Figure 17.11 An infinite CFL.

Note that there are loops in the graph, therefore, the corresponding CFL is infinite.

17.14 Solved Problems

PROBLEM 17.1 Design NFA-ε for the following regular expressions:

(i) $r = 0^*12$

(ii) $(0 \mid 1)^*a \mid 0$

Solution

The NFA-ε are given in Figure 17.12 and Figure 17.13.

(i) The automaton for regular expression $r = 0^*12$:

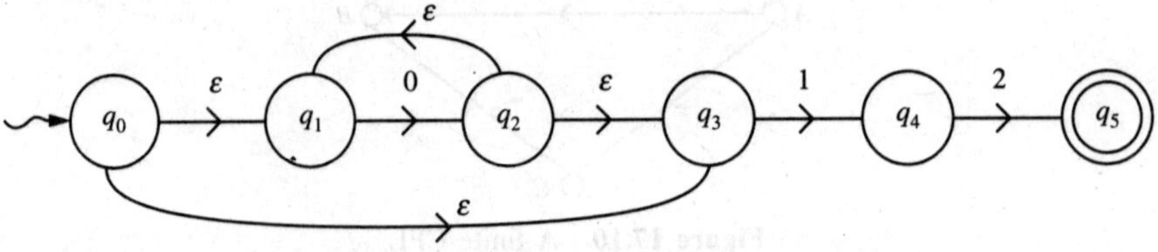

Figure 17.12 $r = 0^*12$.

(ii) The automaton for regular expression $r = (0 \mid 1)^* a \mid 0$:

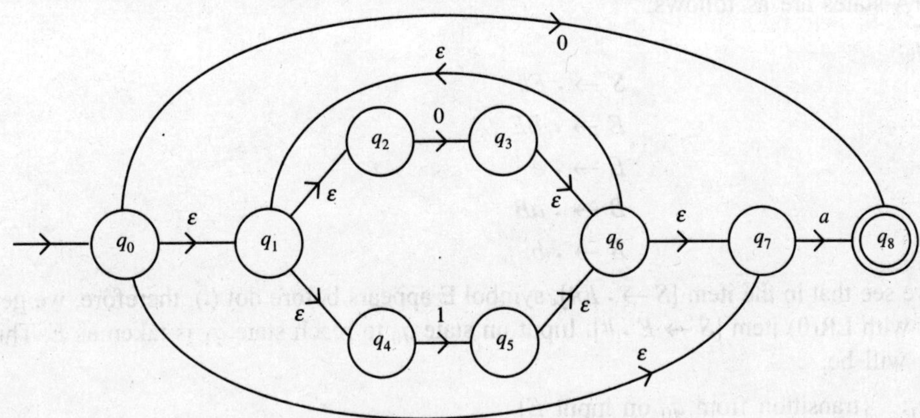

Figure 17.13 $(0 \mid 1)^* a \mid 0$.

PROBLEM 17.2 Design a context-free grammar for the language consisting of all strings over $\Sigma = \{0, 1, 2\}$, containing an unequal number of 0's and 1's, and ending with 2.

Solution

A context-free grammar generating strings of unequal number of 0's and 1's, and ending with 2 is given below.

$$S \to X2 \mid Y2$$
$$X \to Z0X \mid Z0Z$$
$$Y \to Z1Y \mid Z1Z$$
$$Z \to 0Z1Z \mid 1Z0Z \mid \varepsilon$$

PROBLEM 17.3 Find the language generated by the following grammar:

$$S \to aS$$
$$S \to bS$$
$$S \to b$$

or in short, $S \to aS \mid bS \mid b.$

Solution

Here S is the starting symbol. The language generated by this grammar contains all non-empty strings of symbols a and b that end in b.

PROBLEM 17.4 Design an LR(0) parser for the following grammar:

$$\textbf{G: } S \to E\#$$
$$E \to BE \mid e$$
$$B \to aB \mid b$$

Solution

The DFA states are as follows:

State q_0:

$$S \rightarrow .\, E\#$$
$$E \rightarrow .\, BE$$
$$E \rightarrow .\, e$$
$$B \rightarrow .\, aB$$
$$B \rightarrow .\, b$$

Now, we see that in the item $[S \rightarrow .\, E\#]$, symbol E appears before dot (.), therefore, we get next state q_1 with LR(0) item $[S \rightarrow E\, .\, \#]$. Input on state q_0 to reach state q_1 is taken as E. Thus the state q_1 will be:

State q_1: {transition from q_0 on input E}

$$S \rightarrow E\, .\, \#$$

Similarly, we can continue to get other states.

State q_2: {transition from q_0 on input B}

$$E \rightarrow B\, .\, E$$
$$E \rightarrow .\, BE$$
$$E \rightarrow .\, e$$
$$B \rightarrow .\, aB$$
$$B \rightarrow .\, b$$

State q_3: {transition from q_0 on input a}

$$B \rightarrow a\, .\, B$$
$$B \rightarrow .\, aB$$
$$B \rightarrow .\, b$$

State q_4: {transition from q_0 on input b}

$$B \rightarrow b\, .$$

State q_5: {transition from q_2 on input E}

$$E \rightarrow BE\, .$$

State q_6: {transition from q_3 on input B}

$$B \rightarrow aB\, .$$

State q_7: {transition from q_0 on input e}

$$E \rightarrow e\, .$$

These DFA states, which we have computed, are the only ones possible at this stage. The resulting DFA is given in Figure 17.14. Its transition table is given in Table 17.8.

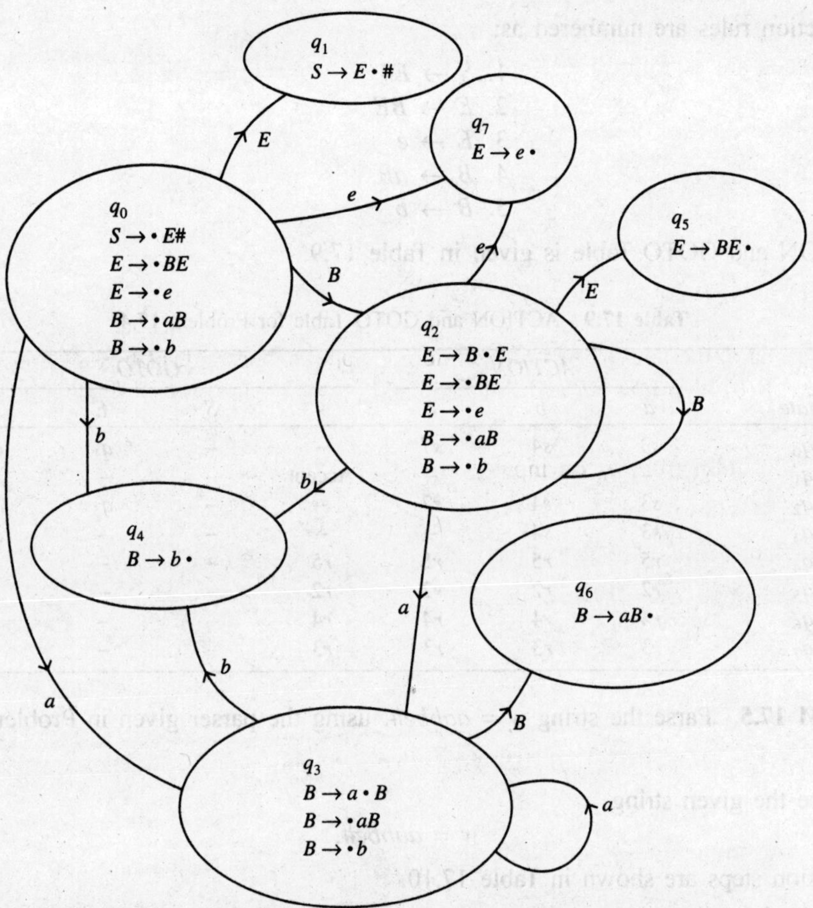

Figure 17.14 Transition diagram.

Table 17.8 Transition Table of DFA for Problem 17.4

States	Input					
	a	*b*	e	#	E	B
q_0	q_3	q_4	q_7	–	q_1	q_2
q_1	–	–	–	–	–	–
q_2	q_3	q_4	q_7	–	q_5	q_2
q_3	q_3	q_4	–	–	–	q_6
q_4	–	–	–	–	–	–
q_5	–	–	–	–	–	–
q_6	–	–	–	–	–	–
q_7	–	–	–	–	–	–

The ACTION and GOTO table can be constructed using the procedure described in Section 17.11.8.

The production rules are numbered as:

1. $S \to E\#$
2. $E \to BE$
3. $E \to e$
4. $B \to aB$
5. $B \to b$

The ACTION and GOTO Table is given in Table 17.9.

Table 17.9 ACTION and GOTO Table for Problem 17.4

State	*ACTION*				*GOTO*		
	a	*b*	*e*	#	*S*	*E*	*B*
q_0	*s*3	*s*4	*s*7	–	–	q_1	q_2
q_1	–	–	–	Accept	–	–	–
q_2	*s*3	*s*4	*s*7	–	–	q_5	q_2
q_3	*s*3	*s*4	–	–	–	–	q_6
q_4	*r*5	*r*5	*r*5	*r*5	–	–	–
q_5	*r*2	*r*2	*r*2	*r*2	–	–	–
q_6	*r*4	*r*4	*r*4	*r*4	–	–	–
q_7	*r*3	*r*3	*r*3	*r*3	–	–	–

PROBLEM 17.5 Parse the string $w = aabbe\#$, using the parser given in Problem 17.4.

Solution

Let us parse the given string

$$w = aabbe\#.$$

The derivation steps are shown in Table 17.10.

Table 17.10 Derivation Steps of String $w = aabbe\#$

S. No.	*Stack*	*Input stream*	*Action*
1	q_0	*aabbe#*	*s*3: shift state q_3 on the stack
2	q_0, q_3	*abbe#*	*s*3: shift state q_3 on the stack
3	q_0, q_3, q_3	*bbe#*	*s*4: shift state q_4 on the stack
4	q_0, q_3, q_3, q_4	*be#*	*r*5: reduce by production rule 5
5	q_0, q_3, q_3, q_6	*be#*	*r*4: reduce by production rule 4
6	q_0, q_3, q_6	*be#*	*r*4: reduce by production rule 4
7	q_0, q_2	*be#*	*s*4: shift state q_4 on the stack
8	q_0, q_2, q_4	*e#*	*r*5: reduce by production rule 5
9	q_0, q_2, q_2	*e#*	*s*7: shift state q_7 on the stack
10	q_0, q_2, q_2, q_7	#	*r*3: reduce by production rule 3
11	q_0, q_2, q_2, q_5	#	*r*2: reduce by production rule 2
12	q_0, q_2, q_5	#	*r*2: reduce by production rule 2
13	q_0, q_1	#	accept

Exercises

17.1 Construct DFA for the following regular expressions:

(i) $0 \mid 1 \mid 01+$

(ii) $abb \mid ab^* \mid a^*b$

17.2 Obtain a regular expression for the following DFA.

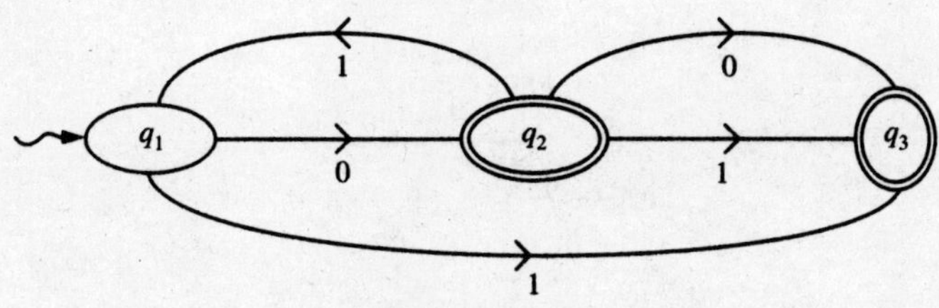

Figure 17.15

17.3 List all the LR(0) items and construct an automaton for the following grammar:

$$S \rightarrow AS \mid b$$
$$A \rightarrow SA \mid a$$

17.4 Construct LR(0) parsers for the following grammars.

(i) $E \rightarrow E + T \mid T$

$T \rightarrow T^* F \mid F$

$F \rightarrow a$

(ii) $S \rightarrow SA \mid A$

$A \rightarrow aSb \mid ab$

17.5 Construct LR(1) parsers for the following grammars.

(i) $E \rightarrow E + T \mid T$

$T \rightarrow T^* F \mid F$

$F \rightarrow (a)$

(ii) $S \rightarrow AA$

$A \rightarrow aA \mid b$

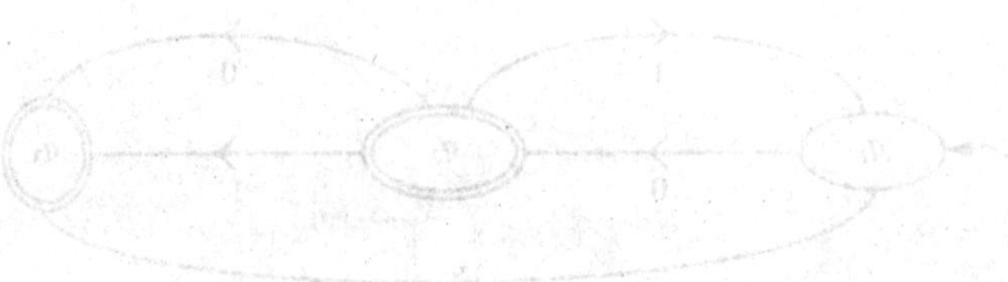

Part F

NUMERIC FUNCTIONS AND RECURRENCE RELATIONS

18

Discrete Numeric Functions and Recurrence Relations

18.1 Introduction

In this chapter, we will discuss discrete numeric functions and recurrence relations. The discrete numeric functions and recurrence relations frequently occur in mathematical computations. We will also describe generating function, which is another alternative representation of a discrete numeric function.

18.2 Discrete Numeric Function

The function, whose domain is the set of positive integers, and range, the set of real numbers, is called **discrete numeric function** or **numeric function**. A discrete numeric function may be written by listing all its values, exhaustively. There is, however, a concise way for it. For instance, $a_r = 2r^2 + 3r + 2$, $r \geq 0$ is a discrete numeric function. We can list all its values, exhaustively. Six values are given in Table 18.1.

Table 18.1 Discrete Numeric Function

S. No.	r	$a_r = 2r^2 + 3r + 1$
1	0	$a_0 = 1$
2	1	$a_1 = 6$
3	2	$a_2 = 15$
4	3	$a_3 = 28$
5	4	$a_4 = 45$
6	5	$a_5 = 66$

Some other examples of the numeric functions are:

$$b_r = 6r^3 + 3r^2 + 1,\ 0 \le r \le 100$$

$$c_r = 2r^2 + 3r + 1,\ r \ge 0$$

$$d_r = r^4 + 2r + 3,\ r \ge 0$$

$$e_r = \begin{cases} 2r, & r \ge 5 \\ 0, & \text{otherwise,} \end{cases}$$

$$f_r = \begin{cases} 2r, & r \ge 2 \\ 2r^3 + 3r + 1, & r = 1 \\ 0, & \text{otherwise.} \end{cases}$$

18.3 Operations on Discrete Numeric Functions

We can get a new numeric function by performing algebraic operations on the given numeric functions. Some algebraic operations performed on numeric functions are described in the following subsection.

18.3.1 Addition of Numeric Functions

Two numeric functions e_r and f_r, given above, can be added to get a new numeric function g_r. That is,

$$g_r = e_r + f_r.$$

The values of g_r has been computed up to $r = 5$.

The computation details of g_r are given in Table 18.2.

Table 18.2 Addition of Numeric Functions

S. No.	r	$e_r = \begin{cases} 2r, & r \ge 5 \\ 0, & \text{otherwise} \end{cases}$	$f_r = \begin{cases} 2r, & r \ge 2 \\ 2r^3 + 3r + 1, & r = 1 \\ 0, & \text{otherwise} \end{cases}$	$g_r = e_r + f_r$
1	0	0	0	0
2	1	0	$2r^3 + 3r + 1$	$2r^3 + 3r + 1$
3	2	0	$2r$	$2r$
4	3	0	$2r$	$2r$
5	4	0	$2r$	$2r$
6	5	$2r$	$2r$	$4r$

From Table 18.2, g_r can be written concisely as:

$$g_r = \begin{cases} 2r, & r \geq 5 \\ 0, & \text{otherwise} \end{cases} + \begin{cases} 2r, & r \geq 2 \\ 2r^3 + 3r + 1, & r = 1 \\ 0, & \text{otherwise} \end{cases}$$

$$= \begin{cases} 0, & r = 0 \\ 2r^3 + 3r + 1, & r = 1 \\ 2r, & 2 \leq r \leq 4 \\ 4r, & r \geq 5 \end{cases}$$

18.3.2 Subtraction of Numeric Functions

Let us perform subtraction of numeric functions e_r and f_r. That is

$$g_r = e_r - f_r$$

The values of e_r and f_r, which are listed in Table 18.2, are used here. The computation details of g_r are given in Table 18.3.

Table 18.3 Subtraction of Numeric Functions

S. No.	r	$e_r = \begin{cases} 2r, & r \geq 5 \\ 0, & \text{otherwise} \end{cases}$	$f_r = \begin{cases} 2r, & r \geq 2 \\ 2r^3 + 3r + 1, & r = 1 \\ 0, & \text{otherwise} \end{cases}$	$g_r = e_r - f_r$
1	0	0	0	0
2	1	0	$2r^3 + 3r + 1$	$-2r^3 - 3r - 1$
3	2	0	$2r$	$-2r$
4	3	0	$2r$	$-2r$
5	4	0	$2r$	$-2r$
6	5	$2r$	$2r$	0

Thus, concisely, g_r can be written as:

$$g_r = \begin{cases} 2r, & r \geq 5 \\ 0, & \text{otherwise} \end{cases} - \begin{cases} 2r, & r \geq 2 \\ 2r^3 + 3r + 1, & r = 1 \\ 0, & \text{otherwise.} \end{cases}$$

$$= \begin{cases} -2r^3 - 3r - 1, & r = 1 \\ -2r, & 2 \leq r \leq 4 \\ 0, & \text{otherwise.} \end{cases}$$

18.3.3 Multiplication of Numeric Functions

Like addition and subtraction, multiplication is also performed term by term for different values of r. Let us multiply functions e_r and f_r given above, such that

$$g_r = e_r \times f_r$$

The complete computation details of g_r are listed in Table 18.4. The values of e_r and f_r have been taken from Table 18.2.

Table 18.4 Multiplication of Numeric Functions

S. No.	r	$e_r = \begin{cases} 2r, & r \geq 5 \\ 0, & \text{otherwise} \end{cases}$	$f_r = \begin{cases} 2r, & r \geq 2 \\ 2r^3 + 3r + 1, & r = 1 \\ 0, & \text{otherwise} \end{cases}$	$g_r = e_r \times f_r$
1	0	0	0	0
2	1	0	$2r^3 + 3r + 1$	0
3	2	0	$2r$	0
4	3	0	$2r$	0
5	4	0	$2r$	0
6	5	$2r$	$2r$	$4r^2$

Therefore, concisely, g_r can be written as:

$$g_r = \begin{cases} 2r, & r \geq 5 \\ 0, & \text{otherwise} \end{cases} \times \begin{cases} 2r, & r \geq 2 \\ 2r^3 + 3r + 1, & r = 1 \\ 0, & \text{otherwise.} \end{cases}$$

$$= \begin{cases} 4r^2, & r \geq 5 \\ 0, & \text{otherwise.} \end{cases}$$

18.3.4 Scaling of a Numeric Function

The value of a discrete numeric function can be scaled up, or scaled down by a factor, say k. Let,

$$a_r = 2r^2 + 3r + 2, \; r \geq 0$$

be a discrete numeric function and k be a real number. Assume $k = 10$

Then,
$$ka_r = 10\{2r^2 + 3r + 2\},$$
$$= 20r^2 + 30r + 20.$$

Similarly,
$$\frac{1}{k}a_r = \frac{1}{10}\{2r^2 + 3r + 2\},$$
$$= 0.2r^2 + 0.3r + 0.2.$$

18.3.5 Convolution of Numeric Functions

Convolution of two discrete numeric functions is defined as:

$c_r = a_r \circ b_r$, where, symbol '$\circ$' is termed as convolution operator.

$$= \sum_{i=0}^{r} a_i\, b_{r-i}$$

$$= a_0 b_r + a_1 b_{r-1} + a_2 b_{r-2} + \cdots + a_r b_0$$

For instance, let

$$a_r = 5^r, \text{ and } b_r = 7^r, \text{ for } r \geq 0.$$

Then

$$c_r = a_r \circ b_r$$

$$= \sum_{i=0}^{r} 5^i \cdot 7^{r-i}, \quad r \geq 0.$$

18.4 Generating Function

A generating function is an alternative representation of a discrete numeric function. It is written in the form of an infinite series. The advantage of this representation is the convenience in manipulation, particularly addition and subtraction.

Let

$$a_r = (a_0, a_1, a_2, \ldots, a_r, \ldots)$$

be a discrete numeric function.

Then an infinite series written as:

$$a(z) = a_0 + a_1 z^1 + a_2 z^2 + \cdots + a_r z^r + \cdots$$

is called the generating function of a_r.

For example, let

$$a_r = (5^0\ 5, 5^2, \ldots, 5^r, \ldots)$$

then the generating function of a_r is written as:

$$a(z) = 5^0 + 5^1 z^1 + 5^2 z^2 + \cdots + 5^r z^r + \cdots$$

Series $a(z)$ can be written in a compact form as:

$$a(z) = \frac{1}{1 - 5z} \qquad \text{\{sum of an infinite geometric progression\}}$$

18.5 Operations on Generating Functions

In this section, we will discuss some operations performed on generating functions.

18.5.1 Addition of Generating Functions

Let us consider two numeric functions a_r and b_r, such that

$$a_r = 5 \times 7^r, \quad r \geq 0$$

and

$$b_r = 5^{r+3}, \qquad r \geq 0.$$

Then the generating functions of a_r and b_r will be as follows:
Generating function of a_r,

$$a(z) = \frac{5}{1-7z}$$

Generating function of b_r,

$$b(z) = \frac{125}{1-5z}$$

First, we will compute the sum by the method used for finding the sum of numeric functions. Let the sum of numeric functions a_r and b_r be c_r.

Then
$$c_r = a_r + b_r$$
$$= 5 \times 7^r + 5^{r+3}, \quad r \geq 0.$$

The values of c_r, for $r = 0$ to $r = 5$, are given in Table 18.5.

Table 18.5 Computation Details of c_r

S. No.	r	$a_r = 5 \times 7^r$	$b_r = 5^{r+3}$	$c_r = a_r + b_r$
1	0	5	125	130
2	1	35	625	660
3	2	245	3125	3370
4	3	1715	15625	17340
5	4	12005	78125	90130
6	5	84035	390625	474660

Thus, from Table 18.5, we can see that

$$c_r = (130, 660, 3370, 17340, \ldots).$$

Now, let us compute the sum by using generating functions. We will see that it is convenient and relatively easy. Let $d(z)$ be the sum.

Then $d(z) = a(z) + b(z)$

$$= \frac{5}{1-7z} + \frac{125}{1-5z}$$
$$= 5 \cdot (7^0 + 7^1z^1 + 7^2z^2 + \cdots + 7^rz^r + \cdots) + 125 \cdot (5^0 + 5^1z^1 + 5^2z^2 + \cdots + 5^rz^r + \cdots)$$
$$= (5 \cdot 7^0 + 125 \cdot 5^0) + (5 \cdot 7^1 + 125 \cdot 5^1)z^1 + (5 \cdot 7^2 + 125 \cdot 5^2)\, z^2 + \cdots$$
$$= 130 + 660z + 3370z^2 + \cdots.$$

Note that, this is the generating function of numeric function

$$c_r = (130, 660, 3370, \ldots).$$

Thus, $d(z)$ is the generating function of numeric function c_r.

Hence, we conclude that the **sum** of two numeric functions can be obtained easily by directly adding their corresponding generating functions.

The **subtraction** of generating functions, for instance

$$c(z) = a(z) - b(z)$$

is analogous to addition. Therefore, it is needless to describe.

We also exclude the **multiplication** of two generating functions due to its computational difficulties.

18.5.2 Scaling the Generating Function

Suppose, a_r is a discrete numeric function, and $a(z)$ is its generating function. Consider a numeric function b_r, such that

$$b_r = k^r\, a_r$$

where k is some scaling constant.

Then the generating function of b_r can be written as:

$$\begin{aligned} b(z) &= a_0 + ka_1z^1 + k^2\, a_2z^2 + \cdots + k^r\, a_rz^r + \cdots \\ &= a_0 + a_1\,(kz)^1 + a_2\,(kz)^2 + \cdots + a_r\,(kz)^r + \cdots \\ &= a(kz). \end{aligned}$$

EXAMPLE 18.1 Let $a_r = 1$, $r \geq 0$. Then the generating function of a_r is given as:

$$a(z) = \frac{1}{1-z}$$

Consequently, the generating function of $b_r = k^r\, a_r$, $r \geq 0$, will be

$$b(z) = \frac{1}{1-kz}.$$

18.5.3 Convolution of Generating Function

Consider two numeric functions a_r and b_r. Let

$$c_r = a_r \circ b_r$$

be the convolution of a_r and b_r. Further, let $a(z)$, $b(z)$ and $c(z)$ be the generating functions of a_r, b_r and c_r respectively. Then generating function $c(z)$ is given by

$$c(z) = a(z)\, b(z).$$

EXAMPLE 18.2 Suppose, $a_r = 4^r$ and $b_r = 3^r$ are two numeric functions. Then the corresponding generating functions will be:

$$a(z) = \frac{1}{1-4z}$$

and

$$b(z) = \frac{1}{1-3z}$$

Therefore, the convolution of $a(z)$ and $b(z)$ will be given by

$$c(z) = \frac{1}{(1-4z)} \cdot \frac{1}{(1-3z)}$$

$$= \frac{4}{1-4z} - \frac{3}{1-3z}$$

Therefore,

$$c_r = 4(4^r) - 3(3^r)$$

$$= 4^{r+1} - 3^{r+1}.$$

18.6 Recurrence Relation

Suppose, a_r is a discrete numeric function. Then a relation, among a_r and one or more a_i's, such that

$$i < r$$

is called a recurrence relation. A discrete numeric function is also referred to as the solution of a recurrence relation. Some authors use the name **difference equation** in place of recurrence relation.

EXAMPLE 18.3 Consider a relation

$$a_r = a_{r-1} + a_{r-2}.$$

It is an example of recurrence relation. In this recurrence relation, if we assume $a_0 = 1$ and $a_1 = 1$, which are called the **boundary conditions**, then we can compute values a_2, a_3, and a_4 etc.

For example,

$$a_2 = a_1 + a_0$$
$$= 1 + 1$$
$$= 2$$
$$a_3 = a_2 + a_1$$
$$= 2 + 1$$
$$= 3$$
$$a_4 = a_3 + a_2$$
$$= 3 + 2$$
$$= 5$$

and so on.

Sequence of these values is called **Fibonacci sequence**. Note that the collection of these values $(a_0, a_1, a_2, a_3, a_4, ..., a_r, ...)$, is the solution of the given recurrence relation.

18.7 Linear Recurrence Relation with Constant Coefficient

Consider the following recurrence relation,

$$c_0 a_r + c_1 a_{r-1} + c_2 a_{r-2} + \cdots + c_k a_{r-k} = f(r) \tag{18.1}$$

If each c_i in recurrence relation (18.1) is constant, then it is called linear recurrence relation with constant coefficients. It is linear because **degree** of each a_i is one. Further, if c_0 and c_k are non-zero, then the recurrence relation given in Eq. (18.1) is of **order** k.

Thus,
$$a_r + a_{r-1} = 2r$$
is a first order linear recurrence relation with constant coefficients.

Similarly,
$$a_r + 2a_{r-1} - 3a_{r-2} = r^3$$
is a second order linear recurrence relation with constant coefficients.

If right hand side of the recurrence relation is equal to zero,

that is,
$$f(r) = 0$$
then the recurrence relation is called **homogeneous**.

In this chapter, we will keep our discussion limited to the solution of linear recurrence relation, with constant coefficients.

18.8 Solution of Recurrence Relation

Solution of a recurrence relation consists of two parts, homogeneous solution $a_r(h)$ and particular solution $a_r(p)$. In other words, the numeric function, which is the solution of a recurrence relation, is the sum of two parts:

1. a numeric function, which is the homogeneous solution and
2. a numeric function, which is the particular solution of the recurrence relation.

These two parts are discussed in the following subsections.

18.8.1 Homogeneous Solution

Homogeneous solution, $a_r(h)$, is obtained by substituting $A\alpha^r$ in place of a_r in a recurrence relation, and setting the right hand side of the recurrence relation equal to zero. Here α is called the **characteristic root**, and A is an arbitrary constant. We will understand the procedure of finding homogeneous solution with the help of an example.

EXAMPLE 18.4 Consider the recurrence relation
$$a_r = a_{r-1} + a_{r-2}. \tag{18.2}$$
Let the boundary conditions be
$$a_1 = 1$$
and
$$a_2 = 1.$$
It is a homogeneous recurrence relation with constant coefficients. Let us find its solution. Equation (18.2) can be rewritten as:
$$a_r - a_{r-1} - a_{r-2} = 0 \tag{18.3}$$
Replacing a_r in Eq. (18.3) by $A\alpha^r$ we get
$$A\alpha^r - A\alpha^{r-1} - A\alpha^{r-2} = 0$$
$$\Rightarrow \quad \alpha^r - \alpha^{r-1} - \alpha^{r-2} = 0, \quad \text{since } A \neq 0$$
$$\Rightarrow \quad \alpha^2 - \alpha - 1 = 0. \tag{18.4}$$
Equation (18.4) is called the **characteristic equation**.

There are two roots of this equation, say α_1 and α_2.

Then, $$\alpha_1 = \frac{1+\sqrt{5}}{2}$$

and $$\alpha_2 = \frac{1-\sqrt{5}}{2}.$$

Thus, the homogeneous solution is

$$a_r(h) = A_1\left(\frac{1+\sqrt{5}}{2}\right)^r + A_2\left(\frac{1-\sqrt{5}}{2}\right)^r.$$

Constants A_1 and A_2 can be determined by using boundary conditions in the following way.

$$A_1\left(\frac{1+\sqrt{5}}{2}\right)^0 + A_2\left(\frac{1-\sqrt{5}}{2}\right)^0 = 1 \tag{18.5}$$

and $$A_1\left(\frac{1+\sqrt{5}}{2}\right)^1 + A_2\left(\frac{1-\sqrt{5}}{2}\right)^1 = 1. \tag{18.6}$$

Equations (18.5) and (18.6) can be rewritten as:

$$A_1 + A_2 = 1 \tag{18.7}$$

and $$A_1\left(\frac{1+\sqrt{5}}{2}\right) + A_2\left(\frac{1-\sqrt{5}}{2}\right) = 1. \tag{18.8}$$

Solving Eqs. (18.7) and (18.8), we get

$$A_1 = \frac{1+\sqrt{5}}{2\sqrt{5}}$$

and $$A_2 = -\frac{1-\sqrt{5}}{2\sqrt{5}}.$$

Hence, $$a_r(h) = \left(\frac{1+\sqrt{5}}{2\sqrt{5}}\right)\left(\frac{1+\sqrt{5}}{2}\right)^r - \left(\frac{1-\sqrt{5}}{2\sqrt{5}}\right)\left(\frac{1-\sqrt{5}}{2}\right)^r.$$

which is the required solution.

Note: If a root of the characteristic equation is repeated, then the corresponding homogeneous solution will be of the form

$$(A_1 r^{m-1} + A_2 r^{m-2} + \cdots + A_m) \cdot \beta^r$$

where β is a repeated root of **multiplicity *m*.**

18.8.2 Particular Solution

Determining the particular solution, $a_r(p)$, of a recurrence relation is a tricky process. In fact, there is no general procedure. However, in some cases the solution can be obtained by method of inspection. The solution depends on the form of $f(r)$, which is the right hand side of the recurrence relation. The following are some cases, which help in determining the particular solution.

Case 1. If $f(r)$ is a constant, then particular solution shall also be a constant.

Case 2. If $f(r)$ is a polynomial of degree k, and is of the form

$$F_1 r^k + F_2 r^{k-1} + \cdots F_k r + F_{k+1}$$

where $F_1, F_2, \ldots, F_{k+1}$ are all constants,
then the particular solution shall be of the form,

$$P_1 r^k + P_2 r^{k-1} + \cdots P_k r + P_{k+1}$$

with $P_1, P_2, \ldots, P_{k+1}$ being all constants.

Case 3. If $f(r)$ is of the form β^r, where β is a constant but not the characteristic root of the recurrence relation, then the particular solution shall be of the form $P\beta^r$. In general, when $f(r)$ is of the form

$$(F_1r^k + F_2r^{k-1} + \cdots F_kr + F_{k+1})\beta^r$$

then the particular solution shall assume the form

$$(P_1r^k + P_2r^{k-1} + \cdots P_kr + P_{k+1})\beta^r$$

provided β is not a characteristic root of the recurrence relation.

Case 4. In case, β is a characteristic root of multiplicity m, and $f(r)$ is of the form

$$(F_1r^k + F_2r^{k-1} + \cdots F_kr + F_{k+1})\beta^r$$

then the particular solution shall assume the form

$$r^m(P_1r^k + P_2r^{k-1} + \cdots P_kr + P_{k+1})\beta^r.$$

The following examples will help in understanding the procedure to find the particular solution of a recurrence relation.

EXAMPLE 18.5 Consider the following recurrence relation:

$$a_r + 3a_{r-1} + 6a_{r-2} = 1 \qquad (18.9)$$

Since the right hand side of Eq. (18.9) is a constant, therefore, the particular solution will also be a constant, say P.

Further, if P is the solution of given recurrence relation, then it must satisfy Eq. (18.9). Therefore, substituting P in Eq. (18.9), we get

$$P + 3P + 6P = 1.$$

Therefore, $$P = \frac{1}{10}.$$

Thus, the particular solution is

$$a_r(p) = \frac{1}{10}.$$

EXAMPLE 18.6 Consider the following recurrence relation:

$$a_r + 3a_{r-1} + 6a_{r-2} = 4r^2 - 2r + 1 \tag{18.10}$$

Obviously, the particular solution will be of the form:

$$P_1\, r^2 + P_2\, r + P_3. \tag{18.11}$$

Further, if Eq. (18.11) is the solution of the given recurrence relation, then it must satisfy Eq. (18.10).

Therefore, substituting Eq. (18.11) in Eq. (18.10), we get

$$P_1\, r^2 + P_2\, r + P_3 + 3\, P_1(r-1)^2 + 3P_2\,(r-1)$$
$$+\, 3P_3 + 6P_1\,(r-2)^2 + 6P_2\,(r-2) + 6P_3 = 4r^2 - 2r + 1$$
$$\Rightarrow \quad 10P_1r^2 + (10P_2 - 30P_1)r + 27P_1 - 15P_2 + 10P_3 = 4r^2 - 2r + 1$$

Comparing the coefficients both sides, we get

$$10P_1 = 4$$
$$10P_2 - 30P_1 = -2$$
$$27P_1 - 15P_2 + 10P_3 = 1$$

Solving these, we get

$$P_1 = \frac{2}{5},\; P_2 = 1, \text{ and } P_3 = \frac{13}{25}.$$

Thus, the particular solution is

$$a_r(p) = \frac{2}{5}\, r^2 + r + \frac{13}{25}.$$

EXAMPLE 18.7 Consider the following recurrence relation:

$$a_r + 3a_{r-1} + 6a_{r-2} = 6 \cdot 4^r \tag{18.12}$$

In this relation, 4 is not a characteristic root, therefore, the particular solution will be of the form: $P4^r$ {using case 3 of particular solution described in Subsection 18.8.2}.
Then, substituting $P4^r$ in Eq. (18.12), we get

$$P4^r + 3P4^{r-1} + 6P4^{r-2} = 6 \cdot 4^r.$$

$$\Rightarrow \quad \frac{17}{8}P4^r = 6 \cdot 4^r$$

$$\Rightarrow \quad P = \frac{48}{17}.$$

Thus, the particular solution is

$$a_r(p) = \frac{48}{17} \cdot 4^r$$

EXAMPLE 18.8 Consider the following recurrence relation:

$$a_r - 3a_{r-1} = 6 \cdot 3^r \tag{18.13}$$

In this recurrence relation, 3 is a characteristic root of multiplicity 1, therefore, the particular solution will be of the form: $Pr3^r$ {using case 4 of particular solution described in Subsection 18.8.2}.

Therefore, substituting $Pr3^r$ in Eq. (18.13), we get

$$Pr3^r - 3P(r-1)3^{r-1} = 6 \cdot 3^r.$$

$$\Rightarrow \quad P3^r = 6 \cdot 3^r$$

$$\Rightarrow \quad P = 6.$$

Thus, the particular solution is

$$a_r(p) = 6r3^r.$$

18.8.3 Total Solution

Total solution is obtained by combining homogeneous solution and particular solution. The arbitrary constants of homogeneous solution that appear in total solution are determined with the help of boundary conditions. If there are k arbitrary constants, then k boundary conditions are needed. However, if boundary conditions are not given then the solution is called **general solution**.

For example,

Let

$$a_r = A_1(-3)^r + A_2(-4)^r + 5 \cdot 2^r \tag{18.14}$$

be a total solution of a recurrence relation. Note that there are two arbitrary constants A_1 and A_2 in Eq. (18.14). Therefore, two boundary conditions are needed to determine these values.

Let $a_0 = 10$ and $a_1 = 20$ be two given values.

Then, we get two equations,

$$5 = A_1 + A_2$$

and

$$10 = -3A_1 - 4A_2.$$

On solving, we get

$$A_1 = 30 \text{ and } A_2 = -25.$$

Thus, the total solution becomes

$$a_r = 30(-3)^r - 25(-4)^r + 5 \cdot 2^r.$$

18.9 Solved Problems

PROBLEM 18.1 If $a_r = \begin{cases} 0, & 0 \le r \le 4 \\ 2^{-r} + 7, & r \ge 5 \end{cases}$ and $b_r = \begin{cases} 3 - 2^r, & 0 \le r \le 3 \\ r + 2, & r \ge 4 \end{cases}$, then find $a_r + b_r$.

Solution

Let $c_r = a_r + b_r$

$$\Rightarrow \quad c_r = \begin{cases} 0, & 0 \le r \le 4 \\ 2^{-r} + 7, & r \ge 5 \end{cases} + \begin{cases} 3 - 2^r, & 0 \le r \le 3 \\ r + 2, & r \ge 4 \end{cases}$$

$$= \begin{cases} 3 - 2^r, & 0 \le r \le 3 \\ r + 2, & r = 4 \\ 2^{-r} + r + 9, & r \ge 5. \end{cases}$$

PROBLEM 18.2 Let $a_r = (-1)^r \cdot \dfrac{5}{r}, r \ge 4$. Find $|a_r|$ and $5|a_r|$.

Solution

Obviously, $$|a_r| = \frac{5}{r}, \quad r \ge 4$$

and

let $b_r = 5|a_r|$, then

$$b_r = \frac{25}{r}, \quad r \ge 4.$$

PROBLEM 18.3 Let a_r and b_r be two numeric functions, such that

$$a_r = 3 \times 7^r, r \ge 0$$

and $$b_r = 5^{r+2}, r \ge 0.$$

Find the corresponding generating functions. Also, find their sum.

Solution

Generating function of a_r,

$$a(z) = \frac{3}{1 - 7z}$$

Generating function of b_r,

$$b(z) = \frac{25}{1 - 5z}$$

Let the sum of the given numeric functions be c_r.

Then, $$c_r = a_r + b_r.$$

Further, if $c(z)$ be the generating function of c_r, then

$$c(z) = a(z) + b(z)$$

$$= \frac{3}{1 - 7z} + \frac{25}{1 - 5z}$$

$$= 3 \cdot (7^0 + 7^1z^1 + 7^2z^2 + \cdots + 7^rz^r + \cdots) + 25 \cdot (5^0 + 5^1z^1 + 5^2z^2 + \cdots + 5^rz^r + \cdots)$$
$$= (3 \cdot 7^0 + 25 \cdot 5^0) + (3 \cdot 7^1 + 25 \cdot 5^1)z^1 + (3 \cdot 7^2 + 25 \cdot 5^2)z^2 + \cdots$$
$$= 28 + 146z^1 + 772z^2 + \cdots .$$

Thus, the generating function of numeric function

$$c_r = (28, 146, 772, ...).$$

PROBLEM 18.4 Find the general solution of the recurrence relation

$$a_r - 5a_{r-1} + 6a_{r-2} = 0.$$

Solution

It is a homogeneous recurrence relation. Its characteristic equation is:

$$\alpha^2 - 5\alpha - 6 = 0.$$

There are two roots of this equation,

$$\alpha_1 = 2$$

and

$$\alpha_2 = 3.$$

Thus, the homogeneous solution is

$$a_r(h) = A_1 2^r + A_2 3^r.$$

PROBLEM 18.5 Find the particular solution of the following recurrence relation:

$$a_r + 3a_{r-1} = 4r^2 - 2r + 1.$$

Solution

The particular solution will be of the form:

$$P_1\, r^2 + P_2\, r + P_3.$$

Substituting it in the given recurrence relation, we get

$$P_1\, r^2 + P_2\, r + P_3 + 3P_1(r-1)^2 + 3P_2\,(r-1) + 3P_3 = 4r^2 - 2r + 1$$

$$\Rightarrow \quad 4P_1r^2 + (4P_2 - 6P_1)r + 3P_1 - 3P_2 + 4P_3 = 4r^2 - 2r + 1.$$

Comparing the coefficients on both sides, we get

$$4P_1 = 4,$$
$$4P_2 - 6P_1 = -2$$

and

$$3P_1 - 3P_2 + 4P_3 = 1.$$

Solving these, we get

$$P_1 = 1,\ P_2 = 1, \text{ and } P_3 = \frac{1}{4}.$$

Thus, the particular solution is

$$a_r(p) = r^2 + r + \frac{1}{4}.$$

PROBLEM 18.6 Find the total solution of the recurrence relation

$$a_r - 6a_{r-1} + 9a_{r-2} = 6 \cdot 3^r$$

The boundary conditions are, $a_1 = 2$ and $a_2 = 3$.

Solution

There will be two parts of the solution: homogeneous solution and particular solution. These solutions are described below.

(i) **Homogeneous solution:** For homogeneous solution, equate right hand side of given recurrence relation to 0.

That is,
$$a_r - 6a_{r-1} + 9a_{r-2} = 0$$

Then the characteristic equation is
$$\alpha^2 - 6\alpha + 9 = 0$$

This gives $\alpha = 3, 3$. It is a repeated root of multiplicity 2.

Thus, homogeneous solution is
$$a_r(h) = (A_1 + A_2\, r)3^r$$

(ii) **Particular solution:** There is a repeated root, $\alpha = 3$, of multiplicity 2. Therefore, particular solution will be of the form $Pr^2\, 3^r$.

Substituting $Pr^2 3^r$ in the given recurrence relation, we get
$$Pr^2 3^r - 6P(r-1)^2\, 3^{r-1} + 9P(r-2)^2\, 3^{r-2} = 6 \cdot 3^r$$

$$\Rightarrow \qquad P = 3$$

Thus, the particular solution is
$$a_r(p) = 3r^2 3^r$$
$$= r^2 3^{r+1}$$

(iii) **Total solution:** Total solution is given by the sum of homogeneous solution and particular solution.

Therefore,
$$a_r = (A_1 + A_2\, r)3^r + r^2 3^{r+1}.$$

Using the given boundary conditions, we get
$$-7 = 3(A_1 + A_2)$$
$$-35 = 3(A_1 + 2A_2).$$

On solving these, we get
$$A_1 = \frac{66}{3} \text{ and } A_2 = \frac{-73}{3}.$$

Thus, the total solution is
$$a_r = \left(\frac{66}{3} + \frac{-73}{3} r\right) 3^r + r^2 3^{r+1}.$$

Exercises

18.1 Find the sum, difference and multiplication of the following discrete numeric functions:

$$a_r = \begin{cases} 0, & 0 \le r \le 4 \\ 2^{-r} + 7, & r > 4 \end{cases}$$

$$b_r = \begin{cases} 3 - 2^r, & 0 \le r \le 2 \\ r + 3, & r > 2 \end{cases}$$

18.2 Find the solution of the following recurrence relation

$$a_r + 6a_{r-1} + 12a_{r-2} + 8a_{r-3} = 0$$

18.3 Find the particular solution of the following recurrence relation

$$a_r + 6a_{r-1} + 12a_{r-2} + 8a_{r-3} = 6 \cdot 3^r$$

18.4 Find the total solution of the following recurrence relation

$$a_r + 9a_{r-1} + 27a_{r-2} + 27a_{r-3} = 5 \cdot 3^r$$

The boundary conditions are: $a_0 = 0$, $a_1 = 2$ and $a_2 = 3$.

18.5 Find the general solution of the following recurrence relation.

$$a_r + 5a_{r-1} + 6a_{r-2} = 4 \cdot r^2$$

18.6 Find the total solution of the following recurrence relation

$$a_r + 9a_{r-1} + 27a_{r-2} + 27a_{r-3} = 8r^2 - 4r + 5$$

The boundary conditions are: $a_1 = 3$, $a_2 = 6$, and $a_3 = 10$.

18.7 Find the general solution of the following recurrence relation

$$a_r + 5a_{r-1} + 6a_{r-2} = 3r^2 - 2r + 1$$

Bibliography

Algebraic Structure

Clifford, A.H. and Preston, G.B., *The Algebraic Theory of Semigroups* Vol. II, 2nd ed., American Mathematical Society, USA, 1967.

Grillet, P.A., *Semigroups: An Introduction to the Structure Theory*, CRC Press, USA, 1995.

Howie, J.M., *Fundamentals of Semigroup Theory*, Oxford University Press, London, UK, 1991.

Nagy, Attila, *Special Classes of Semigroups*, Springer, Germany, 2001.

Mathematical Logic

Barwise, Jon, *Handbook of Mathematical Logic*, Studies in Logic and the Foundations of Mathematics, Elsevier Science Publishers, North Holland, 1982.

Hamilton, A.G., *Logic for Mathematicians*, 2nd ed., Cambridge University Press, UK, 1988.

Katz, J. Victor, *A History of Mathematics*, Addison-Wesley, USA, 1998.

Katz, Robert, *Axiomatic Analysis*, D.C. Heath and Company, Boston, USA, 1964.

Mendelson, Elliott, *Introduction to Mathematical Logic*, 4th ed., Chapman & Hall, London, UK, 1997.

Boolean Algebra and Digital Logic

Arnold, B.H, *Logic and Boolean Algebra*, Prentice Hall, Englewood, N.J., 1962.

Bartee, Thomas, *Digital Computer Fundamentals*, McGraw-Hill, NY, 1962.

Caldwell, S.H., *Switching Circuits and Logic Design*, John Wiley & Sons, NY, 1958.

Katz, R.H., *Contemporary Logic Design*, The Benjamin/Cummings Publishing Company, USA, 1994.

Koppelberg, Sabine, *General Theory of Boolean Algebras*. Handbook of Boolean Algebras, Vol. 1, Ed. J. Donald Monk with Robert Bonnet, Amsterdam, North Holland, 1989.

Sikorski, Roman, *Boolean Algebras,* 3rd ed., Springer-Verlag, Berlin, 1969.

Graph Theory

Biggs, N., Lloyd, E. and Wilson, R., *Graph Theory*, Oxford University Press, UK, pp. 1736–1936, 1986.

Chartrand, Gary, *Introductory Graph Theory*, Dover, USA, 1985.

Gibbons, Alan, *Algorithmic Graph Theory*, Cambridge University Press, UK, 1985.

Golumbic, Martin, *Algorithmic Graph Theory and Perfect Graphs*, Academic Press, USA, 1980.

Harary, Frank, *Graph Theory*, Reading MA, Addison-Wesley, USA, 1969.

Automata Theory and Languages

Aho, A.V. and Ullman, J.D., *The Theory of Parsing, Translation and Computing*, Vol. 1, Prentice Hall, Inc., Englewood, NJ, 1972.

John, E., Hopcroft, Rajeev Motwani, and Jeffrey D. Ullman, *Introduction to Automata Theory, Languages, and Computation*, 2nd ed., Pearson Education, USA, 2000.

Kain, Richard Y., *Automata Theory, Machines and Languages*, McGraw-Hill, NY, 1972.

Nelson, Raymond J., *Introduction to Automata Theory*, John Wiley & Sons, NY, 1972.

Recurrence Relations

Levy H. and Lessman, F., *Finite Difference Equations*, Macmillan, NY, USA, 1961.

Liu, C.L., *Introduction to Combinatorial Mathematics*, McGraw-Hill, NY, USA, 1975.

Wilf., Herbert S., *Generating functionology*, Academic Press, USA, 1990.

Answers to Exercises

Chapter 1

1.2 (i) 0 (ii) 2 (iii) 4 (iv) 3

1.4 $\{(a, e), (b, e), (c, e), (d, e), (e, a), (e, b), (e, c), (e, d)\}$

1.5 $\{(a, a), (a, b), (b, a), (b, b), ((c, d), a), ((c, d), b), (a, (c, d)), (b, (c, d))\}$

1.6 $\{ (0, a), (0, b), (1, a), (1, b), (a, 0), (a, 1), (b, 0), (b, 1)\}$

Chapter 2

2.1 (i) $\{((1, 3), (1, 6), (3, 3), (3, 6), (4, 6), (5, 6)\}$

(ii) $\{(3, 3), (4, 3), (5, 3)\}$

(iii) $\{(1, 3), (1, 6)\}$

(iv) φ

2.2 (i) not a function (ii) function

(iii) function (iv) not a function

2.4 (i) not a function (ii) yes, a function

2.6 [0], [1], [2], [3], [4].

2.7 $fog = (x^2 + 3x + 7)^3$

$gof = x^6 + 3x^3 + 7$

Chapter 4

4.2 0

4.4 $\langle 1, 1\rangle$

4.6 0

Chapter 5

5.3 (i) cyclic, $\{-1, 1\}$ (ii) noncyclic
(iii) cyclic, $\{-6, 6\}$ (iv) noncyclic.

Chapter 6

6.2 (i) $\{-1, 1\}$ (ii) $\{(-1, 1), (1, -1), (1, 1), (-1, -1)\}$
(iii) $\{1, 3\}$ (iv) $\{(1, q, -1), (-1, q, 1), (-1, q, -1), (1, q, 1)\}$

6.6 **Hint:** Z_{12} has elements 3 and 4, and $3 \cdot 4 = 12$, i.e., $3 \cdot 4 = 0$.

6.9 (i) 0 (ii) 0
(iii) 12 (iv) 30

Chapter 7

7.1 (i) $(\sim R) \wedge H$ (ii) $(\sim R) \wedge (\sim H)$
(iii) $R \vee ((\sim R) \wedge H)$ (iv) $(\sim R) \rightarrow (\sim H)$
(v) $(<R) \rightleftharpoons (\sim H)$ (vi) $(\sim(R \wedge (\sim H)))$

7.4 (i) valid (ii) invalid

7.5 (i) valid (ii) valid

7.6 (i) invalid (ii) invalid

7.7 valid

7.8 invalid

7.9 invalid

Chapter 8

8.3 invalid

8.4 valid

8.5 valid

8.6 invalid

Chapter 9

9.1

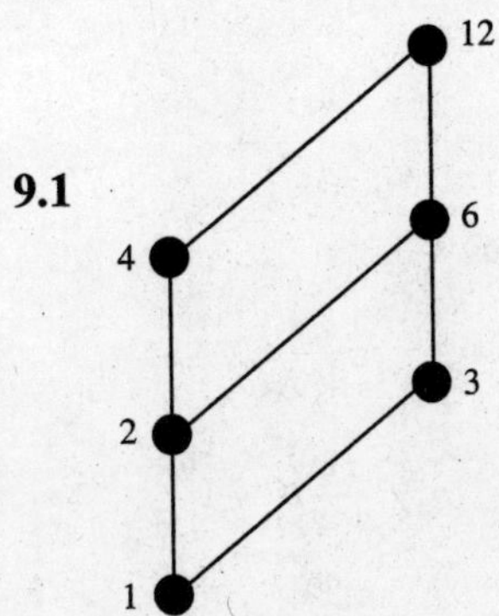

9.4 **Hint:** $a * (b + c) \neq (a * b) + (a * c)$.

Chapter 10

10.3 (i) $x' + (y \cdot z) = (x' + y) \cdot (x' + z)$
(ii) $(x \cdot 1) + (x \cdot 0) = x$
(iii) $(x + y) \cdot (x' + y) \cdot 0 = 0$

Chapter 11

11.1 (i) $(x + y + z)(x' + y + z)(x + y + z')(x' + y + z')(x' + y' + z')$
(ii) $(x + y + z + w)(x + y + z + w')(x + y + z' + w)(x + y + z' + w')(x' + y + z + w)$
$(x' + y + z + w') \cdot (x' + y + z' + w)(x' + y + z' + w')(x' + y' + z' + w)$
$(x' + y' + z' + w')(x + y' + z + w)$

11.2 (i) $xy'zt + xy'zt' + x'yzt + x'yzt' + xyz't + x'yz't' + xyzt + xy'z't + x'yz't + x'y'zt + x'y'zt'$
(iii) $xyz't + xyz't' + x'yzt + x'yzt' + x'yz't + x'yz't' + xyzt + xy'zt + xy'z't$

11.3 (i) $z(x'yw + xy') + y(z' + zt)$ (iii) $x + z' + x'yz$
(v) $y'z + (x'yz + x)t + z'$ (vii) $(x + x'yz)t + xy'z + x'y + yz'$

11.4 (i) $xy + y'zw' + x'z'w'$ (iii) $x'y + yzt + xy'z'w' + y'z'w't'$

11.5 (i) $x'z' + xy + xz + y'z$ (iii) $x'y'z'w' + x'ywt' + ywt + xz't' + xy'wt' + xzwt$

Chapter 12

12.2 (i) 101101 (ii) 10100101
(iii) 1111111

12.5 (i) $x'y + xy' + xy$ (ii) $(x'y + xy') \cdot xy$

12.9 $A'B'C + A'BC' + AB'C' + AB'C + ABC$

Chapter 13

13.2 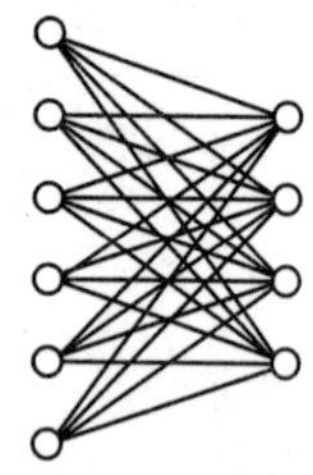 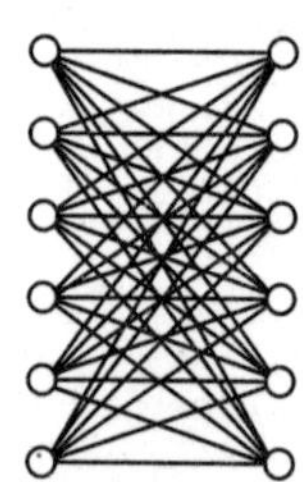

Chapter 14

14.2 (a) planar (b) non-planar

14.5 (a) minimum weight = 19 (b) minimum weight = 18

14.6 (a) minimum weight = 22 (b) minimum weight = 18

14.8 (a) value = 65 (b) value = 25

14.9 (a) nonplanar (b) nonplanar

14.11 (a) chromatic no = 3 (b) chromatic no = 3

Chapter 15

15.1 (a) *a-c-e-f-d-e-b-c-d-b-a* (b) *a-f-g-a-e-g-d-e-b-d-c-b-a*

15.2 (a) *a-c-d-f* (b) *a-d-g-f*

15.3 (a) strongly connected (b) weakly connected

15.4 (a) strongly connected (b) strongly connected

15.7 max flow = 15

15.8 max flow = 21

15.9 (a) each maximum matching contains 3 edges
(b) each maximum matching contains 4 edges
(c) each maximum matching contains 4 edges

15.10 (a) $\{x_3, x_5, x_6\}$ (b) $\{x_1, x_4, x_5, x_6, x_8\}$
(c) $\{x_1, x_2, x_3, x_4\}$

Chapter 16

16.1 Hint: Initial state = (*A*-3, *B*-0, *C*-0); final state = (*A*-0, *B*-0, *C*-3). Take shifting of a disc from one peg to other as input.

16.2 Hint: Assign names to containers of sizes 8, 5 and 3 liters as *A*, *B* and *C*, respectively. Then initial state = (*A*-8, *B*-0, *C*-0); final state = (*A*-4, *B*-4, *C*-0). Take transferring of oil from one container to other as input.

Chapter 17

17.2 $r = 1 \mid 0 \mid 0(0 \mid 1) \mid 0(10)^*(\varepsilon + 0 \mid 1)$

Chapter 18

18.1 (a) addition: $a_r + b_r = \begin{cases} 3-2^r, & 0 \le r \le 2 \\ r+3, & 2 < r \le 4 \\ 2^{-r}+r+10, & r > 4 \end{cases}$

(b) subtraction: $a_r - b_r = \begin{cases} -3+2^r, & 0 \le r \le 2 \\ -(r+3), & 3 \le r \le 4 \\ 2^{-r}-r+4, & r > 4 \end{cases}$

(c) multiplication: $a_r \cdot b_r = \begin{cases} 0, & 0 \le r \le 4 \\ (2^{-r}+7)(r+3) & r > 4 \end{cases}$

18.2 $a_r = (A_1r^2 + A_2r + A_3)(-2)^r$

18.3 $a_r = (A_1r^2 + A_2r + A_3)(-2)^r - \dfrac{162}{125}\,3^r$

18.4 $a_r = \left(-\dfrac{5}{12}r^2 + r - \dfrac{5}{8}\right)(-3)^r + \dfrac{5}{8}3^r$

18.5 $a_r = A(-3)^r + B(-2)^r + \dfrac{1}{3}r^2 + \dfrac{17}{18}r + \dfrac{115}{216}$

18.6 $a_r = \left(-\dfrac{191}{216}r^2 + \dfrac{67}{18}r - \dfrac{187}{54}\right)(-3)^r + \dfrac{1}{8}r^2 + \dfrac{1}{2}r + \dfrac{1}{2}$

18.7 $a_r = A(-3)^r + B(-2)^r + \dfrac{1}{4}r^2 + \dfrac{13}{24}r + \dfrac{71}{288}$

Index